SpringerBriefs in Statistics

For further volumes:
http://www.springer.com/series/8921

Photograph 1 Tiritiri Island, Auckland, New Zealand. (Photo: SP)

Simo Puntanen
George P. H. Styan
Jarkko Isotalo

Formulas Useful for Linear Regression Analysis and Related Matrix Theory

It's Only Formulas But We Like Them

Springer

Simo Puntanen
School of Information Sciences
University of Tampere
Tampere, Finland

Jarkko Isotalo
Department of Forest Sciences
University of Helsinki
Helsinki, Finland

George P. H. Styan
Department of Mathematics
and Statistics
McGill University
Montréal, QC, Canada

ISSN 2191-544X ISSN 2191-5458 (electronic)
ISBN 978-3-642-32930-2 ISBN 978-3-642-32931-9 (eBook)
DOI 10.1007/978-3-642-32931-9
Springer Heidelberg New York Dordrecht London

Library of Congress Control Number: 2012948184

Printed on acid-free paper

Springer is part of Springer Science+Business Media (www.springer.com)

Preface

Lie la lie, lie la la-lie lie la-lie.
There must be fifty ways to leave your lover.
Oh, still crazy after all these years.

PAUL SIMON[1]

Think about going to a lonely island for some substantial time and that you are supposed to decide what books to take with you. This book is then a serious alternative: it does not only guarantee a good night's sleep (reading in the late evening) but also offers you a survival kit in your urgent regression problems (definitely met at the day time on any lonely island, see for example Photograph 1, p. ii).

Our experience is that even though a huge amount of the formulas related to linear models is available in the statistical literature, it is not always so easy to catch them when needed. The purpose of this book is to collect together a good bunch of helpful rules—within a limited number of pages, however. They all exist in literature but are pretty much scattered. The first version (technical report) of the *Formulas* appeared in 1996 (54 pages) and the fourth one in 2008. Since those days, the authors have never left home without the *Formulas*.

This book is not a regular textbook—this is supporting material for courses given in linear regression (and also in multivariate statistical analysis); such courses are extremely common in universities providing teaching in quantitative statistical analysis. We assume that the reader is somewhat familiar with linear algebra, matrix calculus, linear statistical models, and multivariate statistical analysis, although a thorough knowledge is not needed, one year of undergraduate study of linear algebra and statistics is expected. A short course in regression would also be necessary before traveling with our book. Here are some examples of smooth introductions to regression: Chatterjee & Hadi (2012) (first ed. 1977), Draper & Smith (1998) (first ed. 1966), Seber & Lee (2003) (first ed. 1977), and Weisberg (2005) (first ed. 1980).

The term *regression* itself has an exceptionally interesting history: see the excellent chapter entitled *Regression towards Mean* in Stigler (1999), where (on p. 177) he says that the story of Francis Galton's (1822–1911) discovery of regression is "an exciting one, involving science, experiment, mathematics, simulation, and one of the great thought experiments of all time".

[1] From (1) *The Boxer*, a folk rock ballad written by Paul Simon in 1968 and first recorded by Simon & Garfunkel, (2) *50 Ways to Leave Your Lover*, a 1975 song by Paul Simon, from his album "Still Crazy After All These Years", (3) *Still Crazy After All These Years*, a 1975 song by Paul Simon and title track from his album "Still Crazy After All These Years".

This book is neither a real handbook: by a handbook we understand a thorough representation of a particular area. There are some recent handbook-type books dealing with matrix algebra helpful for statistics. The book by Seber (2008) should be mentioned in particular. Some further books are, for example, by Abadir & Magnus (2005) and Bernstein (2009). Quick visits to matrices in linear models and multivariate analysis appear in Puntanen, Seber & Styan (2013) and in Puntanen & Styan (2013).

We do not provide any proofs nor references. The book by Puntanen, Styan & Isotalo (2011) offers many proofs for the formulas. The website http://www.sis.uta.fi/tilasto/matrixtricks supports both these books by additional material.

Sincere thanks go to Götz Trenkler, Oskar Maria Baksalary, Stephen J. Haslett, and Kimmo Vehkalahti for helpful comments. We give special thanks to Jarmo Niemelä for his outstanding LaTeX assistance. The Figure 1 (p. xii) was prepared using the Survo software, online at http://www.survo.fi (thanks go to Kimmo Vehkalahti) and the Figure 2 (p. xii) using PSTricks (thanks again going to Jarmo Niemelä).

We are most grateful to Alice Blanck, Ulrike Stricker-Komba, and to Niels Peter Thomas of Springer for advice and encouragement.

This research has been supported in part by the Natural Sciences and Engineering Research Council of Canada.

SP, GPHS & JI
June 7, 2012

MSC 2000: 15-01, 15-02, 15A09, 15A42, 15A99, 62H12, 62J05.

Key words and phrases: Best linear unbiased estimation, Cauchy–Schwarz inequality, column space, eigenvalue decomposition, estimability, Gauss–Markov model, generalized inverse, idempotent matrix, linear model, linear regression, Löwner ordering, matrix inequalities, oblique projector, ordinary least squares, orthogonal projector, partitioned linear model, partitioned matrix, rank cancellation rule, reduced linear model, Schur complement, singular value decomposition.

References

Abadir, K. M. & Magnus, J. R. (2005). *Matrix Algebra*. Cambridge University Press.

Bernstein, D. S. (2009). *Matrix Mathematics: Theory, Facts, and Formulas*. Princeton University Press.

Chatterjee, S. & Hadi, A. S. (2012). *Regression Analysis by Example*, 5th Edition. Wiley.

Draper, N. R. & Smith, H. (1998). *Applied Regression Analysis*, 3rd Edition. Wiley.

Puntanen, S., Styan, G. P. H. & Isotalo, J. (2011). *Matrix Tricks for Linear Statistical Models: Our Personal Top Twenty*. Springer.

Puntanen, S. & Styan, G. P. H. (2013). Chapter 52: Random Vectors and Linear Statistical Models. *Handbook of Linear Algebra*, 2nd Edition (Leslie Hogben, ed.), Chapman & Hall, in press.

Puntanen, S., Seber, G. A. F. & Styan, G. P. H. (2013). Chapter 53: Multivariate Statistical Analysis. *Handbook of Linear Algebra*, 2nd Edition (Leslie Hogben, ed.), Chapman & Hall, in press.

Seber, G. A. F. (2008). *A Matrix Handbook for Statisticians*. Wiley.

Seber, G. A. F. & Lee, A. J. (2006). *Linear Regression Analysis*, 2nd Edition. Wiley.

Stigler, S. M. (1999). *Statistics on the Table: The History of Statistical Concepts and Methods*. Harvard University Press.

Weisberg, S. (2005). *Applied Linear Regression*, 3rd Edition. Wiley.

Contents

Notation

$\mathbb{R}^{n\times m}$	the set of $n \times m$ real matrices: all matrices considered in this book are real
$\mathbb{R}^{n\times m}_r$	the subset of $\mathbb{R}^{n\times m}$ consisting of matrices with rank r
NND_n	the subset of symmetric $n \times n$ matrices consisting of nonnegative definite (nnd) matrices
PD_n	the subset of NND_n consisting of positive definite (pd) matrices
$\mathbf{0}$	null vector, null matrix
$\mathbf{1}_n$	column vector of ones, shortened $\mathbf{1}$
$\mathbf{I}_n$	identity matrix, shortened $\mathbf{I}$
$\mathbf{i}_j$	the jth column of $\mathbf{I}$; jth standard basis vector
$\mathbf{A}_{n\times m} = \{a_{ij}\}$	$n \times m$ matrix $\mathbf{A}$ with its elements a_{ij}, $\mathbf{A} = (\mathbf{a}_1 : \ldots : \mathbf{a}_m)$ presented columnwise, $\mathbf{A} = (\mathbf{a}_{(1)} : \ldots : \mathbf{a}_{(n)})'$ presented row-wise
$\mathbf{a}$	column vector $\mathbf{a} \in \mathbb{R}^n$
$\mathbf{A}'$	transpose of matrix $\mathbf{A}$; $\mathbf{A}$ is symmetric if $\mathbf{A}' = \mathbf{A}$, skew-symmetric if $\mathbf{A}' = -\mathbf{A}$
$(\mathbf{A} : \mathbf{B})$	partitioned (augmented) matrix
$\mathbf{A}^{-1}$	inverse of matrix $\mathbf{A}_{n\times n}$: $\mathbf{AB} = \mathbf{BA} = \mathbf{I}_n \implies \mathbf{B} = \mathbf{A}^{-1}$
$\mathbf{A}^{-}$	generalized inverse of matrix $\mathbf{A}$: $\mathbf{AA}^{-}\mathbf{A} = \mathbf{A}$
$\mathbf{A}^{+}$	the Moore–Penrose inverse of matrix $\mathbf{A}$: $\mathbf{AA}^{+}\mathbf{A} = \mathbf{A}$, $\mathbf{A}^{+}\mathbf{AA}^{+} = \mathbf{A}^{+}$, $(\mathbf{AA}^{+})' = \mathbf{AA}^{+}$, $(\mathbf{A}^{+}\mathbf{A})' = \mathbf{A}^{+}\mathbf{A}$
$\mathbf{A}^{1/2}$	nonnegative definite square root of $\mathbf{A} \in \mathrm{NND}_n$
$\mathbf{A}^{+1/2}$	nonnegative definite square square root of $\mathbf{A}^{+} \in \mathrm{NND}_n$
$\langle \mathbf{a}, \mathbf{b} \rangle$	standard inner product in $\mathbb{R}^n$: $\langle \mathbf{a}, \mathbf{b} \rangle = \mathbf{a}'\mathbf{b}$
$\langle \mathbf{a}, \mathbf{b} \rangle_{\mathbf{V}}$	inner product $\mathbf{a}'\mathbf{V}\mathbf{b}$; $\mathbf{V}$ is a positive definite inner product matrix (ipm)

$\|\mathbf{a}\|$	Euclidean norm (standard norm, 2-norm) of vector $\mathbf{a}$: $\|\mathbf{a}\|^2 = \mathbf{a}'\mathbf{a}$, also denoted as $\|\mathbf{a}\|_2$
$\|\mathbf{a}\|_{\mathbf{V}}$	$\|\mathbf{a}\|_{\mathbf{V}}^2 = \mathbf{a}'\mathbf{V}\mathbf{a}$, norm when the ipm is positive definite $\mathbf{V}$
$\|\mathbf{A}\|_F$	Euclidean (Frobenius) norm of matrix $\mathbf{A}$: $\|\mathbf{A}\|_F^2 = \mathrm{tr}(\mathbf{A}'\mathbf{A})$
$\det(\mathbf{A})$	determinant of matrix $\mathbf{A}$, also denoted as $\|\mathbf{A}\|$
$\mathrm{diag}(d_1, \dots, d_n)$	$n \times n$ diagonal matrix with listed diagonal entries
$\mathrm{diag}(\mathbf{A})$	diagonal matrix formed by the diagonal entries of $\mathbf{A}_{n\times n}$, denoted also as $\mathbf{A}_\delta$
$\mathrm{r}(\mathbf{A})$	rank of matrix $\mathbf{A}$, denoted also as $\mathrm{rank}(\mathbf{A})$
$\mathrm{tr}(\mathbf{A})$	trace of matrix $\mathbf{A}_{n\times n}$, denoted also as $\mathrm{trace}(\mathbf{A})$: $\mathrm{tr}(\mathbf{A}) = a_{11} + a_{22} + \cdots + a_{nn}$
$\mathbf{A} \geq_{\mathsf{L}} \mathbf{0}$	$\mathbf{A}$ is nonnegative definite: $\mathbf{A} = \mathbf{L}\mathbf{L}'$ for some $\mathbf{L}$
$\mathbf{A} >_{\mathsf{L}} \mathbf{0}$	$\mathbf{A}$ is positive definite: $\mathbf{A} = \mathbf{L}\mathbf{L}'$ for some invertible $\mathbf{L}$
$\mathbf{A} \geq_{\mathsf{L}} \mathbf{B}$	$\mathbf{A} - \mathbf{B}$ is nonnegative definite, Löwner partial ordering
$\mathbf{A} >_{\mathsf{L}} \mathbf{B}$	$\mathbf{A} - \mathbf{B}$ is positive definite
$\cos(\mathbf{a}, \mathbf{b})$	the cosine of the angle, θ, between the nonzero vectors $\mathbf{a}$ and $\mathbf{b}$: $\cos(\mathbf{a}, \mathbf{b}) = \cos\theta = \langle \mathbf{a}, \mathbf{b}\rangle / (\|\mathbf{a}\|\,\|\mathbf{b}\|)$
$\mathrm{vec}(\mathbf{A})$	the vector of columns of $\mathbf{A}$, $\mathrm{vec}(\mathbf{A}_{n\times m}) = (\mathbf{a}_1', \dots, \mathbf{a}_m')' \in \mathbb{R}^{nm}$
$\mathbf{A} \otimes \mathbf{B}$	Kronecker product of $\mathbf{A}_{n\times m}$ and $\mathbf{B}_{p\times q}$: $$\mathbf{A} \otimes \mathbf{B} = \begin{pmatrix} a_{11}\mathbf{B} & \dots & a_{1m}\mathbf{B} \\ \vdots & \vdots & \vdots \\ a_{n1}\mathbf{B} & \dots & a_{nm}\mathbf{B} \end{pmatrix} \in \mathbb{R}^{np\times mq}$$
$\mathbf{A}_{22\cdot 1}$	Schur complement of $\mathbf{A}_{11}$ in $\mathbf{A} = \begin{pmatrix} \mathbf{A}_{11} & \mathbf{A}_{12} \\ \mathbf{A}_{21} & \mathbf{A}_{22} \end{pmatrix}$: $\mathbf{A}_{22\cdot 1} = \mathbf{A}_{22} - \mathbf{A}_{21}\mathbf{A}_{11}^{-}\mathbf{A}_{12} = \mathbf{A}/\mathbf{A}_{11}$
$\mathbf{P}_{\mathbf{A}}$	orthogonal projector onto $\mathscr{C}(\mathbf{A})$ w.r.t. ipm $\mathbf{I}$: $\mathbf{P}_{\mathbf{A}} = \mathbf{A}(\mathbf{A}'\mathbf{A})^{-}\mathbf{A}' = \mathbf{A}\mathbf{A}^{+}$
$\mathbf{P}_{\mathbf{A};\mathbf{V}}$	orthogonal projector onto $\mathscr{C}(\mathbf{A})$ w.r.t. ipm $\mathbf{V}$: $\mathbf{P}_{\mathbf{A};\mathbf{V}} = \mathbf{A}(\mathbf{A}'\mathbf{V}\mathbf{A})^{-}\mathbf{A}'\mathbf{V}$
$\mathbf{P}_{\mathbf{A}\|\mathbf{B}}$	projector onto $\mathscr{C}(\mathbf{A})$ along $\mathscr{C}(\mathbf{B})$: $\mathbf{P}_{\mathbf{A}\|\mathbf{B}}(\mathbf{A} : \mathbf{B}) = (\mathbf{A} : \mathbf{0})$
$\mathscr{C}(\mathbf{A})$	column space of matrix $\mathbf{A}_{n\times m}$: $\mathscr{C}(\mathbf{A}) = \{\, \mathbf{y} \in \mathbb{R}^n : \mathbf{y} = \mathbf{A}\mathbf{x}$ for some $\mathbf{x} \in \mathbb{R}^m \,\}$
$\mathscr{N}(\mathbf{A})$	null space of matrix $\mathbf{A}_{n\times m}$: $\mathscr{N}(\mathbf{A}) = \{\, \mathbf{x} \in \mathbb{R}^m : \mathbf{A}\mathbf{x} = \mathbf{0} \,\}$
$\mathscr{C}(\mathbf{A})^{\perp}$	orthocomplement of $\mathscr{C}(\mathbf{A})$ w.r.t. ipm $\mathbf{I}$: $\mathscr{C}(\mathbf{A})^{\perp} = \mathscr{N}(\mathbf{A}')$
$\mathbf{A}^{\perp}$	matrix whose column space is $\mathscr{C}(\mathbf{A}^{\perp}) = \mathscr{C}(\mathbf{A})^{\perp} = \mathscr{N}(\mathbf{A}')$
$\mathscr{C}(\mathbf{A})^{\perp}_{\mathbf{V}}$	orthocomplement of $\mathscr{C}(\mathbf{A})$ w.r.t. ipm $\mathbf{V}$
$\mathbf{A}^{\perp}_{\mathbf{V}}$	matrix whose column space is $\mathscr{C}(\mathbf{A})^{\perp}_{\mathbf{V}}$: $\mathbf{A}^{\perp}_{\mathbf{V}} = (\mathbf{V}\mathbf{A})^{\perp} = \mathbf{V}^{-1}\mathbf{A}^{\perp}$

$\mathrm{ch}_i(\mathbf{A}) = \lambda_i$	the ith largest eigenvalue of $\mathbf{A}_{n\times n}$ (all eigenvalues being real): $(\lambda_i, \mathbf{t}_i)$ is the ith eigenpair of $\mathbf{A}$: $\mathbf{A}\mathbf{t}_i = \lambda_i \mathbf{t}_i,\ \mathbf{t}_i \neq \mathbf{0}$
$\mathrm{ch}(\mathbf{A})$	set of all n eigenvalues of $\mathbf{A}_{n\times n}$, including multiplicities
$\mathrm{nzch}(\mathbf{A})$	set of nonzero eigenvalues of $\mathbf{A}_{n\times n}$, including multiplicities
$\mathrm{sg}_i(\mathbf{A}) = \delta_i$	the ith largest singular value of $\mathbf{A}_{n\times m}$
$\mathrm{sg}(\mathbf{A})$	set of singular values of $\mathbf{A}_{n\times m}$
$\mathcal{U} + \mathcal{V}$	sum of vector spaces $\mathcal{U}$ and $\mathcal{V}$
$\mathcal{U} \oplus \mathcal{V}$	direct sum of vector spaces $\mathcal{U}$ and $\mathcal{V}$; here $\mathcal{U} \cap \mathcal{V} = \{\mathbf{0}\}$
$\mathcal{U} \boxplus \mathcal{V}$	direct sum of orthogonal vector spaces $\mathcal{U}$ and $\mathcal{U}$
$\mathrm{var_d}(\mathbf{y}) = s_y^2$	sample variance: argument is variable vector $\mathbf{y} \in \mathbb{R}^n$: $$\mathrm{var_d}(\mathbf{y}) = \tfrac{1}{n-1}\mathbf{y}'\mathbf{C}\mathbf{y}$$
$\mathrm{cov_d}(\mathbf{x},\mathbf{y}) = s_{xy}$	sample covariance: $$\mathrm{cov_d}(\mathbf{x},\mathbf{y}) = \tfrac{1}{n-1}\mathbf{x}'\mathbf{C}\mathbf{y} = \tfrac{1}{n-1}\sum_{i=1}^{n}(x_i - \bar{x}_i)(y_i - \bar{y})$$
$\mathrm{cor_d}(\mathbf{x},\mathbf{y}) = r_{xy}$	sample correlation: $r_{xy} = \mathbf{x}'\mathbf{C}\mathbf{y}/\sqrt{\mathbf{x}'\mathbf{C}\mathbf{x}\cdot\mathbf{y}'\mathbf{C}\mathbf{y}}$, $\mathbf{C}$ is the centering matrix
$\mathrm{E}(\mathbf{x})$	expectation of a p-dimensional random vector $\mathbf{x}$: $\mathrm{E}(\mathbf{x}) = \boldsymbol{\mu}_\mathbf{x} = (\mu_1, \dots, \mu_p)' \in \mathbb{R}^p$
$\mathrm{cov}(\mathbf{x}) = \boldsymbol{\Sigma}$	covariance matrix ($p \times p$) of a p-dimensional random vector $\mathbf{x}$: $$\mathrm{cov}(\mathbf{x}) = \boldsymbol{\Sigma} = \{\sigma_{ij}\} = \mathrm{E}(\mathbf{x}-\boldsymbol{\mu}_\mathbf{x})(\mathbf{x}-\boldsymbol{\mu}_\mathbf{x})',$$ $$\mathrm{cov}(x_i, x_j) = \sigma_{ij} = \mathrm{E}(x_i - \mu_i)(x_j - \mu_j),$$ $$\mathrm{var}(x_i) = \sigma_{ii} = \sigma_i^2$$
$\mathrm{cor}(\mathbf{x})$	correlation matrix of a p-dimensional random vector $\mathbf{x}$: $$\mathrm{cor}(\mathbf{x}) = \{\varrho_{ij}\} = \left\{\frac{\sigma_{ij}}{\sigma_i \sigma_j}\right\}$$

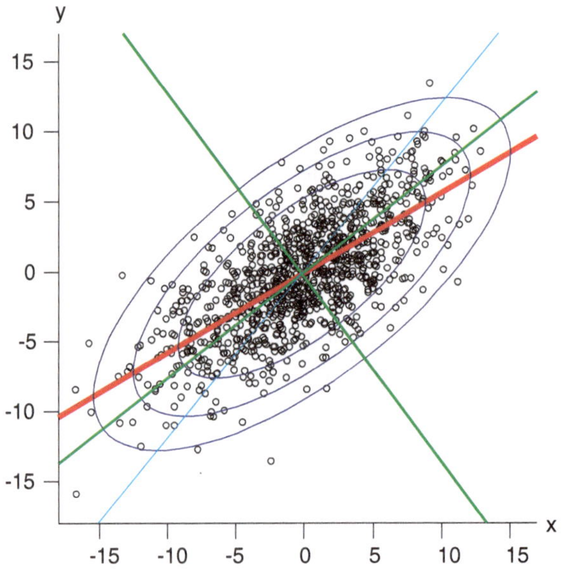

Figure 1 Observations from $N_2(\mathbf{0}, \mathbf{\Sigma})$; $\sigma_x = 5$, $\sigma_y = 4$, $\varrho_{xy} = 0.7$. Regression line has the slope $\hat{\beta}_1 \approx \varrho_{xy}\sigma_y/\sigma_x$. Also the regression line of x on y is drawn. The direction of the first major axis of the contour ellipse is determined by $\mathbf{t}_1$, the first eigenvector of $\mathbf{\Sigma}$.

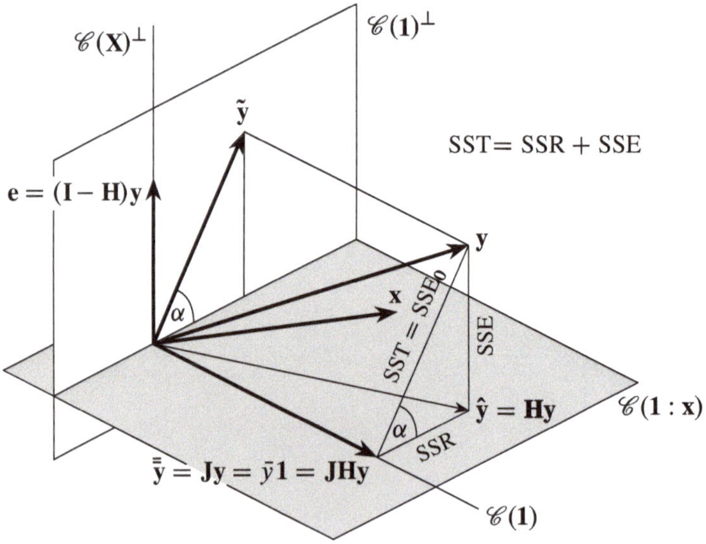

Figure 2 Illustration of SST = SSR + SSE.

Formulas Useful for Linear Regression Analysis and Related Matrix Theory

1 The model matrix & other preliminaries

1.1 Linear model. By $\mathscr{M} = \{\mathbf{y}, \mathbf{X}\boldsymbol{\beta}, \sigma^2\mathbf{V}\}$ we mean that we have the model $\mathbf{y} = \mathbf{X}\boldsymbol{\beta} + \boldsymbol{\varepsilon}$, where $\mathrm{E}(\mathbf{y}) = \mathbf{X}\boldsymbol{\beta} \in \mathbb{R}^n$ and $\mathrm{cov}(\mathbf{y}) = \sigma^2\mathbf{V}$, i.e., $\mathrm{E}(\boldsymbol{\varepsilon}) = \mathbf{0}$ and $\mathrm{cov}(\boldsymbol{\varepsilon}) = \sigma^2\mathbf{V}$; $\mathscr{M}$ is often called the Gauss–Markov model.

- $\mathbf{y}$ is an observable random vector, $\boldsymbol{\varepsilon}$ is unobservable random error vector, $\mathbf{X} = (\mathbf{1} : \mathbf{X}_0)$ is a given $n \times p$ ($p = k + 1$) model (design) matrix, the vector $\boldsymbol{\beta} = (\beta_0, \beta_1, \dots, \beta_k)' = (\beta_0, \boldsymbol{\beta}_{\mathrm{x}}')'$ and scalar $\sigma^2 > 0$ are unknown
- $y_i = \mathrm{E}(y_i) + \varepsilon_i = \beta_0 + \mathbf{x}_{(i)}'\boldsymbol{\beta}_{\mathrm{x}} + \varepsilon_i = \beta_0 + \beta_1 x_{i1} + \dots + \beta_k x_{ik} + \varepsilon_i$, where $\mathbf{x}_{(i)}'$ = the ith row of $\mathbf{X}_0$
- from the context it is apparent when $\mathbf{X}$ has full column rank; when distributional properties are considered, we assume that $\mathbf{y} \sim \mathrm{N}_n(\mathbf{X}\boldsymbol{\beta}, \sigma^2\mathbf{V})$
- according to the model, we believe that $\mathrm{E}(\mathbf{y}) \in \mathscr{C}(\mathbf{X})$, i.e., $\mathrm{E}(\mathbf{y})$ is a linear combination of the columns of $\mathbf{X}$ but we do not know which linear combination
- from the context it is clear which formulas require that the model has the intercept term β_0; p refers to the number of columns of $\mathbf{X}$ and hence in the no-intercept model $p = k$
- if the explanatory variables x_i are random variables, then the model $\mathscr{M}$ may be interpreted as the conditional model of $\mathbf{y}$ given $\mathbf{X}$: $\mathrm{E}(\mathbf{y} \mid \mathbf{X}) = \mathbf{X}\boldsymbol{\beta}$, $\mathrm{cov}(\mathbf{y} \mid \mathbf{X}) = \sigma^2\mathbf{V}$, and the error term is difference $\mathbf{y} - \mathrm{E}(\mathbf{y} \mid \mathbf{X})$. In short, regression is the study of how the conditional distribution of y, when x is given, changes with the value of x.

S. Puntanen et al., *Formulas Useful for Linear Regression Analysis and Related Matrix Theory*, SpringerBriefs in Statistics,
DOI: 10.1007/978-3-642-32931-9_1, © The Author(s) 2013

1.2 $\mathbf{X} = (\mathbf{1} : \mathbf{X}_0) = (\mathbf{1} : \mathbf{x}_1 : \ldots : \mathbf{x}_k) = \begin{pmatrix} \mathbf{x}'_{(1*)} \\ \vdots \\ \mathbf{x}'_{(n*)} \end{pmatrix} \in \mathbb{R}^{n\times(k+1)}$ model matrix $n \times p,\ p = k+1$

1.3 $\mathbf{X}_0 = (\mathbf{x}_1 : \ldots : \mathbf{x}_k) = \begin{pmatrix} \mathbf{x}'_{(1)} \\ \vdots \\ \mathbf{x}'_{(n)} \end{pmatrix} = \begin{pmatrix} x_{11} & x_{12} & \ldots & x_{1k} \\ \vdots & \vdots & & \vdots \\ x_{n1} & x_{n2} & \ldots & x_{nk} \end{pmatrix} \in \mathbb{R}^{n\times k}$ data matrix of $x_1, \ldots, x_k$

1.4 $\mathbf{1} = (1, \ldots, 1)' \in \mathbb{R}^n, \quad \mathbf{i}_i = (0, \ldots, 1\ (i\text{th}) \ldots, 0)' \in \mathbb{R}^n,$
$\mathbf{i}'_i\mathbf{X} = \mathbf{x}'_{(i*)} = (1, \mathbf{x}'_{(i)}) = \text{the } i\text{th row of } \mathbf{X},$
$\mathbf{i}'_i\mathbf{X}_0 = \mathbf{x}'_{(i)} = \text{the } i\text{th row of } \mathbf{X}_0$

1.5 $\mathbf{x}_1, \ldots, \mathbf{x}_k$ "variable vectors" in "variable space" $\mathbb{R}^n$
$\mathbf{x}_{(1)}, \ldots, \mathbf{x}_{(n)}$ "observation vectors" in "observation space" $\mathbb{R}^k$

1.6 $\mathbf{X}_y = (\mathbf{X}_0 : \mathbf{y}) \in \mathbb{R}^{n\times(k+1)}$ joint data matrix of $x_1, \ldots, x_k$ and response y

1.7 $\mathbf{J} = \mathbf{1}(\mathbf{1}'\mathbf{1})^{-1}\mathbf{1}' = \frac{1}{n}\mathbf{1}\mathbf{1}' = \mathbf{P_1} = \mathbf{J}_n = \text{orthogonal projector onto } \mathscr{C}(\mathbf{1}_n)$
$\mathbf{J}\mathbf{y} = \bar{y}\mathbf{1} = \bar{\bar{\mathbf{y}}} = (\bar{y}, \bar{y}, \ldots, \bar{y})' \in \mathbb{R}^n$

1.8 $\mathbf{I} - \mathbf{J} = \mathbf{C}$ $\quad \mathbf{C} = \mathbf{C}_n = \text{orthogonal projector onto } \mathscr{C}(\mathbf{1}_n)^\perp,$ centering matrix

1.9 $(\mathbf{I} - \mathbf{J})\mathbf{y} = \mathbf{C}\mathbf{y} = \mathbf{y} - \bar{y}\mathbf{1}_n$
$= \mathbf{y} - \bar{\bar{\mathbf{y}}} = \tilde{\mathbf{y}} = (y_1 - \bar{y}, \ldots, y_n - \bar{y})'$ centered $\mathbf{y}$

1.10 $\bar{\mathbf{x}} = (\bar{x}_1, \ldots, \bar{x}_k)' = \frac{1}{n}\mathbf{X}'_0\mathbf{1}_n$
$= \frac{1}{n}(\mathbf{x}_{(1)} + \cdots + \mathbf{x}_{(n)}) \in \mathbb{R}^k$ vector of x-means

1.11 $\mathbf{J}\mathbf{X}_y = (\mathbf{J}\mathbf{X}_0 : \mathbf{J}\mathbf{y}) = (\bar{x}_1\mathbf{1} : \ldots : \bar{x}_k\mathbf{1} : \bar{y}\mathbf{1})$
$= (\bar{\bar{\mathbf{x}}}_1 : \ldots : \bar{\bar{\mathbf{x}}}_k : \bar{\bar{\mathbf{y}}}) = \mathbf{1}(\bar{\mathbf{x}}', \bar{y}) = \begin{pmatrix} \bar{\mathbf{x}}' & \bar{y} \\ \vdots & \vdots \\ \bar{\mathbf{x}}' & \bar{y} \end{pmatrix} \in \mathbb{R}^{n\times(k+1)}$

1.12 $\tilde{\mathbf{X}}_0 = (\mathbf{I} - \mathbf{J})\mathbf{X}_0 = \mathbf{C}\mathbf{X}_0 = (\mathbf{x}_1 - \bar{\bar{\mathbf{x}}}_1 : \ldots : \mathbf{x}_k - \bar{\bar{\mathbf{x}}}_k) = (\tilde{\mathbf{x}}_1 : \ldots : \tilde{\mathbf{x}}_k)$
$= \begin{pmatrix} \mathbf{x}'_{(1)} - \bar{\mathbf{x}}' \\ \vdots \\ \mathbf{x}'_{(n)} - \bar{\mathbf{x}}' \end{pmatrix} = \begin{pmatrix} \tilde{\mathbf{x}}'_{(1)} \\ \vdots \\ \tilde{\mathbf{x}}'_{(n)} \end{pmatrix} \in \mathbb{R}^{n\times k}$ centered $\mathbf{X}_0$

1.13 $\tilde{\mathbf{X}}_y = (\mathbf{I} - \mathbf{J})\mathbf{X}_y$

$$= \mathbf{C}\mathbf{X}_y = (\mathbf{C}\mathbf{X}_0 : \mathbf{C}\mathbf{y}) = (\tilde{\mathbf{X}}_0 : \mathbf{y} - \bar{\bar{\mathbf{y}}}) = (\tilde{\mathbf{X}}_0 : \tilde{\mathbf{y}}) \qquad \text{centered } \mathbf{X}_y$$

1.14 $\mathbf{T} = \mathbf{X}_y'(\mathbf{I} - \mathbf{J})\mathbf{X}_y = \tilde{\mathbf{X}}_y'\tilde{\mathbf{X}}_y = \begin{pmatrix} \mathbf{X}_0'\mathbf{C}\mathbf{X}_0 & \mathbf{X}_0'\mathbf{C}\mathbf{y} \\ \mathbf{y}'\mathbf{C}\mathbf{X}_0 & \mathbf{y}'\mathbf{C}\mathbf{y} \end{pmatrix} = \begin{pmatrix} \tilde{\mathbf{X}}_0'\tilde{\mathbf{X}}_0 & \tilde{\mathbf{X}}_0'\tilde{\mathbf{y}} \\ \tilde{\mathbf{y}}'\tilde{\mathbf{X}}_0 & \tilde{\mathbf{y}}'\tilde{\mathbf{y}} \end{pmatrix}$

$$= \begin{pmatrix} \mathbf{T}_{\mathbf{xx}} & \mathbf{t}_{\mathbf{x}y} \\ \mathbf{t}_{\mathbf{x}y}' & t_{yy} \end{pmatrix} = \begin{pmatrix} \mathrm{ssp}(\mathbf{X}_0) & \mathrm{ssp}(\mathbf{X}_0, \mathbf{y}) \\ \mathrm{ssp}(\mathbf{y}, \mathbf{X}_0) & \mathrm{ssp}(\mathbf{y}) \end{pmatrix} = \begin{pmatrix} t_{11} & t_{12} & \dots & t_{1k} & t_{1y} \\ \vdots & \vdots & & \vdots & \vdots \\ t_{k1} & t_{k2} & \dots & t_{kk} & t_{ky} \\ t_{y1} & t_{y2} & \dots & t_{yk} & t_{yy} \end{pmatrix}$$

$$= \mathrm{ssp}(\mathbf{X}_0 : \mathbf{y}) = \mathrm{ssp}(\mathbf{X}_y) \quad \text{“corrected” sums of squares and cross products}$$

1.15 $\mathbf{T}_{\mathbf{xx}} = \tilde{\mathbf{X}}_0'\tilde{\mathbf{X}}_0 = \mathbf{X}_0'\mathbf{C}\mathbf{X}_0 = \sum_{i=1}^n \tilde{\mathbf{x}}_{(i)}\tilde{\mathbf{x}}_{(i)}' = \sum_{i=1}^n (\mathbf{x}_{(i)} - \bar{\mathbf{x}})(\mathbf{x}_{(i)} - \bar{\mathbf{x}})'$

$$= \sum_{i=1}^n \mathbf{x}_{(i)}\mathbf{x}_{(i)}' - n\bar{\mathbf{x}}\bar{\mathbf{x}}' = \{\mathbf{x}_i'\mathbf{C}\mathbf{x}_j\} = \{t_{ij}\} = \mathrm{ssp}(\mathbf{X}_0)$$

1.16 $\mathbf{S} = \mathrm{cov_d}(\mathbf{X}_y) = \mathrm{cov_d}(\mathbf{X}_0 : \mathbf{y}) = \frac{1}{n-1}\mathbf{T} = \begin{pmatrix} \mathbf{S}_{\mathbf{xx}} & \mathbf{s}_{\mathbf{x}y} \\ \mathbf{s}_{\mathbf{x}y}' & s_y^2 \end{pmatrix}$

$$= \begin{pmatrix} \mathrm{cov_d}(\mathbf{X}_0) & \mathrm{cov_d}(\mathbf{X}_0, \mathbf{y}) \\ \mathrm{cov_d}(\mathbf{y}, \mathbf{X}_0) & \mathrm{var_d}(\mathbf{y}) \end{pmatrix} \quad \text{sample covariance matrix of } x_i\text{'s and } y$$

$$= \mathrm{cov_s}\begin{pmatrix} \mathbf{x} \\ y \end{pmatrix} = \begin{pmatrix} \mathrm{cov_s}(\mathbf{x}) & \mathrm{cov_s}(\mathbf{x}, y) \\ \mathrm{cov_s}(y, \mathbf{x}) & \mathrm{var_s}(y) \end{pmatrix}$$

here $\mathbf{x}$ is the vector of x's to be observed

1.17 $\mathbf{T}_\delta = \mathrm{diag}(\mathbf{T}) = \mathrm{diag}(t_{11}, \dots, t_{kk}, t_{yy})$,
$\mathbf{S}_\delta = \mathrm{diag}(\mathbf{S}) = \mathrm{diag}(s_1^2, \dots, s_k^2, s_y^2)$

1.18 $\tilde{\tilde{\mathbf{X}}}_y = (\tilde{\tilde{\mathbf{x}}}_1 : \dots : \tilde{\tilde{\mathbf{x}}}_k : \tilde{\tilde{\mathbf{y}}})$

$$= \tilde{\mathbf{X}}_y\mathbf{T}_\delta^{-1/2}, \text{ centering \& scaling: } \mathrm{diag}(\tilde{\tilde{\mathbf{X}}}_y'\tilde{\tilde{\mathbf{X}}}_y) = \mathbf{I}_{k+1}$$

1.19 While calculating the correlations, we assume that all variables have nonzero variances, that is, the matrix $\mathrm{diag}(\mathbf{T})$ is positive definite, or in other words: $\mathbf{x}_i \notin \mathscr{C}(\mathbf{1}),\ i = 1, \dots, k,\ \mathbf{y} \notin \mathscr{C}(\mathbf{1})$.

1.20 $\mathbf{R} = \mathrm{cor_d}(\mathbf{X}_0 : \mathbf{y}) = \mathrm{cor_d}(\mathbf{X}_y)$ sample correlation matrix of x_i's and y

$$= \mathbf{S}_\delta^{-1/2}\mathbf{S}\mathbf{S}_\delta^{-1/2} = \mathbf{T}_\delta^{-1/2}\mathbf{T}\mathbf{T}_\delta^{-1/2}$$

$$= \tilde{\tilde{\mathbf{X}}}_y'\tilde{\tilde{\mathbf{X}}}_y = \begin{pmatrix} \tilde{\tilde{\mathbf{X}}}_0'\tilde{\tilde{\mathbf{X}}}_0 & \tilde{\tilde{\mathbf{X}}}_0'\tilde{\tilde{\mathbf{y}}} \\ \tilde{\tilde{\mathbf{y}}}'\tilde{\tilde{\mathbf{X}}}_0 & \tilde{\tilde{\mathbf{y}}}'\tilde{\tilde{\mathbf{y}}} \end{pmatrix} = \begin{pmatrix} \mathbf{R}_{\mathbf{xx}} & \mathbf{r}_{\mathbf{x}y} \\ \mathbf{r}_{\mathbf{x}y}' & 1 \end{pmatrix} = \begin{pmatrix} \mathrm{cor_d}(\mathbf{X}_0) & \mathrm{cor_d}(\mathbf{X}_0, \mathbf{y}) \\ \mathrm{cor_d}(\mathbf{y}, \mathbf{X}_0) & 1 \end{pmatrix}$$

1.21 $\mathbf{T} = \begin{pmatrix} \mathbf{T}_{xx} & \mathbf{t}_{xy} \\ \mathbf{t}'_{xy} & t_{yy} \end{pmatrix}, \quad \mathbf{S} = \begin{pmatrix} \mathbf{S}_{xx} & \mathbf{s}_{xy} \\ \mathbf{s}'_{xy} & s_y^2 \end{pmatrix} = \frac{1}{n-1}\mathbf{T}, \quad \mathbf{R} = \begin{pmatrix} \mathbf{R}_{xx} & \mathbf{r}_{xy} \\ \mathbf{r}'_{xy} & 1 \end{pmatrix}$

1.22 $\mathbf{t}_{xy} = \begin{pmatrix} t_{1y} \\ \vdots \\ t_{ky} \end{pmatrix}, \quad \mathbf{s}_{xy} = \begin{pmatrix} s_{1y} \\ \vdots \\ s_{ky} \end{pmatrix}, \quad \mathbf{r}_{xy} = \begin{pmatrix} r_{1y} \\ \vdots \\ r_{ky} \end{pmatrix}$

1.23 $$\begin{aligned} \mathrm{SS}_y &= \sum_{i=1}^n (y_i - \bar{y})^2 = \sum_{i=1}^n y_i^2 - \tfrac{1}{n}\Big(\sum_{i=1}^n y_i\Big)^2 \\ &= \sum_{i=1}^n y_i^2 - n\bar{y}^2 = \mathbf{y}'\mathbf{C}\mathbf{y} = \mathbf{y}'\mathbf{y} - \mathbf{y}'\mathbf{J}\mathbf{y} \end{aligned}$$

1.24 $$\begin{aligned} \mathrm{SP}_{xy} &= \sum_{i=1}^n (x_i - \bar{x})(y_i - \bar{y}) = \sum_{i=1}^n x_i y_i - \tfrac{1}{n}\Big(\sum_{i=1}^n x_i\Big)\Big(\sum_{i=1}^n y_i\Big) \\ &= \sum_{i=1}^n x_i y_i - n\bar{x}\bar{y} = \mathbf{x}'\mathbf{C}\mathbf{y} = \mathbf{x}'\mathbf{y} - \mathbf{x}'\mathbf{J}\mathbf{y} \end{aligned}$$

1.25 $$\begin{aligned} s_y^2 &= \mathrm{var_s}(y) = \mathrm{var_d}(\mathbf{y}) = \tfrac{1}{n-1}\mathbf{y}'\mathbf{C}\mathbf{y}, \\ s_{xy} &= \mathrm{cov_s}(x, y) = \mathrm{cov_d}(\mathbf{x}, \mathbf{y}) = \tfrac{1}{n-1}\mathbf{x}'\mathbf{C}\mathbf{y} \end{aligned}$$

1.26 $$\begin{aligned} r_{ij} &= \mathrm{cor_s}(x_i, x_j) = \mathrm{cor_d}(\mathbf{x}_i, \mathbf{x}_j) = \cos(\mathbf{C}\mathbf{x}_i, \mathbf{C}\mathbf{x}_j) = \cos(\tilde{\mathbf{x}}_i, \tilde{\mathbf{x}}_j) = \tilde{\tilde{\mathbf{x}}}_i'\tilde{\tilde{\mathbf{x}}}_j \\ &= \frac{t_{ij}}{\sqrt{t_{ii}t_{jj}}} = \frac{s_{ij}}{s_i s_j} = \frac{\mathbf{x}_i'\mathbf{C}\mathbf{x}_j}{\sqrt{\mathbf{x}_i'\mathbf{C}\mathbf{x}_i \cdot \mathbf{x}_j'\mathbf{C}\mathbf{x}_j}} = \frac{\mathrm{SP}_{ij}}{\sqrt{\mathrm{SS}_i \mathrm{SS}_j}} \end{aligned}$$ sample correlation

1.27 If $\mathbf{x}$ and $\mathbf{y}$ are centered then $\mathrm{cor_d}(\mathbf{x}, \mathbf{y}) = \cos(\mathbf{x}, \mathbf{y})$.

1.28 Keeping observed data as a theoretical distribution. Let $u_1, \dots, u_n$ be the observed values of some empirical variable u, and let u_* be a discrete random variable whose values are $u_1, \dots, u_n$, each with $\mathrm{P}(u_* = u_i) = \frac{1}{n}$. Then $\mathrm{E}(u_*) = \bar{u}$ and $\mathrm{var}(u_*) = \frac{n-1}{n}s_u^2$. More generally, consider a data matrix $\mathbf{U} = (\mathbf{u}_{(1)} : \dots : \mathbf{u}_{(n)})'$ and define a discrete random vector $\mathbf{u}_*$ with probability function $\mathrm{P}(\mathbf{u}_* = \mathbf{u}_{(i)}) = \frac{1}{n}$, $i = 1, \dots, n$. Then

$$\mathrm{E}(\mathbf{u}_*) = \bar{\mathbf{u}}, \quad \mathrm{cov}(\mathbf{u}_*) = \tfrac{1}{n}\mathbf{U}'\mathbf{C}\mathbf{U} = \tfrac{n-1}{n}\,\mathrm{cov_d}(\mathbf{U}) = \tfrac{n-1}{n}\mathbf{S}.$$

Moreover, the sample correlation matrix of data matrix $\mathbf{U}$ is the same as the (theoretical, population) correlation matrix of $\mathbf{u}_*$. Therefore, any property shown for population statistics, holds for sample statistics and vice versa.

1.29 Mahalanobis distance. Consider a data matrix $\mathbf{U}_{n \times p} = (\mathbf{u}_{(1)} : \dots : \mathbf{u}_{(n)})'$, where $\mathrm{cov_d}(\mathbf{U}) = \mathbf{S} \in \mathrm{PD}_p$. The (squared) sample Mahalanobis distance of the observation $\mathbf{u}_{(i)}$ from the mean vector $\bar{\mathbf{u}}$ is defined as

$$\mathrm{MHLN}^2(\mathbf{u}_{(i)}, \bar{\mathbf{u}}, \mathbf{S}) = (\mathbf{u}_{(i)} - \bar{\mathbf{u}})'\mathbf{S}^{-1}(\mathbf{u}_{(i)} - \bar{\mathbf{u}}) = \|\mathbf{S}^{-1/2}(\mathbf{u}_{(i)} - \bar{\mathbf{u}})\|^2.$$

Moreover, $\mathrm{MHLN}^2(\mathbf{u}_{(i)}, \bar{\mathbf{u}}, \mathbf{S}) = (n-1)\tilde{h}_{ii}$, where $\tilde{\mathbf{H}} = \mathbf{P}_{\mathbf{CU}}$. If $\mathbf{x}$ is a random vector with $\mathrm{E}(\mathbf{x}) = \boldsymbol{\mu} \in \mathbb{R}^p$ and $\mathrm{cov}(\mathbf{x}) = \boldsymbol{\Sigma} \in \mathrm{PD}_p$, then the (squared) Mahalanobis distance between $\mathbf{x}$ and $\boldsymbol{\mu}$ is the random variable

$$\mathrm{MHLN}^2(\mathbf{x}, \boldsymbol{\mu}, \boldsymbol{\Sigma}) = (\mathbf{x}-\boldsymbol{\mu})'\boldsymbol{\Sigma}^{-1}(\mathbf{x}-\boldsymbol{\mu}) = \mathbf{z}'\mathbf{z}; \quad \mathbf{z} = \boldsymbol{\Sigma}^{-1/2}(\mathbf{x}-\boldsymbol{\mu}).$$

1.30 Statistical distance. The squared Euclidean distance of the ith observation $\mathbf{u}_{(i)}$ from the mean $\bar{\mathbf{u}}$ is of course $\|\mathbf{u}_{(i)} - \bar{\mathbf{u}}\|^2$. Given the data matrix $\mathbf{U}_{n\times p}$, one may wonder if there is a more informative way, in statistical sense, to measure the distance between $\mathbf{u}_{(i)}$ and $\bar{\mathbf{u}}$. Consider a new variable $z = \mathbf{a}'\mathbf{u}$ so that the n values of z are in the variable vector $\mathbf{z} = \mathbf{U}\mathbf{a}$. Then $z_i = \mathbf{a}'\mathbf{u}_{(i)}$ and $\bar{z} = \mathbf{a}'\bar{\mathbf{u}}$, and we may define

$$D_i(\mathbf{a}) = \frac{|z_i - \bar{z}|}{\sqrt{\mathrm{var}_s(z)}} = \frac{|\mathbf{a}'(\mathbf{u}_{(i)} - \bar{\mathbf{u}})|}{\sqrt{\mathbf{a}'\mathbf{S}\mathbf{a}}}, \quad \text{where } \mathbf{S} = \mathrm{cov}_d(\mathbf{U}).$$

Let us find a vector $\mathbf{a}_*$ which maximizes $D_i(\mathbf{a})$. In view of 22.24c (p. 102),

$$\max_{\mathbf{a}\neq\mathbf{0}} D_i^2(\mathbf{a}) = (\mathbf{u}_{(i)} - \bar{\mathbf{u}})'\mathbf{S}^{-1}(\mathbf{u}_{(i)} - \bar{\mathbf{u}}) = \mathrm{MHLN}^2(\mathbf{u}_{(i)}, \bar{\mathbf{u}}, \mathbf{S}).$$

The maximum is attained for any vector $\mathbf{a}_*$ proportional to $\mathbf{S}^{-1}(\mathbf{u}_{(i)} - \bar{\mathbf{u}})$.

1.31 $$\begin{aligned}\mathscr{C}(\mathbf{A}) &= \text{the column space of } \mathbf{A}_{n\times m} = (\mathbf{a}_1 : \ldots : \mathbf{a}_m) \\ &= \{\, \mathbf{z} \in \mathbb{R}^n : \mathbf{z} = \mathbf{A}\mathbf{t} = \mathbf{a}_1 t_1 + \cdots + \mathbf{a}_m t_m \text{ for some } \mathbf{t} \in \mathbb{R}^m \,\} \subset \mathbb{R}^n\end{aligned}$$

1.32 $$\begin{aligned}\mathscr{C}(\mathbf{A})^\perp &= \text{the orthocomplement of } \mathscr{C}(\mathbf{A}) \\ &= \text{the set of vectors which are orthogonal (w.r.t. the standard inner product } \mathbf{u}'\mathbf{v}\text{) to every vector in } \mathscr{C}(\mathbf{A}) \\ &= \{\, \mathbf{u} \in \mathbb{R}^n : \mathbf{u}'\mathbf{A}\mathbf{t} = 0 \text{ for all } \mathbf{t} \,\} = \{\, \mathbf{u} \in \mathbb{R}^n : \mathbf{A}'\mathbf{u} = \mathbf{0} \,\} \\ &= \mathscr{N}(\mathbf{A}') = \text{the null space of } \mathbf{A}'\end{aligned}$$

1.33 Linear independence and $\mathrm{rank}(\mathbf{A})$. The columns of $\mathbf{A}_{n\times m}$ are linearly independent iff $\mathscr{N}(\mathbf{A}) = \{\mathbf{0}\}$. The rank of $\mathbf{A}$, $\mathrm{r}(\mathbf{A})$, is the maximal number of linearly independent columns (equivalently, rows) of $\mathbf{A}$; $\mathrm{r}(\mathbf{A}) = \dim \mathscr{C}(\mathbf{A})$.

1.34 $\mathbf{A}^\perp$ = a matrix whose column space is $\mathscr{C}(\mathbf{A}^\perp) = \mathscr{C}(\mathbf{A})^\perp = \mathscr{N}(\mathbf{A}')$:

$$\mathbf{Z} \in \{\mathbf{A}^\perp\} \iff \mathbf{A}'\mathbf{Z} = \mathbf{0} \text{ and } \mathrm{r}(\mathbf{Z}) = n - \mathrm{r}(\mathbf{A}) = \dim \mathscr{C}(\mathbf{A})^\perp.$$

1.35 The rank of the model matrix $\mathbf{X} = (\mathbf{1} : \mathbf{X}_0)$ can be expressed as

$$\begin{aligned}\mathrm{r}(\mathbf{X}) &= 1 + \mathrm{r}(\mathbf{X}_0) - \dim \mathscr{C}(\mathbf{1}) \cap \mathscr{C}(\mathbf{X}_0) = \mathrm{r}(\mathbf{1} : \mathbf{C}\mathbf{X}_0) \\ &= 1 + \mathrm{r}(\mathbf{C}\mathbf{X}_0) = 1 + \mathrm{r}(\mathbf{T}_{\mathbf{xx}}) = 1 + \mathrm{r}(\mathbf{S}_{\mathbf{xx}}),\end{aligned}$$

and thereby

$$\mathrm{r}(\mathbf{S}_{\mathbf{xx}}) = \mathrm{r}(\mathbf{X}) - 1 = \mathrm{r}(\mathbf{C}\mathbf{X}_0) = \mathrm{r}(\mathbf{X}_0) - \dim \mathscr{C}(\mathbf{1}) \cap \mathscr{C}(\mathbf{X}_0).$$

If all x-variables have nonzero variances, i.e., the correlation matrix $\mathbf{R}_{\mathbf{xx}}$ is properly defined, then $\mathrm{r}(\mathbf{R}_{\mathbf{xx}}) = \mathrm{r}(\mathbf{S}_{\mathbf{xx}}) = \mathrm{r}(\mathbf{T}_{\mathbf{xx}})$. Moreover,

$$\mathbf{S}_{\mathbf{xx}} \text{ is pd} \iff \mathrm{r}(\mathbf{X}) = k + 1 \iff \mathrm{r}(\mathbf{X}_0) = k \text{ and } \mathbf{1} \notin \mathscr{C}(\mathbf{X}_0).$$

1.36
- In 1.1 vector $\mathbf{y}$ is a random vector but for example in 1.6 $\mathbf{y}$ is an observed sample value.
- $\mathrm{cov}(\mathbf{x}) = \mathbf{\Sigma} = \{\sigma_{ij}\}$ refers to the covariance matrix ($p \times p$) of a random vector $\mathbf{x}$ (with p elements), $\mathrm{cov}(x_i, x_j) = \sigma_{ij}$, $\mathrm{var}(x_i) = \sigma_{ii} = \sigma_i^2$:

$$\begin{aligned} \mathrm{cov}(\mathbf{x}) = \mathbf{\Sigma} &= \mathrm{E}(\mathbf{x} - \boldsymbol{\mu}_{\mathbf{x}})(\mathbf{x} - \boldsymbol{\mu}_{\mathbf{x}})' \\ &= \mathrm{E}(\mathbf{x}\mathbf{x}') - \boldsymbol{\mu}_{\mathbf{x}}\boldsymbol{\mu}_{\mathbf{x}}', \quad \boldsymbol{\mu}_{\mathbf{x}} = \mathrm{E}(\mathbf{x}). \end{aligned}$$

- notation $\mathbf{x} \sim (\boldsymbol{\mu}, \mathbf{\Sigma})$ indicates that $\mathrm{E}(\mathbf{x}) = \boldsymbol{\mu}$ and $\mathrm{cov}(\mathbf{x}) = \mathbf{\Sigma}$
- the determinant $\det(\mathbf{\Sigma})$ is called the (population) generalized variance
- $\mathrm{cor}(\mathbf{x}) = \boldsymbol{\varrho} = \left\{\frac{\sigma_{ij}}{\sigma_i\sigma_j}\right\}$ refers to the correlation matrix of a random vector $\mathbf{x}$:

$$\mathrm{cor}(\mathbf{x}) = \boldsymbol{\varrho} = \mathbf{\Sigma}_\delta^{-1/2}\mathbf{\Sigma}\mathbf{\Sigma}_\delta^{-1/2}, \quad \mathbf{\Sigma} = \mathbf{\Sigma}_\delta^{1/2}\boldsymbol{\varrho}\mathbf{\Sigma}_\delta^{1/2}$$

- $\mathrm{cov}(\mathbf{A}\mathbf{x}) = \mathbf{A}\,\mathrm{cov}(\mathbf{x})\mathbf{A}' = \mathbf{A}\mathbf{\Sigma}\mathbf{A}'$, $\mathbf{A} \in \mathbb{R}^{a \times p}$, $\quad \mathrm{E}(\mathbf{A}\mathbf{x}) = \mathbf{A}\boldsymbol{\mu}_{\mathbf{x}}$
- $\mathrm{E}[\mathbf{\Sigma}^{-1/2}(\mathbf{x} - \boldsymbol{\mu})] = \mathbf{0}$, $\quad \mathrm{cov}[\mathbf{\Sigma}^{-1/2}(\mathbf{x} - \boldsymbol{\mu})] = \mathbf{I}_p$, when $\mathbf{x} \sim (\boldsymbol{\mu}, \mathbf{\Sigma})$
- $\mathrm{cov}(\mathbf{T}'\mathbf{x}) = \mathbf{\Lambda}$, if $\mathbf{\Sigma} = \mathbf{T}\mathbf{\Lambda}\mathbf{T}'$ is the eigenvalue decomposition of $\mathbf{\Sigma}$
- $\mathrm{var}(\mathbf{a}'\mathbf{x}) = \mathbf{a}'\,\mathrm{cov}(\mathbf{x})\mathbf{a} = \mathbf{a}'\mathbf{\Sigma}\mathbf{a} \geq 0$ for all $\mathbf{a} \in \mathbb{R}^p$ and hence every covariance matrix is nonnegative definite; $\mathbf{\Sigma}$ is singular iff there exists a nonzero $\mathbf{a} \in \mathbb{R}^p$ such that $\mathbf{a}'\mathbf{x}$ = a constant with probability 1
- $\mathrm{var}(x_1 \pm x_2) = \sigma_1^2 + \sigma_2^2 \pm 2\sigma_{12}$
- $\mathrm{cov}\begin{pmatrix}\mathbf{x}\\ \mathbf{y}\end{pmatrix} = \begin{pmatrix}\mathrm{cov}(\mathbf{x}) & \mathrm{cov}(\mathbf{x},\mathbf{y})\\ \mathrm{cov}(\mathbf{y},\mathbf{x}) & \mathrm{cov}(\mathbf{y})\end{pmatrix} = \begin{pmatrix}\mathbf{\Sigma}_{\mathbf{xx}} & \mathbf{\Sigma}_{\mathbf{xy}}\\ \mathbf{\Sigma}_{\mathbf{yx}} & \mathbf{\Sigma}_{\mathbf{yy}}\end{pmatrix}$,

  $\mathrm{cov}\begin{pmatrix}\mathbf{x}\\ y\end{pmatrix} = \begin{pmatrix}\mathbf{\Sigma}_{\mathbf{xx}} & \boldsymbol{\sigma}_{\mathbf{x}y}\\ \boldsymbol{\sigma}_{\mathbf{x}y}' & \sigma_y^2\end{pmatrix}$
- $\mathrm{cov}(\mathbf{x}, \mathbf{y})$ refers to the covariance matrix between random vectors $\mathbf{x}$ and $\mathbf{y}$:

$$\mathrm{cov}(\mathbf{x}, \mathbf{y}) = \mathrm{E}(\mathbf{x} - \boldsymbol{\mu}_{\mathbf{x}})(\mathbf{y} - \boldsymbol{\mu}_{\mathbf{y}})' = \mathrm{E}(\mathbf{x}\mathbf{y}') - \boldsymbol{\mu}_{\mathbf{x}}\boldsymbol{\mu}_{\mathbf{y}}' = \mathbf{\Sigma}_{\mathbf{xy}}$$

- $\mathrm{cov}(\mathbf{x}, \mathbf{x}) = \mathrm{cov}(\mathbf{x})$
- $\mathrm{cov}(\mathbf{A}\mathbf{x}, \mathbf{B}\mathbf{y}) = \mathbf{A}\,\mathrm{cov}(\mathbf{x}, \mathbf{y})\mathbf{B}' = \mathbf{A}\mathbf{\Sigma}_{\mathbf{xy}}\mathbf{B}'$
- $\mathrm{cov}(\mathbf{A}\mathbf{x} + \mathbf{B}\mathbf{y}) = \mathbf{A}\mathbf{\Sigma}_{\mathbf{xx}}\mathbf{A}' + \mathbf{B}\mathbf{\Sigma}_{\mathbf{yy}}\mathbf{B}' + \mathbf{A}\mathbf{\Sigma}_{\mathbf{xy}}\mathbf{B}' + \mathbf{B}\mathbf{\Sigma}_{\mathbf{yx}}\mathbf{A}'$

- $\operatorname{cov}(\mathbf{Ax}, \mathbf{By} + \mathbf{Cz}) = \operatorname{cov}(\mathbf{Ax}, \mathbf{By}) + \operatorname{cov}(\mathbf{Ax}, \mathbf{Cz})$
- $\operatorname{cov}(\mathbf{a}'\mathbf{x}, y) = \mathbf{a}' \operatorname{cov}(\mathbf{x}, y) = \mathbf{a}'\boldsymbol{\sigma}_{\mathbf{x}y} = a_1\sigma_{1y} + \cdots + a_p\sigma_{py}$
- $\operatorname{cov}(\mathbf{z}) = \mathbf{I}_2$,

$$\mathbf{A} = \begin{pmatrix} \sigma_x & 0 \\ \sigma_y \varrho & \sigma_y\sqrt{1-\varrho^2} \end{pmatrix} \implies \operatorname{cov}(\mathbf{Az}) = \begin{pmatrix} \sigma_x^2 & \sigma_{xy} \\ \sigma_{xy} & \sigma_y^2 \end{pmatrix},$$

i.e.,

$$\operatorname{cov}(\mathbf{z}) = \operatorname{cov}\begin{pmatrix} u \\ v \end{pmatrix} = \mathbf{I}_2$$
$$\implies \operatorname{cov}\begin{pmatrix} \sigma_x u \\ \sigma_y(\varrho u + \sqrt{1-\varrho^2}\, v) \end{pmatrix} = \begin{pmatrix} \sigma_x^2 & \sigma_{xy} \\ \sigma_{xy} & \sigma_y^2 \end{pmatrix}$$

- $\operatorname{cov}\left[\begin{pmatrix} 1 & 0 \\ -\sigma_{xy}/\sigma_x^2 & 1 \end{pmatrix}\begin{pmatrix} x \\ y \end{pmatrix}\right] = \operatorname{cov}\begin{pmatrix} x \\ y - \frac{\sigma_{xy}}{\sigma_x^2}x \end{pmatrix} = \begin{pmatrix} \sigma_x^2 & 0 \\ 0 & \sigma_y^2(1-\varrho^2) \end{pmatrix}$
- $\operatorname{cov}_{\mathrm{d}}(\mathbf{U}_{n\times p}) = \operatorname{cov}_{\mathrm{s}}(\mathbf{u})$ refers to the sample covariance matrix

$$\operatorname{cov}_{\mathrm{d}}(\mathbf{U}) = \tfrac{1}{n-1}\mathbf{U}'\mathbf{CU} = \tfrac{1}{n-1}\sum_{i=1}^{n}(\mathbf{u}_{(i)} - \bar{\mathbf{u}})(\mathbf{u}_{(i)} - \bar{\mathbf{u}})' = \mathbf{S}$$

- the determinant $\det(\mathbf{S})$ is called the (sample) generalized variance
- $\operatorname{cov}_{\mathrm{d}}(\mathbf{UA}) = \mathbf{A}' \operatorname{cov}_{\mathrm{d}}(\mathbf{U})\mathbf{A} = \mathbf{A}'\mathbf{SA} = \operatorname{cov}_{\mathrm{s}}(\mathbf{A}'\mathbf{u})$
- $\operatorname{cov}_{\mathrm{d}}(\mathbf{US}^{-1/2}) = \mathbf{I}_p = \operatorname{cov}_{\mathrm{d}}(\mathbf{CUS}^{-1/2})$
- $\underset{\sim}{\mathbf{U}} = \mathbf{CUS}^{-1/2}$: $\underset{\sim}{\mathbf{U}}$ is centered and transformed so that the new variables are uncorrelated and each has variance 1. Moreover, $\operatorname{diag}(\underset{\sim}{\mathbf{U}}\underset{\sim}{\mathbf{U}}') = \operatorname{diag}(d_1^2, \dots, d_n^2)$, where $d_i^2 = \mathrm{MHLN}^2(\mathbf{u}_{(i)}, \bar{\mathbf{u}}, \mathbf{S})$.
- $\mathbf{U}_* = \mathbf{CU}[\operatorname{diag}(\mathbf{S})]^{-1/2}$: $\mathbf{U}_*$ is centered and scaled so that each variable has variance 1 (and the squared length $n-1$)
- $\tilde{\tilde{\mathbf{U}}} = \mathbf{CU}[\operatorname{diag}(\mathbf{U}'\mathbf{CU})]^{-1/2}$: $\tilde{\tilde{\mathbf{U}}}$ is centered and scaled so that each variable has length 1 (and variance $\frac{1}{n-1}$)
- Denote $\mathbf{U}_\# = \mathbf{CUT}$, where $\mathbf{T}_{p\times p}$ comprises the orthonormal eigenvectors of $\mathbf{S}$: $\mathbf{S} = \mathbf{T\Lambda T}'$, $\boldsymbol{\Lambda} = \operatorname{diag}(\lambda_1, \dots, \lambda_p)$. Then $\mathbf{U}_\#$ is centered and transformed so that the new variables are uncorrelated and the ith variable has variance λ_i: $\operatorname{cov}_{\mathrm{d}}(\mathbf{U}_\#) = \boldsymbol{\Lambda}$.
- $\operatorname{var}_{\mathrm{s}}(u_1 + u_2) = \operatorname{cov}_{\mathrm{d}}(\mathbf{U}_{n\times 2}\mathbf{1}_2) = \mathbf{1}' \operatorname{cov}_{\mathrm{d}}(\mathbf{U})\mathbf{1} = \mathbf{1}'\mathbf{S}_{2\times 2}\mathbf{1} = s_1^2 + s_2^2 + 2s_{12}$
- $\operatorname{var}_{\mathrm{s}}(u_1 \pm u_2) = s_1^2 + s_2^2 \pm 2s_{12}$

2 Fitted values and residuals

2.1 $\mathbf{H} = \mathbf{X}(\mathbf{X}'\mathbf{X})^{-}\mathbf{X}' = \mathbf{X}\mathbf{X}^{+} = \mathbf{P}_{\mathbf{X}}$ orthogonal projector onto $\mathscr{C}(\mathbf{X})$

$= \mathbf{X}(\mathbf{X}'\mathbf{X})^{-1}\mathbf{X}', \quad \text{when } \mathrm{r}(\mathbf{X}) = p$

2.2 $\mathbf{H} = \mathbf{P}_{\mathbf{1}} + \mathbf{P}_{(\mathbf{I}-\mathbf{J})\mathbf{X}_0} = \mathbf{P}_{\mathbf{1}} + \mathbf{P}_{\tilde{\mathbf{X}}_0}$ $\quad \mathbf{X} = (\mathbf{1} : \mathbf{X}_0)$

$= \mathbf{J} + \tilde{\mathbf{X}}_0(\tilde{\mathbf{X}}_0'\tilde{\mathbf{X}}_0)^{-}\tilde{\mathbf{X}}_0' = \mathbf{J} + \tilde{\mathbf{X}}_0\mathbf{T}_{\mathbf{xx}}^{-}\tilde{\mathbf{X}}_0'$ $\quad \tilde{\mathbf{X}}_0 = (\mathbf{I} - \mathbf{J})\mathbf{X}_0 = \mathbf{C}\mathbf{X}_0$

2.3 $\mathbf{H} - \mathbf{J} = \mathbf{P}_{\tilde{\mathbf{X}}_0} = \tilde{\mathbf{X}}_0(\tilde{\mathbf{X}}_0'\tilde{\mathbf{X}}_0)^{-}\tilde{\mathbf{X}}_0' = \mathbf{P}_{\mathscr{C}(\mathbf{X})\cap\mathscr{C}(\mathbf{1})^{\perp}}$

2.4 $\mathbf{H} = \mathbf{P}_{\mathbf{X}_1} + \mathbf{P}_{\mathbf{M}_1\mathbf{X}_2};$

$\mathbf{M}_1 = \mathbf{I} - \mathbf{P}_{\mathbf{X}_1}, \quad \mathbf{X} = (\mathbf{X}_1 : \mathbf{X}_2), \quad \mathbf{X}_1 \in \mathbb{R}^{n\times p_1}, \quad \mathbf{X}_2 \in \mathbb{R}^{n\times p_2}$

2.5 $\mathbf{H} - \mathbf{P}_{\mathbf{X}_1} = \mathbf{P}_{\mathbf{M}_1\mathbf{X}_2} = \mathbf{P}_{\mathscr{C}(\mathbf{X})\cap\mathscr{C}(\mathbf{M}_1)} = \mathbf{M}_1\mathbf{X}_2(\mathbf{X}_2'\mathbf{M}_1\mathbf{X}_2)^{-}\mathbf{X}_2'\mathbf{M}_1$

2.6 $\mathscr{C}(\mathbf{M}_1\mathbf{X}_2) = \mathscr{C}(\mathbf{X}) \cap \mathscr{C}(\mathbf{M}_1),$

$\mathscr{C}(\mathbf{M}_1\mathbf{X}_2)^{\perp} = \mathscr{N}(\mathbf{X}_2'\mathbf{M}_1) = \mathscr{C}(\mathbf{X})^{\perp} \boxplus \mathscr{C}(\mathbf{X}_1)$

2.7 $\mathscr{C}(\mathbf{C}\mathbf{x}) = \mathscr{C}(\mathbf{1} : \mathbf{x}) \cap \mathscr{C}(\mathbf{1})^{\perp}, \quad \mathscr{C}(\mathbf{C}\mathbf{x})^{\perp} = \mathscr{C}(\mathbf{1} : \mathbf{x})^{\perp} \boxplus \mathscr{C}(\mathbf{1})$

2.8 $\mathrm{r}(\mathbf{X}_2'\mathbf{M}_1\mathbf{X}_2) = \mathrm{r}(\mathbf{X}_2'\mathbf{M}_1) = \mathrm{r}(\mathbf{X}_2) - \dim \mathscr{C}(\mathbf{X}_1) \cap \mathscr{C}(\mathbf{X}_2)$

2.9 The matrix $\mathbf{X}_2'\mathbf{M}_1\mathbf{X}_2$ is pd iff $\mathrm{r}(\mathbf{M}_1\mathbf{X}_2) = p_2$, i.e., iff $\mathscr{C}(\mathbf{X}_1) \cap \mathscr{C}(\mathbf{X}_2) = \{\mathbf{0}\}$ and $\mathbf{X}_2$ has full column rank. In particular, $\mathbf{T}_{\mathbf{xx}} = \mathbf{X}_0'\mathbf{C}\mathbf{X}_0$ is pd iff $\mathrm{r}(\mathbf{C}\mathbf{X}_0) = k$ iff $\mathrm{r}(\mathbf{X}) = k + 1$ iff

$\mathscr{C}(\mathbf{X}_0) \cap \mathscr{C}(\mathbf{1}) = \{\mathbf{0}\}$ and $\mathbf{X}_0$ has full column rank.

2.10 $\mathbf{H} = \mathbf{X}_1(\mathbf{X}_1'\mathbf{M}_2\mathbf{X}_1)^{-}\mathbf{X}_1'\mathbf{M}_2 + \mathbf{X}_2(\mathbf{X}_2'\mathbf{M}_1\mathbf{X}_2)^{-}\mathbf{X}_2'\mathbf{M}_1$

iff $\mathscr{C}(\mathbf{X}_1) \cap \mathscr{C}(\mathbf{X}_2) = \{\mathbf{0}\}$

2.11 $\mathbf{M} = \mathbf{I} - \mathbf{H}$ orthogonal projector onto $\mathscr{C}(\mathbf{X})^{\perp} = \mathscr{N}(\mathbf{X}')$

2.12 $\mathbf{M} = \mathbf{I} - (\mathbf{P}_{\mathbf{X}_1} + \mathbf{P}_{\mathbf{M}_1\mathbf{X}_2}) = \mathbf{M}_1 - \mathbf{P}_{\mathbf{M}_1\mathbf{X}_2} = \mathbf{M}_1(\mathbf{I} - \mathbf{P}_{\mathbf{M}_1\mathbf{X}_2})$

2.13 $\hat{\mathbf{y}} = \mathbf{H}\mathbf{y} = \mathbf{X}\hat{\boldsymbol{\beta}} = \widehat{\mathbf{X}\boldsymbol{\beta}} = \mathrm{OLSE}(\mathbf{X}\boldsymbol{\beta})$ OLSE of $\mathbf{X}\boldsymbol{\beta}$, the fitted values

2.14 Because $\hat{\mathbf{y}}$ is the projection of $\mathbf{y}$ onto $\mathscr{C}(\mathbf{X})$, it depends only on $\mathscr{C}(\mathbf{X})$, not on a particular choice of $\mathbf{X} = \mathbf{X}_*$, as long as $\mathscr{C}(\mathbf{X}) = \mathscr{C}(\mathbf{X}_*)$. The coordinates of $\hat{\mathbf{y}}$ with respect to $\mathbf{X}$, i.e., $\hat{\boldsymbol{\beta}}$, depend on the choice of $\mathbf{X}$.

2.15 $\hat{\mathbf{y}} = \mathbf{H}\mathbf{y} = \mathbf{X}\hat{\boldsymbol\beta} = \mathbf{1}\hat\beta_0 + \mathbf{X}_0\hat{\boldsymbol\beta}_{\mathrm{x}} = \hat\beta_0\mathbf{1} + \hat\beta_1\mathbf{x}_1 + \dots + \hat\beta_k\mathbf{x}_k$
$= (\mathbf{J} + \mathbf{P}_{\tilde{\mathbf{X}}_0})\mathbf{y} = \mathbf{J}\mathbf{y} + (\mathbf{I}-\mathbf{J})\mathbf{X}_0\mathbf{T}_{\mathbf{xx}}^{-1}\mathbf{t}_{\mathbf{x}y} = (\bar{y} - \bar{\mathbf{x}}'\mathbf{T}_{\mathbf{xx}}^{-1}\mathbf{t}_{\mathbf{x}y})\mathbf{1} + \mathbf{X}_0\mathbf{T}_{\mathbf{xx}}^{-1}\mathbf{t}_{\mathbf{x}y}$

2.16 $\hat{\mathbf{y}} = \mathbf{X}_1\hat{\boldsymbol\beta}_1 + \mathbf{X}_2\hat{\boldsymbol\beta}_2 = \mathbf{X}_1(\mathbf{X}_1'\mathbf{M}_2\mathbf{X}_1)^{-1}\mathbf{X}_1'\mathbf{M}_2\mathbf{y} + \mathbf{X}_2(\mathbf{X}_2'\mathbf{M}_1\mathbf{X}_2)^{-1}\mathbf{X}_2'\mathbf{M}_1\mathbf{y}$
$= (\mathbf{P}_{\mathbf{X}_1} + \mathbf{P}_{\mathbf{M}_1\mathbf{X}_2})\mathbf{y} = \mathbf{X}_1(\mathbf{X}_1'\mathbf{X}_1)^{-1}\mathbf{X}_1'\mathbf{y} + \mathbf{M}_1\mathbf{X}_2(\mathbf{X}_2'\mathbf{M}_1\mathbf{X}_2)^{-1}\mathbf{X}_2'\mathbf{M}_1\mathbf{y}$
$= \mathbf{P}_{\mathbf{X}_1}\mathbf{y} + \mathbf{M}_1\mathbf{X}_2\hat{\boldsymbol\beta}_2$ here and in 2.15 $\mathrm{r}(\mathbf{X}) = p$

2.17 OLS criterion. Let $\hat{\boldsymbol\beta}$ be any vector minimizing $\|\mathbf{y} - \mathbf{X}\boldsymbol\beta\|^2$. Then $\mathbf{X}\hat{\boldsymbol\beta}$ is $\mathrm{OLSE}(\mathbf{X}\boldsymbol\beta)$. Vector $\mathbf{X}\hat{\boldsymbol\beta}$ is always unique but $\hat{\boldsymbol\beta}$ is unique iff $\mathrm{r}(\mathbf{X}) = p$. Even if $\mathrm{r}(\mathbf{X}) < p$, $\hat{\boldsymbol\beta}$ is called the OLSE of $\boldsymbol\beta$ even though it is not an ordinary estimator because of its nonuniqueness; it is merely a solution to the minimizing problem. The OLSE of $\mathbf{K}'\boldsymbol\beta$ is $\mathbf{K}'\hat{\boldsymbol\beta}$ which is unique iff $\mathbf{K}'\boldsymbol\beta$ is estimable.

2.18 Normal equations. Let $\hat{\boldsymbol\beta} \in \mathbb{R}^p$ be any solution to normal equation $\mathbf{X}'\mathbf{X}\boldsymbol\beta = \mathbf{X}'\mathbf{y}$. Then $\hat{\boldsymbol\beta}$ minimizes $\|\mathbf{y} - \mathbf{X}\boldsymbol\beta\|$. The general solution to $\mathbf{X}'\mathbf{X}\boldsymbol\beta = \mathbf{X}'\mathbf{y}$ is

$$\hat{\boldsymbol\beta} = (\mathbf{X}'\mathbf{X})^{-}\mathbf{X}'\mathbf{y} + [\mathbf{I}_p - (\mathbf{X}'\mathbf{X})^{-}\mathbf{X}'\mathbf{X}]\mathbf{z},$$

where $\mathbf{z} \in \mathbb{R}^p$ is free to vary and $(\mathbf{X}'\mathbf{X})^{-}$ is an arbitrary (but fixed) generalized inverse of $\mathbf{X}'\mathbf{X}$.

2.19 Generalized normal equations. Let $\tilde{\boldsymbol\beta} \in \mathbb{R}^p$ be any solution to the generalized normal equation $\mathbf{X}'\mathbf{V}^{-1}\mathbf{X}\boldsymbol\beta = \mathbf{X}'\mathbf{V}^{-1}\mathbf{y}$, where $\mathbf{V} \in \mathrm{PD}_n$. Then $\tilde{\boldsymbol\beta}$ minimizes $\|\mathbf{y} - \mathbf{X}\boldsymbol\beta\|_{\mathbf{V}^{-1}}$. The general solution to the equation $\mathbf{X}'\mathbf{V}^{-1}\mathbf{X}\boldsymbol\beta = \mathbf{X}'\mathbf{V}^{-1}\mathbf{y}$ is

$$\tilde{\boldsymbol\beta} = (\mathbf{X}'\mathbf{V}^{-1}\mathbf{X})^{-}\mathbf{X}'\mathbf{V}^{-1}\mathbf{y} + [\mathbf{I}_p - (\mathbf{X}'\mathbf{V}^{-1}\mathbf{X})^{-}\mathbf{X}'\mathbf{V}^{-1}\mathbf{X}]\mathbf{z},$$

where $\mathbf{z} \in \mathbb{R}^p$ is free to vary.

2.20 Under the model $\{\mathbf{y}, \mathbf{X}\boldsymbol\beta, \sigma^2\mathbf{I}\}$ the following holds:

(a) $\mathrm{E}(\hat{\mathbf{y}}) = \mathrm{E}(\mathbf{H}\mathbf{y}) = \mathbf{X}\boldsymbol\beta, \quad \mathrm{cov}(\hat{\mathbf{y}}) = \mathrm{cov}(\mathbf{H}\mathbf{y}) = \sigma^2\mathbf{H}$

(b) $\mathrm{cov}(\hat{\mathbf{y}}, \mathbf{y} - \hat{\mathbf{y}}) = \mathbf{0}, \quad \mathrm{cov}(\mathbf{y}, \mathbf{y} - \hat{\mathbf{y}}) = \sigma^2\mathbf{M}$

(c) $\hat{\mathbf{y}} \sim \mathrm{N}_n(\mathbf{X}\boldsymbol\beta, \sigma^2\mathbf{H})$ under normality

(d) $\hat{y}_i = \mathbf{x}_{(i*)}'\hat{\boldsymbol\beta}$, where $\mathbf{x}_{(i*)}'$ is the ith row of $\mathbf{X}$ $\quad \hat{y}_i \sim \mathrm{N}(\mathbf{x}_{(i*)}'\boldsymbol\beta, \sigma^2 h_{ii})$

(e) $\hat{\boldsymbol\varepsilon} = \mathbf{y} - \hat{\mathbf{y}} = (\mathbf{I}_n - \mathbf{H})\mathbf{y} = \mathbf{M}\mathbf{y} = \mathrm{res}(\mathbf{y}; \mathbf{X})$ $\quad \hat{\boldsymbol\varepsilon}$ = the residual vector

(f) $\boldsymbol\varepsilon = \mathbf{y} - \mathbf{X}\boldsymbol\beta, \quad \mathrm{E}(\boldsymbol\varepsilon) = \mathbf{0}, \quad \mathrm{cov}(\boldsymbol\varepsilon) = \sigma^2\mathbf{I}_n$ $\quad \boldsymbol\varepsilon$ = error vector

(g) $\mathrm{E}(\hat{\boldsymbol\varepsilon}) = \mathbf{0}$, $\mathrm{cov}(\hat{\boldsymbol\varepsilon}) = \sigma^2\mathbf{M}$ and hence the components of the residual vector $\hat{\boldsymbol\varepsilon}$ may be correlated and have unequal variances

(h) $\hat{\boldsymbol\varepsilon} = \mathbf{M}\mathbf{y} \sim \mathrm{N}_n(\mathbf{0}, \sigma^2\mathbf{M})$ when normality is assumed

(i) $\hat\varepsilon_i = y_i - \hat{y}_i = y_i - \mathbf{x}_{(i*)}'\hat{\boldsymbol\beta}, \quad \hat\varepsilon_i \sim \mathrm{N}[0, \sigma^2(1 - h_{ii})],$ $\quad \hat\varepsilon_i$ = the ith residual

(j) $\operatorname{var}(\hat{\varepsilon}_i) = \sigma^2(1 - h_{ii}) = \sigma^2 m_{ii}$

(k) $\operatorname{cor}(\hat{\varepsilon}_i, \hat{\varepsilon}_j) = \dfrac{-h_{ij}}{[(1-h_{ii})(1-h_{jj})]^{1/2}} = \dfrac{m_{ij}}{(m_{ii}m_{jj})^{1/2}}$

(l) $\dfrac{\hat{\varepsilon}_i}{\sigma\sqrt{1-h_{ii}}} \sim \mathrm{N}(0,1), \quad \dfrac{\varepsilon_i}{\sigma} \sim \mathrm{N}(0,1)$

2.21 Under the intercept model $\{\mathbf{y}, (\mathbf{1} : \mathbf{X}_0)\boldsymbol{\beta}, \sigma^2\mathbf{I}\}$ the following holds:

(a) $\operatorname{cor_d}(\mathbf{y}, \hat{\mathbf{y}}) = \dfrac{\mathbf{y}'(\mathbf{I}-\mathbf{J})\mathbf{H}\mathbf{y}}{\sqrt{\mathbf{y}'(\mathbf{I}-\mathbf{J})\mathbf{y} \cdot \mathbf{y}'\mathbf{H}(\mathbf{I}-\mathbf{J})\mathbf{H}\mathbf{y}}} = \left(\dfrac{\mathbf{y}'(\mathbf{H}-\mathbf{J})\mathbf{y}}{\mathbf{y}'(\mathbf{I}-\mathbf{J})\mathbf{y}}\right)^{1/2}$

$= \left(\dfrac{\mathrm{SSR}}{\mathrm{SST}}\right)^{1/2} = R = R_{y\cdot\mathbf{x}} =$ the multiple correlation

(b) $\operatorname{cor_d}(\hat{\mathbf{y}}, \hat{\boldsymbol{\varepsilon}}) = 0$ the fitted values and residuals are uncorrelated

(c) $\operatorname{cor_d}(\mathbf{x}_i, \hat{\boldsymbol{\varepsilon}}) = 0$ each x_i-variable is uncorrelated with the residual vector

(d) $\operatorname{cor_d}(\mathbf{y}, \hat{\boldsymbol{\varepsilon}}) = (+)(1-R^2)^{1/2}\ (\geq 0)$ $\mathbf{y}$ and $\hat{\boldsymbol{\varepsilon}}$ may be positively correlated

(e) $\hat{\boldsymbol{\varepsilon}}'\mathbf{1} = \mathbf{y}'\mathbf{M}\mathbf{1} = 0$, i.e., $\sum_{i=1}^n \hat{\varepsilon}_i = 0$ the residual vector $\hat{\boldsymbol{\varepsilon}}$ is centered

(f) $\hat{\boldsymbol{\varepsilon}}'\hat{\mathbf{y}} = 0, \quad \hat{\boldsymbol{\varepsilon}}'\mathbf{x}_i = 0$

(g) $\hat{\mathbf{y}}'\mathbf{1} = \mathbf{y}'\mathbf{H}\mathbf{1} = \mathbf{y}'\mathbf{1}$, i.e., $\frac{1}{n}\sum_{i=1}^n \hat{y}_i = \bar{y}$ the mean of $\hat{y}_i$-values is $\bar{y}$

2.22 Under the model $\{\mathbf{y}, \mathbf{X}\boldsymbol{\beta}, \sigma^2\mathbf{V}\}$ we have

(a) $\mathrm{E}(\hat{\mathbf{y}}) = \mathbf{X}\boldsymbol{\beta}, \quad \operatorname{cov}(\hat{\mathbf{y}}) = \sigma^2\mathbf{HVH}, \quad \operatorname{cov}(\mathbf{Hy}, \mathbf{My}) = \sigma^2\mathbf{HVM},$

(b) $\mathrm{E}(\hat{\boldsymbol{\beta}}) = \boldsymbol{\beta}, \quad \operatorname{cov}(\hat{\boldsymbol{\beta}}) = \sigma^2(\mathbf{X}'\mathbf{X})^{-1}\mathbf{X}'\mathbf{V}\mathbf{X}(\mathbf{X}'\mathbf{X})^{-1}.$ [if $\mathrm{r}(\mathbf{X}) = p$]

3 Regression coefficients

In this section we consider the model $\{\mathbf{y}, \mathbf{X}\boldsymbol{\beta}, \sigma^2\mathbf{I}\}$, where $\mathrm{r}(\mathbf{X}) = p$ most of the time and $\mathbf{X} = (\mathbf{1} : \mathbf{X}_0)$. As regards distribution, $\mathbf{y} \sim \mathrm{N}_n(\mathbf{X}\boldsymbol{\beta}, \sigma^2\mathbf{I})$.

3.1 $\hat{\boldsymbol{\beta}} = (\mathbf{X}'\mathbf{X})^{-1}\mathbf{X}'\mathbf{y} = \begin{pmatrix} \hat{\beta}_0 \\ \hat{\boldsymbol{\beta}}_{\mathbf{x}} \end{pmatrix} \in \mathbb{R}^{k+1}$ estimated regression coefficients, $\mathrm{OLSE}(\boldsymbol{\beta})$

3.2 $\mathrm{E}(\hat{\boldsymbol{\beta}}) = \boldsymbol{\beta}, \quad \operatorname{cov}(\hat{\boldsymbol{\beta}}) = \sigma^2(\mathbf{X}'\mathbf{X})^{-1}; \quad \hat{\boldsymbol{\beta}} \sim \mathrm{N}_{k+1}[\boldsymbol{\beta}, \sigma^2(\mathbf{X}'\mathbf{X})^{-1}]$

3.3 $\hat{\beta}_0 = \bar{y} - \hat{\boldsymbol{\beta}}_{\mathbf{x}}'\bar{\mathbf{x}} = \bar{y} - (\hat{\beta}_1\bar{x}_1 + \cdots + \hat{\beta}_k\bar{x}_k)$ estimated constant term, intercept

3.4 $$\hat{\boldsymbol\beta}_{\mathbf x} = (\hat\beta_1, \dots, \hat\beta_k)' = (\mathbf{X}_0'\mathbf{C}\mathbf{X}_0)^{-1}\mathbf{X}_0'\mathbf{C}\mathbf{y} = (\tilde{\mathbf{X}}_0'\tilde{\mathbf{X}}_0)^{-1}\tilde{\mathbf{X}}_0'\tilde{\mathbf{y}} = \mathbf{T}_{\mathbf{xx}}^{-1}\mathbf{t}_{\mathbf{x}y} = \mathbf{S}_{\mathbf{xx}}^{-1}\mathbf{s}_{\mathbf{x}y}$$

3.5 $$k = 1, \quad \mathbf{X} = (\mathbf{1} : \mathbf{x}), \quad \hat\beta_0 = \bar y - \hat\beta_1 \bar x, \quad \hat\beta_1 = \frac{\mathrm{SP}_{xy}}{\mathrm{SS}_x} = \frac{s_{xy}}{s_x^2} = r_{xy}\frac{s_y}{s_x}$$

3.6 If the model does not have the intercept term, we denote $p = k$, $\mathbf{X} = \mathbf{X}_0$, and $\hat{\boldsymbol\beta} = (\hat\beta_1, \dots, \hat\beta_p)' = (\mathbf{X}'\mathbf{X})^{-1}\mathbf{X}'\mathbf{y} = (\mathbf{X}_0'\mathbf{X}_0)^{-1}\mathbf{X}_0'\mathbf{y}$.

3.7 If $\mathbf{X} = (\mathbf{X}_1 : \mathbf{X}_2)$, $\mathbf{M}_i = \mathbf{I} - \mathbf{P}_{\mathbf{X}_i}$, $\mathbf{X}_i \in \mathbb{R}^{n \times p_i}$, $i = 1, 2$, then

$$\hat{\boldsymbol\beta} = \begin{pmatrix} \hat{\boldsymbol\beta}_1 \\ \hat{\boldsymbol\beta}_2 \end{pmatrix} = \begin{pmatrix} (\mathbf{X}_1'\mathbf{M}_2\mathbf{X}_1)^{-1}\mathbf{X}_1'\mathbf{M}_2\mathbf{y} \\ (\mathbf{X}_2'\mathbf{M}_1\mathbf{X}_2)^{-1}\mathbf{X}_2'\mathbf{M}_1\mathbf{y} \end{pmatrix} \quad \text{and}$$

3.8 $$\hat{\boldsymbol\beta}_1 = (\mathbf{X}_1'\mathbf{X}_1)^{-1}\mathbf{X}_1'\mathbf{y} - (\mathbf{X}_1'\mathbf{X}_1)^{-1}\mathbf{X}_1'\mathbf{X}_2\hat{\boldsymbol\beta}_2 = (\mathbf{X}_1'\mathbf{X}_1)^{-1}\mathbf{X}_1'(\mathbf{y} - \mathbf{X}_2\hat{\boldsymbol\beta}_2).$$

3.9 Denoting the full model as $\mathscr{M}_{12} = \{\mathbf{y}, (\mathbf{X}_1 : \mathbf{X}_2)\boldsymbol\beta, \sigma^2\mathbf{I}\}$ and

$$\mathscr{M}_1 = \{\mathbf{y}, \mathbf{X}_1\boldsymbol\beta_1, \sigma^2\mathbf{I}\}, \qquad \text{small model}$$

$$\mathscr{M}_{12\cdot 2} = \{\mathbf{M}_2\mathbf{y}, \mathbf{M}_2\mathbf{X}_1\boldsymbol\beta_1, \sigma^2\mathbf{M}_2\}, \text{ with } \mathbf{M}_2 = \mathbf{I} - \mathbf{P}_{\mathbf{X}_2}, \qquad \text{reduced model}$$

$$\hat{\boldsymbol\beta}_i(\mathscr{A}) = \text{OLSE of } \boldsymbol\beta_i \text{ under the model } \mathscr{A},$$

we can write 3.8 as

(a) $\hat{\boldsymbol\beta}_1(\mathscr{M}_{12}) = \hat{\boldsymbol\beta}_1(\mathscr{M}_1) - (\mathbf{X}_1'\mathbf{X}_1)^{-1}\mathbf{X}_1'\mathbf{X}_2\hat{\boldsymbol\beta}_2(\mathscr{M}_{12})$,

and clearly we have

(b) $\hat{\boldsymbol\beta}_1(\mathscr{M}_{12}) = \hat{\boldsymbol\beta}_1(\mathscr{M}_{12\cdot 2})$. (Frisch–Waugh–Lovell theorem)

3.10 Let $\mathscr{M}_{12} = \{\mathbf{y}, (\mathbf{1} : \mathbf{X}_0)\boldsymbol\beta, \sigma^2\mathbf{I}\}$, $\boldsymbol\beta = (\beta_0 : \boldsymbol\beta_{\mathbf x}')'$, $\mathscr{M}_{12\cdot 1} = \{\mathbf{C}\mathbf{y}, \mathbf{C}\mathbf{X}_0\boldsymbol\beta_{\mathbf x}, \sigma^2\mathbf{C}\}$ = centered model. Then 3.9b means that $\boldsymbol\beta_{\mathbf x}$ has the same OLSE in the original model and in the centered model.

3.11 $\hat{\boldsymbol\beta}_1(\mathscr{M}_{12}) = (\mathbf{X}_1'\mathbf{X}_1)^{-1}\mathbf{X}_1'\mathbf{y}$, i.e., the old regression coefficients do not change when the new regressors ($\mathbf{X}_2$) are added iff $\mathbf{X}_1'\mathbf{X}_2\hat{\boldsymbol\beta}_2 = \mathbf{0}$.

3.12 The following statements are equivalent:

(a) $\hat{\boldsymbol\beta}_2 = \mathbf{0}$, (b) $\mathbf{X}_2'\mathbf{M}_1\mathbf{y} = \mathbf{0}$, (c) $\mathbf{y} \in \mathscr{N}(\mathbf{X}_2'\mathbf{M}_1) = \mathscr{C}(\mathbf{M}_1\mathbf{X}_2)^{\perp}$,

(d) $\mathrm{pcor_d}(\mathbf{X}_2, \mathbf{y} \mid \mathbf{X}_{01}) = \mathbf{0}$ or $\mathbf{y} \in \mathscr{C}(\mathbf{X}_1)$; $\mathbf{X} = (\mathbf{1} : \mathbf{X}_{01} : \mathbf{X}_2) = (\mathbf{X}_1 : \mathbf{X}_2)$.

3.13 The old regression coefficients do not change when one new regressor $\mathbf{x}_k$ is added iff $\mathbf{X}_1'\mathbf{x}_k = \mathbf{0}$ or $\hat\beta_k = 0$, with $\hat\beta_k = 0$ being equivalent to $\mathbf{x}_k'\mathbf{M}_1\mathbf{y} = 0$.

3.14 In the intercept model the old regression coefficients $\hat{\beta}_1, \hat{\beta}_2, \dots, \hat{\beta}_{k-p_2}$ (of "real" predictors) do not change when new regressors (whose values are in $\mathbf{X}_2$) are added if $\operatorname{cor}_{\mathrm{d}}(\mathbf{X}_{01}, \mathbf{X}_2) = \mathbf{0}$ or $\hat{\boldsymbol{\beta}}_2 = \mathbf{0}$; here $\mathbf{X} = (\mathbf{1} : \mathbf{X}_{01} : \mathbf{X}_2)$.

3.15 $$\hat{\beta}_k = \frac{\mathbf{x}_k'(\mathbf{I} - \mathbf{P}_{\mathbf{X}_1})\mathbf{y}}{\mathbf{x}_k'(\mathbf{I} - \mathbf{P}_{\mathbf{X}_1})\mathbf{x}_k} = \frac{\mathbf{x}_k'\mathbf{M}_1\mathbf{y}}{\mathbf{x}_k'\mathbf{M}_1\mathbf{x}_k} = \frac{\mathbf{u}'\mathbf{v}}{\mathbf{v}'\mathbf{v}}, \quad \text{where } \mathbf{X} = (\mathbf{X}_1 : \mathbf{x}_k) \text{ and}$$

$\mathbf{u} = \mathbf{M}_1\mathbf{y} = \operatorname{res}(\mathbf{y}; \mathbf{X}_1)$, $\mathbf{v} = \mathbf{M}_1\mathbf{x}_k = \operatorname{res}(\mathbf{x}_k; \mathbf{X}_1)$, i.e., $\hat{\beta}_k$ is the OLSE of β_k in the reduced model $\mathscr{M}_{12\cdot1} = \{\mathbf{M}_1\mathbf{y}, \mathbf{M}_1\mathbf{x}_k\beta_k, \sigma^2\mathbf{M}_1\}$.

3.16 $$\hat{\beta}_k^2 = r^2_{yk\cdot12\dots k-1} \cdot \frac{\mathbf{y}'\mathbf{M}_1\mathbf{y}}{\mathbf{x}_k'\mathbf{M}_1\mathbf{x}_k'} = r^2_{yk\cdot12\dots k-1} \cdot t^{kk} \cdot \operatorname{SSE}(\mathbf{y}; \mathbf{X}_1)$$

3.17 Multiplying $\mathbf{x}_k$ by a means that $\hat{\beta}_k$ will be divided by a.

Multiplying $\mathbf{y}$ by b means that $\hat{\beta}_k$ will be multiplied by b.

3.18 $$\hat{\boldsymbol{\alpha}} = \mathbf{R}_{\mathbf{xx}}^{-1}\mathbf{r}_{\mathbf{x}y}, \quad \hat{\alpha}_i = \hat{\beta}_i \frac{s_i}{s_y}, \quad \hat{\alpha}_0 = 0,$$ standardized regr. coefficients (all variables centered & equal variance)

3.19 $$k = 2\colon\ \hat{\alpha}_1 = \frac{r_{1y} - r_{12}r_{2y}}{1 - r_{12}^2}, \quad \hat{\alpha}_2 = \frac{r_{2y} - r_{12}r_{1y}}{1 - r_{12}^2}, \quad \hat{\alpha}_1 = r_{1y\cdot2}\frac{(1 - r_{2y}^2)^{1/2}}{(1 - r_{12}^2)^{1/2}}$$

3.20 Let $\mathbf{X} = (\mathbf{1} : \mathbf{X}_1 : \mathbf{X}_2) := (\mathbf{Z} : \mathbf{X}_2)$, where $\mathbf{X}_2 \in \mathbb{R}^{n \times p_2}$. Then

$$\begin{aligned}
\operatorname{cov}(\hat{\boldsymbol{\beta}}) &= \sigma^2(\mathbf{X}'\mathbf{X})^{-1} \\
&= \sigma^2 \begin{pmatrix} n & \mathbf{1}'\mathbf{X}_0 \\ \mathbf{X}_0'\mathbf{1} & \mathbf{X}_0'\mathbf{X}_0 \end{pmatrix}^{-1} = \sigma^2 \begin{pmatrix} 1/n + \bar{\mathbf{x}}'\mathbf{T}_{\mathbf{xx}}^{-1}\bar{\mathbf{x}} & -\bar{\mathbf{x}}'\mathbf{T}_{\mathbf{xx}}^{-1} \\ -\mathbf{T}_{\mathbf{xx}}^{-1}\bar{\mathbf{x}} & \mathbf{T}_{\mathbf{xx}}^{-1} \end{pmatrix} \\
&= \begin{pmatrix} \operatorname{var}(\hat{\beta}_0) & \operatorname{cov}(\hat{\beta}_0, \hat{\boldsymbol{\beta}}_{\mathbf{x}}) \\ \operatorname{cov}(\hat{\boldsymbol{\beta}}_{\mathbf{x}}, \hat{\beta}_0) & \operatorname{cov}(\hat{\boldsymbol{\beta}}_{\mathbf{x}}) \end{pmatrix} = \sigma^2 \begin{pmatrix} t^{00} & t^{01} & \dots & t^{0k} \\ t^{10} & t^{11} & \dots & t^{1k} \\ \vdots & \vdots & \ddots & \vdots \\ t^{k0} & t^{k1} & \dots & t^{kk} \end{pmatrix} \\
&= \sigma^2 \begin{pmatrix} \mathbf{Z}'\mathbf{Z} & \mathbf{Z}'\mathbf{X}_2 \\ \mathbf{X}_2'\mathbf{Z} & \mathbf{X}_2'\mathbf{X}_2 \end{pmatrix}^{-1} = \sigma^2 \begin{pmatrix} \mathbf{T}^{zz} & \mathbf{T}^{z2} \\ \mathbf{T}^{2z} & \mathbf{T}^{22} \end{pmatrix} \\
&= \sigma^2 \begin{pmatrix} [\mathbf{Z}'(\mathbf{I} - \mathbf{P}_{\mathbf{X}_2})\mathbf{Z}]^{-1} & \cdot \\ \cdot & [\mathbf{X}_2'(\mathbf{I} - \mathbf{P}_{\mathbf{Z}})\mathbf{X}_2]^{-1} \end{pmatrix}, \quad \text{where}
\end{aligned}$$

(a) $$\begin{aligned}
\mathbf{T}^{22} &= \mathbf{T}_{22\cdot1}^{-1} = [\mathbf{X}_2'(\mathbf{I} - \mathbf{P}_{(\mathbf{1}:\mathbf{X}_1)})\mathbf{X}_2]^{-1} \\
&= (\mathbf{X}_2'\mathbf{M}_1\mathbf{X}_2)^{-1} = [\mathbf{X}_2'(\mathbf{C} - \mathbf{P}_{\mathbf{C}\mathbf{X}_1})\mathbf{X}_2]^{-1} \qquad \mathbf{M}_1 = \mathbf{I} - \mathbf{P}_{(\mathbf{1}:\mathbf{X}_1)} \\
&= [\mathbf{X}_2'\mathbf{C}\mathbf{X}_2 - \mathbf{X}_2'\mathbf{C}\mathbf{X}_1(\mathbf{X}_1'\mathbf{C}\mathbf{X}_1)^{-1}\mathbf{X}_1'\mathbf{C}\mathbf{X}_2]^{-1} \\
&= (\mathbf{T}_{22} - \mathbf{T}_{21}\mathbf{T}_{11}^{-1}\mathbf{T}_{12})^{-1},
\end{aligned}$$

(b) $\mathbf{T_{xx}} = (\mathbf{X}_1 : \mathbf{X}_2)'\mathbf{C}(\mathbf{X}_1 : \mathbf{X}_2) = \begin{pmatrix} \mathbf{T}_{11} & \mathbf{T}_{12} \\ \mathbf{T}_{21} & \mathbf{T}_{22} \end{pmatrix} \in \mathbb{R}^{k\times k}$,

$\mathbf{T_{xx}^{-1}} = \begin{pmatrix} \cdot & \cdot \\ \cdot & \mathbf{T}^{22} \end{pmatrix}$,

(c) $t^{kk} = [\mathbf{x}_k'(\mathbf{I} - \mathbf{P}_{\mathbf{X}_1})\mathbf{x}_k]^{-1}$ $\quad \mathbf{X} = (\mathbf{X}_1 : \mathbf{x}_k),\ \mathbf{X}_1 = (\mathbf{1} : \mathbf{x}_1 : \ldots : \mathbf{x}_{k-1})$

$= 1/\mathrm{SSE}(k) = 1/\mathrm{SSE}(x_k \text{ explained by all other } x\text{'s})$

$= 1/\mathrm{SSE}(\mathbf{x}_k; \mathbf{X}_1) = 1/t_{kk\cdot\mathbf{X}_1}$ $\quad$ corresp. result holds for all t^{ii}

(d) $t^{kk} = (\mathbf{x}_k'\mathbf{C}\mathbf{x}_k)^{-1} = 1/t_{kk}$ iff $\mathbf{X}_1'\mathbf{C}\mathbf{x}_k = \mathbf{0}$ iff $r_{1k} = r_{2k} = \cdots = r_{k-1,k} = 0$,

(e) $t^{00} = \frac{1}{n} + \bar{\mathbf{x}}'\mathbf{T_{xx}^{-1}}\bar{\mathbf{x}} = (n - \mathbf{1}'\mathbf{P}_{\mathbf{X}_0}\mathbf{1})^{-1} = \|(\mathbf{I} - \mathbf{P}_{\mathbf{X}_0})\mathbf{1}\|^{-1}$.

3.21 Under $\{\mathbf{y}, (\mathbf{1} : \mathbf{x})\boldsymbol{\beta}, \sigma^2\mathbf{I}\}$:

$$\mathrm{cov}(\hat{\boldsymbol{\beta}}) = \sigma^2 \begin{pmatrix} 1/n + \bar{x}^2/\mathrm{SS}_x & -\bar{x}/\mathrm{SS}_x \\ -\bar{x}/\mathrm{SS}_x & 1/\mathrm{SS}_x \end{pmatrix} = \frac{\sigma^2}{\mathrm{SS}_x} \begin{pmatrix} \sum x_i^2/n & -\bar{x} \\ -\bar{x} & 1 \end{pmatrix},$$

$$\mathrm{var}(\hat{\beta}_1) = \frac{\sigma^2}{\mathrm{SS}_x}, \quad \mathrm{var}(\hat{\beta}_0) = \sigma^2 \frac{\mathbf{x}'\mathbf{x}}{n\mathrm{SS}_x}, \quad \mathrm{cor}(\hat{\beta}_0, \hat{\beta}_1) = \frac{-\bar{x}}{\sqrt{\mathbf{x}'\mathbf{x}/n}}.$$

3.22 $\mathrm{cov}(\hat{\boldsymbol{\beta}}_2) = \sigma^2(\mathbf{X}_2'\mathbf{M}_1\mathbf{X}_2)^{-1}, \quad \hat{\boldsymbol{\beta}}_2 \in \mathbb{R}^{p_2}, \quad \mathbf{X} = (\mathbf{X}_1 : \mathbf{X}_2), \quad \mathbf{X}_2$ is $n \times p_2$

3.23 $\mathrm{cov}(\hat{\boldsymbol{\beta}}_1 \mid \mathscr{M}_{12}) = \sigma^2(\mathbf{X}_1'\mathbf{M}_2\mathbf{X}_1)^{-1} \geq_{\mathsf{L}} \sigma^2(\mathbf{X}_1'\mathbf{X}_1)^{-1} = \mathrm{cov}(\hat{\boldsymbol{\beta}}_1 \mid \mathscr{M}_1)$: adding new regressors cannot decrease the variances of old regression coefficients.

3.24 $R_i^2 = R^2(x_i \text{ explained by all other } x\text{'s})$

$= R^2(\mathbf{x}_i; \mathbf{X}_{(-i)})$ $\quad \mathbf{X}_{(-i)} = (\mathbf{1} : \mathbf{x}_1, \ldots, \mathbf{x}_{i-1}, \mathbf{x}_{i+1}, \ldots, \mathbf{x}_k)$

$= \dfrac{\mathrm{SSR}(i)}{\mathrm{SST}(i)} = 1 - \dfrac{\mathrm{SSE}(i)}{\mathrm{SST}(i)}$ $\quad \mathrm{SSE}(i) = \mathrm{SSE}(\mathbf{x}_i; \mathbf{X}_{(-i)}),\ \mathrm{SST}(i) = t_{ii}$

3.25 $\mathrm{VIF}_i = \dfrac{1}{1 - R_i^2} = r^{ii}, \quad i = 1, \ldots, k,$ $\quad \mathrm{VIF}_i$ = variance inflation factor

$= \dfrac{\mathrm{SST}(i)}{\mathrm{SSE}(i)} = \dfrac{t_{ii}}{\mathrm{SSE}(i)} = t_{ii}t^{ii}, \quad \mathbf{R_{xx}^{-1}} = \{r^{ij}\},\ \mathbf{T_{xx}^{-1}} = \{t^{ij}\}$

3.26 $\mathrm{VIF}_i \geq 1$ and $\mathrm{VIF}_i = 1$ iff $\mathrm{cor_d}(\mathbf{x}_i, \mathbf{X}_{(-i)}) = \mathbf{0}$

3.27 $\mathrm{var}(\hat{\beta}_i) = \sigma^2 t^{ii}$

$$= \frac{\sigma^2}{\mathrm{SSE}(i)} = \sigma^2 \frac{\mathrm{VIF}_i}{t_{ii}} = \sigma^2 \frac{r^{ii}}{t_{ii}} = \frac{\sigma^2}{(1 - R_i^2)t_{ii}}, \quad i = 1, \ldots, k$$

3.28 $k = 2\colon\ \operatorname{var}(\hat{\beta}_i) = \dfrac{\sigma^2}{(1 - r_{12}^2)t_{ii}}, \quad \operatorname{cor}(\hat{\beta}_1, \hat{\beta}_2) = -r_{12}$

3.29 $\operatorname{cor}(\hat{\beta}_1, \hat{\beta}_2) = -r_{12\cdot34\ldots k} = -\operatorname{pcor_d}(\mathbf{x}_1, \mathbf{x}_2 \mid \mathbf{X}_2), \qquad \mathbf{X}_2 = (\mathbf{x}_3 : \ldots : \mathbf{x}_k)$

3.30 $\widehat{\operatorname{cov}}(\hat{\boldsymbol{\beta}}) = \hat{\sigma}^2(\mathbf{X}'\mathbf{X})^{-1}$ estimated covariance matrix of $\hat{\boldsymbol{\beta}}$

3.31 $\widehat{\operatorname{var}}(\hat{\beta}_i) = \hat{\sigma}^2 t^{ii} = \operatorname{se}^2(\hat{\beta}_i)$ estimated variance of $\hat{\beta}_i$

3.32 $\operatorname{se}(\hat{\beta}_i) = \sqrt{\widehat{\operatorname{var}}(\hat{\beta}_i)} = \hat{\sigma}\sqrt{t^{ii}}$ estimated stdev of $\hat{\beta}_i$, standard error of $\hat{\beta}_i$

3.33 $\hat{\beta}_i \pm t_{\alpha/2;n-k-1} \operatorname{se}(\hat{\beta}_i)$ $(1-\alpha)100\%$ confidence interval for β_i

3.34 Best linear unbiased prediction, BLUP, of y_* under

$$\mathscr{M}_* = \left\{ \begin{pmatrix} \mathbf{y} \\ y_* \end{pmatrix}, \begin{pmatrix} \mathbf{X} \\ \mathbf{x}_\#' \end{pmatrix} \boldsymbol{\beta},\ \sigma^2 \begin{pmatrix} \mathbf{I}_n & \mathbf{0} \\ \mathbf{0} & 1 \end{pmatrix} \right\};$$

a linear model with new future observation; see Section 13 (p. 65). Suppose that $\mathbf{X} = (\mathbf{1} : \mathbf{X}_0)$ and denote $\mathbf{x}_\#' = (1, \mathbf{x}_*') = (1, x_1^*, \ldots, x_k^*)$. Then

(a) $y_* = \mathbf{x}_\#'\boldsymbol{\beta} + \varepsilon_*$ new unobserved value y_* with a given $(1, \mathbf{x}_*')$ under $\mathscr{M}_*$

(b) $\hat{y}_* = \mathbf{x}_\#'\hat{\boldsymbol{\beta}} = \hat{\beta}_0 + \mathbf{x}_*'\hat{\boldsymbol{\beta}}_{\mathbf{x}} = \hat{\beta}_0 + \hat{\beta}_1 x_1^* + \cdots + \hat{\beta}_k x_k^*$
$= (\bar{y} - \hat{\boldsymbol{\beta}}_{\mathbf{x}}'\bar{\mathbf{x}}) + \hat{\boldsymbol{\beta}}_{\mathbf{x}}'\mathbf{x}_* = \bar{y} + \hat{\boldsymbol{\beta}}_{\mathbf{x}}'(\mathbf{x}_* - \bar{\mathbf{x}})$ $\hat{y}_* = \operatorname{BLUP}(y_*)$

(c) $e_* = y_* - \hat{y}_*$ prediction error with a given $\mathbf{x}_*$

(d) $\operatorname{var}(\hat{y}_*) = \operatorname{var}(\mathbf{x}_\#'\hat{\boldsymbol{\beta}}) = \sigma^2 \mathbf{x}_\#'(\mathbf{X}'\mathbf{X})^{-1}\mathbf{x}_\# := \sigma^2 h_\#$
$= \operatorname{var}(\bar{y}) + \operatorname{var}\big[\hat{\boldsymbol{\beta}}_{\mathbf{x}}'(\mathbf{x}_* - \bar{\mathbf{x}})\big]$ Note: $\operatorname{cov}(\hat{\boldsymbol{\beta}}_{\mathbf{x}}, \bar{y}) = \mathbf{0}$
$= \sigma^2\big[\frac{1}{n} + (\mathbf{x}_* - \bar{\mathbf{x}})'\mathbf{T}_{\mathbf{xx}}^{-1}(\mathbf{x}_* - \bar{\mathbf{x}})\big]$
$= \sigma^2\big[\frac{1}{n} + \frac{1}{n-1}(\mathbf{x}_* - \bar{\mathbf{x}})'\mathbf{S}_{\mathbf{xx}}^{-1}(\mathbf{x}_* - \bar{\mathbf{x}})\big]$
$= \sigma^2\big[\frac{1}{n} + \frac{1}{n-1}\operatorname{MHLN}^2(\mathbf{x}_*, \bar{\mathbf{x}}, \mathbf{S}_{\mathbf{xx}})\big]$

(e) $\operatorname{var}(\hat{y}_*) = \sigma^2\left[\dfrac{1}{n} + \dfrac{(x_* - \bar{x})^2}{\mathrm{SS}_x}\right]$, when $k = 1$; $\hat{y}_* = \hat{\beta}_0 + \hat{\beta}_1 x_*$

(f) $\operatorname{var}(e_*) = \operatorname{var}(y_* - \hat{y}_*)$ variance of the prediction error
$= \operatorname{var}(y_*) + \operatorname{var}(\hat{y}_*) = \sigma^2 + \sigma^2 h_\# = \sigma^2[1 + \mathbf{x}_\#'(\mathbf{X}'\mathbf{X})^{-1}\mathbf{x}_\#]$

(g) $\operatorname{var}(e_*) = \sigma^2\left[1 + \dfrac{1}{n} + \dfrac{(x_* - \bar{x})^2}{\mathrm{SS}_x}\right]$, when $k = 1$; $\hat{y}_* = \hat{\beta}_0 + \hat{\beta}_1 x_*$

(h) $\operatorname{se}^2(\hat{y}_*) = \widehat{\operatorname{var}}(\hat{y}_*) = \operatorname{se}^2(\mathbf{x}_\#'\hat{\boldsymbol{\beta}}) = \hat{\sigma}^2 h_\#$ estimated variance of $\hat{y}_*$

(i) $\operatorname{se}^2(e_*) = \widehat{\operatorname{var}}(e_*)$
$= \widehat{\operatorname{var}}(y_* - \hat{y}_*) = \hat{\sigma}^2(1 + h_\#)$ estimated variance of e_*

(j) $\hat{y}_* \pm t_{\alpha/2;n-k-1}\sqrt{\widehat{\operatorname{var}}(\hat{y}_*)} = \mathbf{x}'_\#\hat{\boldsymbol{\beta}} \pm t_{\alpha/2;n-k-1}\operatorname{se}(\mathbf{x}'_\#\hat{\boldsymbol{\beta}})$
$= \hat{y}_* \pm t_{\alpha/2;n-k-1}\hat{\sigma}\sqrt{h_\#}$ confidence interval for $\operatorname{E}(y_*)$

(k) $\hat{y}_* \pm t_{\alpha/2;n-k-1}\sqrt{\widehat{\operatorname{var}}(y_* - \hat{y}_*)} = \hat{y}_* \pm t_{\alpha/2;n-k-1}\hat{\sigma}\sqrt{1 + h_\#}$
prediction interval for the new unobserved y_*

(l) $\hat{y}_* \pm \sqrt{(k+1)F_{\alpha,k+1,n-k-1}}\,\hat{\sigma}\sqrt{h_\#}$
Working–Hotelling confidence band for $\operatorname{E}(y_*)$

4 Decompositions of sums of squares

Unless otherwise stated we assume that $\mathbf{1} \in \mathscr{C}(\mathbf{X})$ holds throughout this section.

4.1 $\operatorname{SST} = \|\mathbf{y} - \bar{\bar{\mathbf{y}}}\|^2 = \|(\mathbf{I} - \mathbf{J})\mathbf{y}\|^2 = \mathbf{y}'(\mathbf{I} - \mathbf{J})\mathbf{y} = \mathbf{y}'\mathbf{y} - n\bar{y}^2 = t_{yy}$ total SS

4.2 $\operatorname{SSR} = \|\hat{\mathbf{y}} - \bar{\bar{\mathbf{y}}}\|^2 = \|(\mathbf{H} - \mathbf{J})\mathbf{y}\|^2 = \|(\mathbf{I} - \mathbf{J})\mathbf{H}\mathbf{y}\|^2 = \mathbf{y}'(\mathbf{H} - \mathbf{J})\mathbf{y}$
$= \mathbf{y}'\mathbf{P}_{\mathbf{C}\mathbf{X}_0}\mathbf{y} = \mathbf{y}'\mathbf{P}_{\tilde{\mathbf{X}}_0}\mathbf{y} = \mathbf{t}'_{\mathbf{x}y}\mathbf{T}_{\mathbf{xx}}^{-1}\mathbf{t}_{\mathbf{x}y}$ SS due to regression; $\mathbf{1} \in \mathscr{C}(\mathbf{X})$

4.3 $\operatorname{SSE} = \|\mathbf{y} - \hat{\mathbf{y}}\|^2 = \|(\mathbf{I} - \mathbf{H})\mathbf{y}\|^2 = \mathbf{y}'(\mathbf{I} - \mathbf{H})\mathbf{y} = \mathbf{y}'\mathbf{M}\mathbf{y} = \mathbf{y}'(\mathbf{C} - \mathbf{P}_{\mathbf{C}\mathbf{X}_0})\mathbf{y}$
$= \mathbf{y}'\mathbf{y} - \mathbf{y}'\mathbf{X}\hat{\boldsymbol{\beta}} = t_{yy} - \mathbf{t}'_{\mathbf{x}y}\mathbf{T}_{\mathbf{xx}}^{-1}\mathbf{t}_{\mathbf{x}y}$ residual sum of squares

4.4 $\operatorname{SST} = \operatorname{SSR} + \operatorname{SSE}$

4.5 (a) $\operatorname{df}(\operatorname{SST}) = \operatorname{r}(\mathbf{I} - \mathbf{J}) = n - 1,\quad s_y^2 = \operatorname{SST}/(n-1),$

(b) $\operatorname{df}(\operatorname{SSR}) = \operatorname{r}(\mathbf{H} - \mathbf{J}) = \operatorname{r}(\mathbf{X}) - 1,\quad \operatorname{MSR} = \operatorname{SSR}/[\operatorname{r}(\mathbf{X}) - 1],$

(c) $\operatorname{df}(\operatorname{SSE}) = \operatorname{r}(\mathbf{I} - \mathbf{H}) = n - \operatorname{r}(\mathbf{X}),\quad \operatorname{MSE} = \operatorname{SSE}/[n - \operatorname{r}(\mathbf{X})] = \hat{\sigma}^2.$

4.6 $\operatorname{SST} = \sum_{i=1}^{n}(y_i - \bar{y})^2,\quad \operatorname{SSR} = \sum_{i=1}^{n}(\hat{y}_i - \bar{y})^2,\quad \operatorname{SSE} = \sum_{i=1}^{n}(y_i - \hat{y}_i)^2$

4.7 $\operatorname{SSE} = \operatorname{SST}\left(1 - \dfrac{\operatorname{SSR}}{\operatorname{SST}}\right) = \operatorname{SST}(1 - R^2) = \operatorname{SST}(1 - R^2_{y\cdot\mathbf{x}})$

4.8 $\operatorname{MSE} = \hat{\sigma}^2 \approx s_y^2(1 - R^2_{y\cdot\mathbf{x}})$ which corresponds to $\sigma^2_{y\cdot\mathbf{x}} = \sigma_y^2(1 - \varrho^2_{y\cdot\mathbf{x}})$

4.9 $\operatorname{MSE} = \hat{\sigma}^2 = \operatorname{SSE}/[n - \operatorname{r}(\mathbf{X})]$ unbiased estimate of σ^2, residual mean square
$= \operatorname{SSE}/(n - k - 1)$, when $\operatorname{r}(\mathbf{X}) = k + 1$

4.10 We always have $(\mathbf{I}-\mathbf{J})\mathbf{y} = (\mathbf{H}-\mathbf{J})\mathbf{y} + (\mathbf{I}-\mathbf{H})\mathbf{y}$ and similarly always $\mathbf{y}'(\mathbf{I}-\mathbf{J})\mathbf{y} = \mathbf{y}'(\mathbf{H}-\mathbf{J})\mathbf{y} + \mathbf{y}'(\mathbf{I}-\mathbf{H})\mathbf{y}$, but the decomposition 4.4,

(a) $\|(\mathbf{I}-\mathbf{J})\mathbf{y}\|^2 = \|(\mathbf{H}-\mathbf{J})\mathbf{y}\|^2 + \|(\mathbf{I}-\mathbf{H})\mathbf{y}\|^2$,

is valid iff $(\hat{\mathbf{y}}-\bar{\mathbf{y}})'(\mathbf{y}-\hat{\mathbf{y}}) = [(\mathbf{H}-\mathbf{J})\mathbf{y}]'(\mathbf{I}-\mathbf{H})\mathbf{y} = \mathbf{y}'\mathbf{J}(\mathbf{H}-\mathbf{I})\mathbf{y} = 0$, which holds for all $\mathbf{y}$ iff $\mathbf{JH} = \mathbf{J}$ which is equivalent to $\mathbf{1} \in \mathscr{C}(\mathbf{X})$, i.e., to $\mathbf{H1} = \mathbf{1}$. Decomposition (a) holds also if $\mathbf{y}$ is centered or $\mathbf{y} \in \mathscr{C}(\mathbf{X})$.

4.11 $\mathbf{1} \in \mathscr{C}(\mathbf{X}) \iff \mathbf{H}-\mathbf{J}$ is orthogonal projector $\iff \mathbf{JH} = \mathbf{HJ} = \mathbf{J}$ in which situation $\mathbf{J}\hat{\mathbf{y}} = \mathbf{JHy} = \mathbf{Jy} = (\bar{y}, \bar{y}, \dots, \bar{y})'$.

4.12 In the intercept model we usually have $\mathbf{X} = (\mathbf{1} : \mathbf{X}_0)$. If $\mathbf{X}$ does not explicitly have $\mathbf{1}$ as a column, but $\mathbf{1} \in \mathscr{C}(\mathbf{X})$, then $\mathscr{C}(\mathbf{X}) = \mathscr{C}(\mathbf{1} : \mathbf{X})$ and we have $\mathbf{H} = \mathbf{P_1} + \mathbf{P_{CX}} = \mathbf{J} + \mathbf{P_{CX}}$, and $\mathbf{H}-\mathbf{J}$ is indeed an orthogonal projector.

4.13 $$\mathrm{SSE} = \min_{\boldsymbol\beta} \|\mathbf{y}-\mathbf{X}\boldsymbol\beta\|^2 = \mathrm{SSE}(\mathbf{y};\mathbf{X}) = \|\mathrm{res}(\mathbf{y};\mathbf{X})\|^2$$

4.14 $$\mathrm{SST} = \min_{\beta} \|\mathbf{y}-\mathbf{1}\beta\|^2 = \mathrm{SSE}(\mathbf{y};\mathbf{1}) = \|\mathrm{res}(\mathbf{y};\mathbf{1})\|^2$$ $\mathbf{y}$ explained only by $\mathbf{1}$

4.15 $$\mathrm{SSR} = \min_{\beta} \|\mathbf{Hy}-\mathbf{1}\beta\|^2 = \mathrm{SSE}(\mathbf{Hy};\mathbf{1}) = \|\mathrm{res}(\mathbf{Hy};\mathbf{1})\|^2$$
$$= \mathrm{SSE}(\mathbf{y};\mathbf{1}) - \mathrm{SSE}(\mathbf{y};\mathbf{X})$$ change in SSE gained adding "real" predictors
$$= \Delta\mathrm{SSE}(\mathbf{X}_0 \mid \mathbf{1})$$ when $\mathbf{1}$ is already in the model

4.16 $$R^2 = R^2_{y\cdot\mathbf{x}} = \frac{\mathrm{SSR}}{\mathrm{SST}} = 1 - \frac{\mathrm{SSE}}{\mathrm{SST}}$$ multiple correlation coefficient squared, coefficient of determination
$$= \frac{\mathrm{SSE}(\mathbf{y};\mathbf{1}) - \mathrm{SSE}(\mathbf{y};\mathbf{X})}{\mathrm{SSE}(\mathbf{y};\mathbf{1})}$$ fraction of $\mathrm{SSE}(\mathbf{y};\mathbf{1}) = \mathrm{SST}$ accounted for by adding predictors $\mathbf{x}_1, \dots, \mathbf{x}_k$
$$= \frac{\mathbf{t}'_{\mathbf{x}y}\mathbf{T}^{-1}_{\mathbf{xx}}\mathbf{t}_{\mathbf{x}y}}{t_{yy}} = \frac{\mathbf{s}'_{\mathbf{x}y}\mathbf{S}^{-1}_{\mathbf{xx}}\mathbf{s}_{\mathbf{x}y}}{s^2_y}$$
$$= \mathbf{r}'_{\mathbf{x}y}\mathbf{R}^{-1}_{\mathbf{xx}}\mathbf{r}_{\mathbf{x}y} = \hat{\boldsymbol\alpha}'\mathbf{r}_{\mathbf{x}y} = \hat\alpha_1 r_{1y} + \dots + \hat\alpha_k r_{ky}$$

4.17 $$R^2 = \max_{\boldsymbol\beta} \mathrm{cor}^2_{\mathrm{d}}(\mathbf{y};\mathbf{X}\boldsymbol\beta) = \mathrm{cor}^2_{\mathrm{d}}(\mathbf{y};\mathbf{X}\hat{\boldsymbol\beta}) = \mathrm{cor}^2_{\mathrm{d}}(\mathbf{y};\hat{\mathbf{y}}) = \cos^2[\mathbf{Cy}, (\mathbf{H}-\mathbf{J})\mathbf{y}]$$

4.18 (a) $$\mathrm{SSE} = \mathbf{y}'[\mathbf{I} - (\mathbf{P}_{\mathbf{X}_1} + \mathbf{P}_{\mathbf{M}_1\mathbf{X}_2})]\mathbf{y}$$ $\mathbf{X} = (\mathbf{X}_1 : \mathbf{X}_2),\ \mathbf{M}_1 = \mathbf{I} - \mathbf{P}_{\mathbf{X}_1}$
$$= \mathbf{y}'\mathbf{M}_1\mathbf{y} - \mathbf{y}'\mathbf{P}_{\mathbf{M}_1\mathbf{X}_2}\mathbf{y} = \mathrm{SSE}(\mathbf{M}_1\mathbf{y};\mathbf{M}_1\mathbf{X}_2)$$
$$= \mathrm{SSE}(\mathbf{y};\mathbf{X}_1) - \mathrm{SSR}(\mathbf{e}_{\mathbf{y}\cdot\mathbf{X}_1};\mathbf{E}_{\mathbf{X}_2\cdot\mathbf{X}_1}),$$ where

(b) $\mathbf{e}_{\mathbf{y}\cdot\mathbf{X}_1} = \mathrm{res}(\mathbf{y};\mathbf{X}_1) = \mathbf{M}_1\mathbf{y} =$ residual of $\mathbf{y}$ after elimination of $\mathbf{X}_1$,

(c) $\mathbf{E}_{\mathbf{X}_2\cdot\mathbf{X}_1} = \mathrm{res}(\mathbf{X}_2;\mathbf{X}_1) = \mathbf{M}_1\mathbf{X}_2$.

4.19 $\mathbf{y}'\mathbf{P}_{\mathbf{M}_1\mathbf{X}_2}\mathbf{y} = \text{SSE}(\mathbf{y};\mathbf{X}_1) - \text{SSE}(\mathbf{y};\mathbf{X}_1,\mathbf{X}_2) = \Delta\text{SSE}(\mathbf{X}_2 \mid \mathbf{X}_1)$
$= \text{reduction in SSE when adding } \mathbf{X}_2 \text{ to the model}$

4.20 Denoting $\mathscr{M}_{12} = \{\mathbf{y}, \mathbf{X}\boldsymbol{\beta}, \sigma^2\mathbf{I}\}$, $\mathscr{M}_1 = \{\mathbf{y}, \mathbf{X}_1\boldsymbol{\beta}_1, \sigma^2\mathbf{I}\}$, and $\mathscr{M}_{12\cdot 1} = \{\mathbf{M}_1\mathbf{y}, \mathbf{M}_1\mathbf{X}_2\boldsymbol{\beta}_2, \sigma^2\mathbf{M}_1\}$, the following holds:

(a) $\text{SSE}(\mathscr{M}_{12\cdot 1}) = \text{SSE}(\mathscr{M}_{12}) = \mathbf{y}'\mathbf{M}\mathbf{y}$,

(b) $\text{SST}(\mathscr{M}_{12\cdot 1}) = \mathbf{y}'\mathbf{M}_1\mathbf{y} = \text{SSE}(\mathscr{M}_1)$,

(c) $\text{SSR}(\mathscr{M}_{12\cdot 1}) = \mathbf{y}'\mathbf{M}_1\mathbf{y} - \mathbf{y}'\mathbf{M}\mathbf{y} = \mathbf{y}'\mathbf{P}_{\mathbf{M}_1\mathbf{X}_2}\mathbf{y}$,

(d) $R^2(\mathscr{M}_{12\cdot 1}) = \dfrac{\text{SSR}(\mathscr{M}_{12\cdot 1})}{\text{SST}(\mathscr{M}_{12\cdot 1})} = \dfrac{\mathbf{y}'\mathbf{P}_{\mathbf{M}_1\mathbf{X}_2}\mathbf{y}}{\mathbf{y}'\mathbf{M}_1\mathbf{y}} = 1 - \dfrac{\mathbf{y}'\mathbf{M}\mathbf{y}}{\mathbf{y}'\mathbf{M}_1\mathbf{y}}$,

(e) $\mathbf{X}_2 = \mathbf{x}_k$: $R^2(\mathscr{M}_{12\cdot 1}) = r^2_{yk\cdot 12\dots k-1}$ and $1 - r^2_{yk\cdot 12\dots k-1} = \dfrac{\mathbf{y}'\mathbf{M}\mathbf{y}}{\mathbf{y}'\mathbf{M}_1\mathbf{y}}$,

(f) $1 - R^2(\mathscr{M}_{12}) = [1 - R^2(\mathscr{M}_1)][1 - R^2(\mathscr{M}_{12\cdot 1})]$
$= [1 - R^2(\mathbf{y};\mathbf{X}_1)][1 - R^2(\mathbf{M}_1\mathbf{y};\mathbf{M}_1\mathbf{X}_2)]$,

(g) $1 - R^2_{y\cdot 12\dots k} = (1 - r^2_{y1})(1 - r^2_{y2\cdot 1})(1 - r^2_{y3\cdot 12})\cdots(1 - r^2_{yk\cdot 12\dots k-1})$.

4.21 If the model does not have the intercept term [or $\mathbf{1} \notin \mathscr{C}(\mathbf{X})$], then the decomposition 4.4 is not valid. In this situation, we consider the decomposition

$$\mathbf{y}'\mathbf{y} = \mathbf{y}'\mathbf{H}\mathbf{y} + \mathbf{y}'(\mathbf{I} - \mathbf{H})\mathbf{y}, \quad \text{SST}_c = \text{SSR}_c + \text{SSE}_c.$$

In the no-intercept model, the coefficient of determination is defined as

$$R_c^2 = \frac{\text{SSR}_c}{\text{SST}_c} = \frac{\mathbf{y}'\mathbf{H}\mathbf{y}}{\mathbf{y}'\mathbf{y}} = 1 - \frac{\text{SSE}}{\mathbf{y}'\mathbf{y}}.$$

In the no-intercept model we may have $R_c^2 = \cos^2(\mathbf{y}, \hat{\mathbf{y}}) \neq \text{cor}_\text{d}^2(\mathbf{y}, \hat{\mathbf{y}})$. However, if both $\mathbf{X}$ and $\mathbf{y}$ are centered (actually meaning that the intercept term is present but not explicitly), then we can use the usual definitions of R^2 and R_i^2. [We can think that $\text{SSR}_c = \text{SSE}(\mathbf{y};\mathbf{0}) - \text{SSE}(\mathbf{y};\mathbf{X})$ = change in SSE gained adding predictors when there are no predictors previously at all.]

4.22 Sample partial correlations. Below we consider the data matrix $(\mathbf{X} : \mathbf{Y}) = (\mathbf{x}_1 : \dots : \mathbf{x}_p : \mathbf{y}_1 : \dots : \mathbf{y}_q)$.

(a) $\mathbf{E}_{\mathbf{Y}\cdot\mathbf{X}} = \text{res}(\mathbf{Y};\mathbf{X}) = (\mathbf{I} - \mathbf{P}_{(\mathbf{1}:\mathbf{X})})\mathbf{Y} = \mathbf{M}\mathbf{Y} = (\mathbf{e}_{\mathbf{y}_1\cdot\mathbf{X}} : \dots : \mathbf{e}_{\mathbf{y}_q\cdot\mathbf{X}})$,

(b) $\mathbf{e}_{\mathbf{y}_i\cdot\mathbf{X}} = (\mathbf{I} - \mathbf{P}_{(\mathbf{1}:\mathbf{X})})\mathbf{y}_i = \mathbf{M}\mathbf{y}_i$,

(c) $\text{pcor}_\text{d}(\mathbf{Y} \mid \mathbf{X}) = \text{cor}_\text{d}[\text{res}(\mathbf{Y};\mathbf{X})] = \text{cor}_\text{d}(\mathbf{E}_{\mathbf{Y}\cdot\mathbf{X}})$
$=$ partial correlations of variables of $\mathbf{Y}$ after elimination of $\mathbf{X}$,

(d) $\text{pcor}_\text{d}(\mathbf{y}_1, \mathbf{y}_2 \mid \mathbf{X}) = \text{cor}_\text{d}(\mathbf{e}_{\mathbf{y}_1\cdot\mathbf{X}}, \mathbf{e}_{\mathbf{y}_2\cdot\mathbf{X}})$,

(e) $\mathbf{T}_{\mathbf{yy}\cdot\mathbf{x}} = \mathbf{E}'_{\mathbf{Y}\cdot\mathbf{X}}\mathbf{E}_{\mathbf{Y}\cdot\mathbf{X}} = \mathbf{Y}'(\mathbf{I} - \mathbf{P}_{(\mathbf{1}:\mathbf{X})})\mathbf{Y} = \mathbf{Y}'\mathbf{M}\mathbf{Y} = \{t_{ij\cdot\mathbf{x}}\} \in \text{PD}_q$,

(f) $t_{ii\cdot\mathbf{x}} = \mathbf{y}_i'(\mathbf{I} - \mathbf{P}_{(\mathbf{1}:\mathbf{X})})\mathbf{y}_i = \mathbf{y}_i'\mathbf{M}\mathbf{y}_i = \mathrm{SSE}(\mathbf{y}_i; \mathbf{1}, \mathbf{X})$.

4.23 Denote $\mathbf{T} = \begin{pmatrix} \mathbf{T}_{\mathbf{xx}} & \mathbf{T}_{\mathbf{xy}} \\ \mathbf{T}_{\mathbf{yx}} & \mathbf{T}_{\mathbf{yy}} \end{pmatrix} = \begin{pmatrix} \mathbf{X}'\mathbf{C}\mathbf{Y} & \mathbf{X}'\mathbf{C}\mathbf{X} \\ \mathbf{Y}'\mathbf{C}\mathbf{X} & \mathbf{Y}'\mathbf{C}\mathbf{Y} \end{pmatrix}$, $\mathbf{C} = \mathbf{I} - \mathbf{J}$, $\mathbf{M} = \mathbf{I} - \mathbf{P}_{(\mathbf{1}:\mathbf{X})}$. Then

(a) $\mathbf{T}^{-1} = \begin{pmatrix} \mathbf{T}^{-1}_{\mathbf{xx}\cdot\mathbf{y}} & \cdot \\ \cdot & \mathbf{T}^{-1}_{\mathbf{yy}\cdot\mathbf{x}} \end{pmatrix}$, $\quad [\mathrm{cor_d}(\mathbf{X} : \mathbf{Y})]^{-1} = \mathbf{R}^{-1} = \begin{pmatrix} \mathbf{R}^{-1}_{\mathbf{xx}\cdot\mathbf{y}} & \cdot \\ \cdot & \mathbf{R}^{-1}_{\mathbf{yy}\cdot\mathbf{x}} \end{pmatrix}$,

(b) $\mathbf{T}_{\mathbf{yy}\cdot\mathbf{x}} = \mathbf{T}_{\mathbf{yy}} - \mathbf{T}_{\mathbf{yx}}\mathbf{T}_{\mathbf{xx}}^{-1}\mathbf{T}_{\mathbf{xy}} = \mathbf{Y}'\mathbf{C}\mathbf{Y} - \mathbf{Y}'\mathbf{C}\mathbf{X}(\mathbf{Y}'\mathbf{C}\mathbf{X})^{-1}\mathbf{Y}'\mathbf{C}\mathbf{X} = \mathbf{Y}'\mathbf{M}\mathbf{Y}$,

(c) $\mathbf{R}_{\mathbf{yy}\cdot\mathbf{x}} = \mathbf{R}_{\mathbf{yy}} - \mathbf{R}_{\mathbf{yx}}\mathbf{R}_{\mathbf{xx}}^{-1}\mathbf{R}_{\mathbf{xy}}$,

(d) $$\begin{aligned} \mathrm{pcor_d}(\mathbf{Y} \mid \mathbf{X}) &= [\mathrm{diag}(\mathbf{T}_{\mathbf{yy}\cdot\mathbf{x}})]^{-1/2}\mathbf{T}_{\mathbf{yy}\cdot\mathbf{x}}[\mathrm{diag}(\mathbf{T}_{\mathbf{yy}\cdot\mathbf{x}})]^{-1/2} \\ &= [\mathrm{diag}(\mathbf{R}_{\mathbf{yy}\cdot\mathbf{x}})]^{-1/2}\mathbf{R}_{\mathbf{yy}\cdot\mathbf{x}}[\mathrm{diag}(\mathbf{R}_{\mathbf{yy}\cdot\mathbf{x}})]^{-1/2}. \end{aligned}$$

4.24 $$\begin{aligned} (\mathbf{Y} - \mathbf{X}\mathbf{B})'\mathbf{C}(\mathbf{Y} - \mathbf{X}\mathbf{B}) &= (\mathbf{C}\mathbf{Y} - \mathbf{C}\mathbf{X}\mathbf{B})'(\mathbf{C}\mathbf{Y} - \mathbf{C}\mathbf{X}\mathbf{B}) \\ &\geq_{\mathsf{L}} (\mathbf{C}\mathbf{Y} - \mathbf{P}_{\mathbf{C}\mathbf{X}}\mathbf{Y})'(\mathbf{C}\mathbf{Y} - \mathbf{P}_{\mathbf{C}\mathbf{X}}\mathbf{Y}) \\ &= \mathbf{Y}'\mathbf{C}(\mathbf{I} - \mathbf{P}_{\mathbf{C}\mathbf{X}})\mathbf{C}\mathbf{Y} = \mathbf{T}_{\mathbf{yy}} - \mathbf{T}_{\mathbf{yx}}\mathbf{T}_{\mathbf{xx}}^{-1}\mathbf{T}_{\mathbf{xy}} \end{aligned}$$

for all $\mathbf{B}$, and hence for all $\mathbf{B}$ we have

$$\mathrm{cov_s}(\mathbf{y} - \mathbf{B}'\mathbf{x}) = \mathrm{cov_d}(\mathbf{Y} - \mathbf{X}\mathbf{B}) \geq_{\mathsf{L}} \mathbf{S}_{\mathbf{yy}} - \mathbf{S}_{\mathbf{yx}}\mathbf{S}_{\mathbf{xx}}^{-1}\mathbf{S}_{\mathbf{xy}},$$

where the equality is attained if $\mathbf{B} = \mathbf{T}_{\mathbf{xx}}^{-1}\mathbf{T}_{\mathbf{xy}} = \mathbf{S}_{\mathbf{xx}}^{-1}\mathbf{S}_{\mathbf{xy}}$; see 6.7 (p. 28).

4.25 $$r_{xy\cdot z} = \frac{r_{xy} - r_{xz}r_{yz}}{\sqrt{(1 - r_{xz}^2)(1 - r_{yz}^2)}}$$ partial correlation

4.26 If $\mathbf{Y} = (\mathbf{x}_1 : \mathbf{x}_2)$, $\mathbf{X} = (\mathbf{x}_3 : \ldots : \mathbf{x}_k)$ and $\mathbf{R}_{\mathbf{yy}\cdot\mathbf{x}} = \{r^{ij}\}$, then

$$\mathrm{cor}(\hat{\beta}_1, \hat{\beta}_2) = -\mathrm{pcor_d}(\mathbf{x}_1, \mathbf{x}_2 \mid \mathbf{x}_3, \ldots, \mathbf{x}_k) = -r_{12\cdot3\ldots k} = \frac{r^{12}}{\sqrt{r^{11}r^{22}}}.$$

4.27 Added variable plot (AVP). Let $\mathbf{X} = (\mathbf{X}_1 : \mathbf{x}_k)$ and denote

$$\begin{aligned} \mathbf{u} &= \mathbf{e}_{\mathbf{y}\cdot\mathbf{X}_1} = \mathbf{M}_1\mathbf{y} = \mathrm{res}(\mathbf{y}; \mathbf{X}_1), \\ \mathbf{v} &= \mathbf{e}_{\mathbf{x}_k\cdot\mathbf{X}_1} = \mathbf{M}_1\mathbf{x}_k = \mathrm{res}(\mathbf{x}_k; \mathbf{X}_1). \end{aligned}$$

The scatterplot of $\mathbf{e}_{\mathbf{x}_k\cdot\mathbf{X}_1}$ versus $\mathbf{e}_{\mathbf{y}\cdot\mathbf{X}_1}$ is an AVP. Moreover, consider the models $\mathscr{M}_{12} = \{\mathbf{y}, (\mathbf{X}_1 : \mathbf{x}_k)\boldsymbol{\beta}, \sigma^2\mathbf{I}\}$, with $\boldsymbol{\beta} = \binom{\boldsymbol{\beta}_1}{\beta_k}$, $\mathscr{M}_1 = \{\mathbf{y}, \mathbf{X}_1\boldsymbol{\beta}_1, \sigma^2\mathbf{I}\}$, and $\mathscr{M}_{12\cdot1} = \{\mathbf{M}_1\mathbf{y}, \mathbf{M}_1\mathbf{x}_k\beta_k, \sigma^2\mathbf{M}_1\} = \{\mathbf{e}_{\mathbf{y}\cdot\mathbf{X}_1}, \mathbf{e}_{\mathbf{x}_k\cdot\mathbf{X}_1}\beta_k, \sigma^2\mathbf{M}_1\}$. Then

(a) $\hat{\beta}_k(\mathscr{M}_{12}) = \hat{\beta}_k(\mathscr{M}_{12\cdot1})$, (Frisch–Waugh–Lovell theorem)

(b) $\mathrm{res}(\mathbf{y}; \mathbf{X}) = \mathrm{res}(\mathbf{M}_1\mathbf{y}; \mathbf{M}_1\mathbf{x}_k)$, $\quad R^2(\mathscr{M}_{12\cdot1}) = r^2_{yk\cdot12\ldots k-1}$,

(c) $1 - R^2(\mathscr{M}_{12}) = [1 - R^2(\mathscr{M}_1)](1 - r^2_{yk\cdot12\ldots k-1})$.

5 Distributions

5.1 Discrete uniform distribution. Let x be a random variable whose values are $1, 2, \dots, N$, each with equal probability $1/N$. Then $\mathrm{E}(x) = \frac{1}{2}(N+1)$, and $\mathrm{var}(x) = \frac{1}{12}(N^2 - 1)$.

5.2 Sum of squares and cubes of integers:

$$\sum_{i=1}^{n} i^2 = \tfrac{1}{6}n(n+1)(2n+1), \quad \sum_{i=1}^{n} i^3 = \tfrac{1}{4}n^2(n+1)^2.$$

5.3 Let $x_1, \dots, x_p$ be a random sample selected without a replacement from $\mathcal{A} = \{1, 2, \dots, N\}$. Denote $y = x_1 + \dots + x_p = \mathbf{1}_p'\mathbf{x}$. Then $\mathrm{var}(x_i) = \frac{N^2-1}{12}$, $\mathrm{cor}(x_i, x_j) = -\frac{1}{N-1} = \varrho$, $i, j = 1, \dots, p$, $\mathrm{cor}^2(x_1, y) = \frac{1}{p} + \big(1 - \frac{1}{p}\big)\varrho$.

5.4 Bernoulli distribution. Let x be a random variable whose values are 0 and 1, with probabilities p and $q = 1 - p$. Then $x \sim \mathrm{Ber}(p)$ and $\mathrm{E}(x) = p$, $\mathrm{var}(x) = pq$. If $y = x_1 + \dots + x_n$, where x_i are independent and each $x_i \sim \mathrm{Ber}(p)$, then y follows the binomial distribution, $y \sim \mathrm{Bin}(n, p)$, and $\mathrm{E}(x) = np$, $\mathrm{var}(x) = npq$.

5.5 Two dichotomous variables. On the basis of the following frequency table:

$$s_x^2 = \frac{1}{n-1}\frac{\gamma\delta}{n} = \frac{n}{n-1} \cdot \frac{\delta}{n}\Big(1 - \frac{\delta}{n}\Big),$$

$$s_{xy} = \frac{1}{n-1}\frac{ad - bc}{n}, \quad r_{xy} = \frac{ad - bc}{\sqrt{\alpha\beta\gamma\delta}},$$

$$\chi^2 = \frac{n(ad - bc)^2}{\alpha\beta\gamma\delta} = n r_{xy}^2.$$

		x: 0	1	total
y	0	a	b	α
	1	c	d	β
	total	γ	δ	n

5.6 Let $\mathbf{z} = \binom{x}{y}$ be a discrete 2-dimensional random vector which is obtained from the frequency table in 5.5 so that each observation has the same probability $1/n$. Then $\mathrm{E}(x) = \frac{\delta}{n}$, $\mathrm{var}(x) = \frac{\delta}{n}\big(1 - \frac{\delta}{n}\big)$, $\mathrm{cov}(x, y) = (ad - bc)/n^2$, and $\mathrm{cor}(x, y) = (ad - bc)/\sqrt{\alpha\beta\gamma\delta}$.

5.7 In terms of the probabilities:

$$\mathrm{var}(x) = p_{\cdot 1} p_{\cdot 2},$$

$$\mathrm{cov}(x, y) = p_{11}p_{22} - p_{12}p_{21},$$

$$\mathrm{cor}(x, y) = \frac{p_{11}p_{22} - p_{12}p_{21}}{\sqrt{p_{\cdot 1}p_{\cdot 2}p_{1\cdot}p_{2\cdot}}} = \varrho_{xy}.$$

		x: 0	1	total
y	0	p_{11}	p_{12}	$p_{1\cdot}$
	1	p_{21}	p_{22}	$p_{2\cdot}$
	total	$p_{\cdot 1}$	$p_{\cdot 2}$	1

5.8 $$\varrho_{xy} = 0 \iff \det\begin{pmatrix} p_{11} & p_{12} \\ p_{21} & p_{22} \end{pmatrix} = \det\begin{pmatrix} a & b \\ c & d \end{pmatrix} = 0$$
$$\iff f\frac{p_{11}}{p_{21}} = \frac{p_{12}}{p_{22}} \iff \frac{a}{c} = \frac{b}{d}$$

5.9 Dichotomous random variables x and y are statistically independent iff $\varrho_{xy} = 0$.

5.10 Independence between random variables means statistical (stochastic) independence: the random vectors $\mathbf{x}$ and $\mathbf{y}$ are statistically independent iff the joint distribution function of $\binom{\mathbf{x}}{\mathbf{y}}$ is the product of the distribution functions of $\mathbf{x}$ and $\mathbf{y}$. For example, if x and y are discrete random variables with values $x_1, \dots, x_r$ and $y_1, \dots, y_c$, then x and y are statistically independent iff
$$\mathrm{P}(x = x_i, y = y_j) = \mathrm{P}(x = x_i)\,\mathrm{P}(y = y_j),\ i = 1, \dots, r,\ j = 1, \dots, c.$$

5.11 Finiteness matters. Throughout this book, we assume that the expectations, variances and covariances that we are dealing with are finite. Then independence of the random variables x and y implies that $\operatorname{cor}(x, y) = 0$. This implication may not be true if the finiteness is not holding.

5.12 Definition N1: A p-dimensional random variable $\mathbf{z}$ is said to have a p-variate normal distribution N_p if every linear function $\mathbf{a}'\mathbf{z}$ has a univariate normal distribution. We denote $\mathbf{z} \sim \mathrm{N}_p(\boldsymbol{\mu}, \boldsymbol{\Sigma})$, where $\boldsymbol{\mu} = \mathrm{E}(\mathbf{z})$ and $\boldsymbol{\Sigma} = \operatorname{cov}(\mathbf{z})$. If $\mathbf{a}'\mathbf{z} = b$, where b is a constant, we define $\mathbf{a}'\mathbf{z} \sim \mathrm{N}(b, 0)$.

5.13 Definition N2: A p-dimensional random variable $\mathbf{z}$, with $\boldsymbol{\mu} = \mathrm{E}(\mathbf{z})$ and $\boldsymbol{\Sigma} = \operatorname{cov}(\mathbf{z})$, is said to have a p-variate normal distribution N_p if it can be expressed as $\mathbf{z} = \boldsymbol{\mu} + \mathbf{F}\mathbf{u}$, where $\mathbf{F}$ is an $p \times r$ matrix of rank r and $\mathbf{u}$ is a random vector of r independent univariate normal random variables.

5.14 If $\mathbf{z} \sim \mathrm{N}_p$ then each element of $\mathbf{z}$ follows N_1. The reverse relation does not necessarily hold.

5.15 If $\mathbf{z} = \binom{\mathbf{x}}{\mathbf{y}}$ is multinormally distributed, then $\mathbf{x}$ and $\mathbf{y}$ are stochastically independent iff they are uncorrelated.

5.16 Let $\mathbf{z} \sim \mathrm{N}_p(\boldsymbol{\mu}, \boldsymbol{\Sigma})$, where $\boldsymbol{\Sigma}$ is positive definite. Then $\mathbf{z}$ has a density
$$n(\mathbf{z}; \boldsymbol{\mu}, \boldsymbol{\Sigma}) = \frac{1}{(2\pi)^{p/2}|\boldsymbol{\Sigma}|^{1/2}} \mathrm{e}^{-\frac{1}{2}(\mathbf{z}-\boldsymbol{\mu})'\boldsymbol{\Sigma}^{-1}(\mathbf{z}-\boldsymbol{\mu})}.$$

5.17 Contours of constant density for $\mathrm{N}_2(\boldsymbol{\mu}, \boldsymbol{\Sigma})$ are ellipses defined by
$$\mathcal{A} = \{\, \mathbf{z} \in \mathbb{R}^2 : (\mathbf{z} - \boldsymbol{\mu})'\boldsymbol{\Sigma}^{-1}(\mathbf{z} - \boldsymbol{\mu}) = c^2 \,\}.$$

These ellipses are centered at $\boldsymbol{\mu}$ and have axes $c\sqrt{\lambda_i}\mathbf{t}_i$, where $\lambda_i = \operatorname{ch}_i(\boldsymbol{\Sigma})$ and $\mathbf{t}_i$ is the corresponding eigenvector. The major axis is the longest diameter (line through $\boldsymbol{\mu}$) of the ellipse, that is, we want to find a point $\mathbf{z}_1$ solving $\max\|\mathbf{z}-\boldsymbol{\mu}\|^2$ subject to $\mathbf{z} \in \mathcal{A}$. Denoting $\mathbf{u} = \mathbf{z}-\boldsymbol{\mu}$, the above task becomes

$$\max \mathbf{u}'\mathbf{u} \quad \text{subject to } \mathbf{u}'\boldsymbol{\Sigma}^{-1}\mathbf{u} = c^2,$$

for which the solution is $\mathbf{u}_1 = \mathbf{z}_1 - \boldsymbol{\mu} = \pm c\sqrt{\lambda_1}\mathbf{t}_1$, and $\mathbf{u}_1'\mathbf{u}_1 = c^2\lambda_1$. Correspondingly, the minor axis is the shortest diameter of the ellipse $\mathcal{A}$.

5.18 The eigenvalues of $\boldsymbol{\Sigma} = \begin{pmatrix} \sigma^2 & \sigma_{12} \\ \sigma_{21} & \sigma^2 \end{pmatrix} = \sigma^2 \begin{pmatrix} 1 & \varrho \\ \varrho & 1 \end{pmatrix}$, where $\sigma_{12} \geq 0$, are

$$\operatorname{ch}_1(\boldsymbol{\Sigma}) = \sigma^2 + \sigma_{12} = \sigma^2(1+\varrho),$$
$$\operatorname{ch}_2(\boldsymbol{\Sigma}) = \sigma^2 - \sigma_{12} = \sigma^2(1-\varrho),$$

and $\mathbf{t}_1 = \frac{1}{\sqrt{2}}\binom{1}{1}$, $\mathbf{t}_2 = \frac{1}{\sqrt{2}}\binom{-1}{1}$. If $\sigma_{12} \leq 0$, then $\mathbf{t}_1 = \frac{1}{\sqrt{2}}\binom{-1}{1}$.

5.19 When $p = 2$ and $\operatorname{cor}(z_1, z_2) = \varrho\ (\neq \pm 1)$, we have

(a) $$\boldsymbol{\Sigma}^{-1} = \begin{pmatrix} \sigma_{11} & \sigma_{12} \\ \sigma_{21} & \sigma_{22} \end{pmatrix}^{-1} = \begin{pmatrix} \sigma_1^2 & \sigma_1\sigma_2\varrho \\ \sigma_1\sigma_2\varrho & \sigma_2^2 \end{pmatrix}^{-1}$$
$$= \frac{1}{\sigma_1^2\sigma_2^2(1-\varrho^2)} \begin{pmatrix} \sigma_2^2 & -\sigma_{12} \\ -\sigma_{12} & \sigma_1^2 \end{pmatrix} = \frac{1}{1-\varrho^2} \begin{pmatrix} \dfrac{1}{\sigma_1^2} & \dfrac{-\varrho}{\sigma_1\sigma_2} \\ \dfrac{-\varrho}{\sigma_1\sigma_2} & \dfrac{1}{\sigma_2^2} \end{pmatrix},$$

(b) $\det(\boldsymbol{\Sigma}) = \sigma_1^2\sigma_2^2(1-\varrho^2) \leq \sigma_1^2\sigma_2^2$,

(c) $$n(\mathbf{z}; \boldsymbol{\mu}, \boldsymbol{\Sigma}) = \frac{1}{2\pi\sigma_1\sigma_2\sqrt{1-\varrho^2}} \cdot \exp\Bigg(\frac{-1}{2(1-\varrho^2)}\Bigg[\frac{(z_1-\mu_1)^2}{\sigma_1^2} - 2\varrho\frac{(z_1-\mu_1)(z_2-\mu_2)}{\sigma_1\sigma_2} + \frac{(z_2-\mu_2)^2}{\sigma_2^2}\Bigg]\Bigg).$$

5.20 Suppose that $\mathbf{z} \sim \mathrm{N}(\boldsymbol{\mu}, \boldsymbol{\Sigma})$, $\mathbf{z} = \binom{\mathbf{x}}{\mathbf{y}}$, $\boldsymbol{\mu} = \binom{\boldsymbol{\mu}_{\mathbf{x}}}{\boldsymbol{\mu}_{\mathbf{y}}}$, $\boldsymbol{\Sigma} = \begin{pmatrix} \boldsymbol{\Sigma}_{\mathbf{xx}} & \boldsymbol{\Sigma}_{\mathbf{xy}} \\ \boldsymbol{\Sigma}_{\mathbf{yx}} & \boldsymbol{\Sigma}_{\mathbf{yy}} \end{pmatrix}$. Then

(a) the conditional distribution of $\mathbf{y}$ given that $\mathbf{x}$ is held fixed at a selected value $\mathbf{x} = \underline{\mathbf{x}}$ is normal with mean

$$\mathrm{E}(\mathbf{y} \mid \underline{\mathbf{x}}) = \boldsymbol{\mu}_{\mathbf{y}} + \boldsymbol{\Sigma}_{\mathbf{yx}}\boldsymbol{\Sigma}_{\mathbf{xx}}^{-1}(\underline{\mathbf{x}} - \boldsymbol{\mu}_{\mathbf{x}})$$
$$= (\boldsymbol{\mu}_{\mathbf{y}} - \boldsymbol{\Sigma}_{\mathbf{yx}}\boldsymbol{\Sigma}_{\mathbf{xx}}^{-1}\boldsymbol{\mu}_{\mathbf{x}}) + \boldsymbol{\Sigma}_{\mathbf{yx}}\boldsymbol{\Sigma}_{\mathbf{xx}}^{-1}\underline{\mathbf{x}},$$

(b) and the covariance matrix (partial covariances)

$$\operatorname{cov}(\mathbf{y} \mid \underline{\mathbf{x}}) = \boldsymbol{\Sigma}_{\mathbf{yy}\cdot\mathbf{x}} = \boldsymbol{\Sigma}_{\mathbf{yy}} - \boldsymbol{\Sigma}_{\mathbf{yx}}\boldsymbol{\Sigma}_{\mathbf{xx}}^{-1}\boldsymbol{\Sigma}_{\mathbf{xy}} = \boldsymbol{\Sigma}/\boldsymbol{\Sigma}_{\mathbf{xx}}.$$

5.21 If $\mathbf{z} = \begin{pmatrix} \mathbf{x} \\ y \end{pmatrix} \sim \mathrm{N}_{p+1}(\boldsymbol{\mu}, \boldsymbol{\Sigma})$, $\boldsymbol{\mu} = \begin{pmatrix} \boldsymbol{\mu}_{\mathbf{x}} \\ \mu_y \end{pmatrix}$,

$$\boldsymbol{\Sigma} = \begin{pmatrix} \boldsymbol{\Sigma}_{\mathbf{xx}} & \boldsymbol{\sigma}_{\mathbf{x}y} \\ \boldsymbol{\sigma}'_{\mathbf{x}y} & \sigma_y^2 \end{pmatrix}, \quad \boldsymbol{\Sigma}^{-1} = \begin{pmatrix} \cdot & \cdot \\ \cdot & \sigma^{yy} \end{pmatrix}, \text{ then}$$

(a) $\mathrm{E}(y \mid \underline{\mathbf{x}}) = \mu_y + \boldsymbol{\sigma}'_{\mathbf{x}y}\boldsymbol{\Sigma}_{\mathbf{xx}}^{-1}(\underline{\mathbf{x}} - \boldsymbol{\mu}_{\mathbf{x}}) = (\mu_y - \boldsymbol{\sigma}'_{\mathbf{x}y}\boldsymbol{\Sigma}_{\mathbf{xx}}^{-1}\boldsymbol{\mu}_{\mathbf{x}}) + \boldsymbol{\sigma}'_{\mathbf{x}y}\boldsymbol{\Sigma}_{\mathbf{xx}}^{-1}\underline{\mathbf{x}}$,

(b) $$\begin{aligned} \mathrm{var}(y \mid \underline{\mathbf{x}}) &= \sigma^2_{y\cdot 12\ldots p} = \sigma^2_{y\cdot\mathbf{x}} = \sigma_y^2 - \boldsymbol{\sigma}'_{\mathbf{x}y}\boldsymbol{\Sigma}_{\mathbf{xx}}^{-1}\boldsymbol{\sigma}_{\mathbf{x}y} \\ &= \sigma_y^2\left(1 - \frac{\boldsymbol{\sigma}'_{\mathbf{x}y}\boldsymbol{\Sigma}_{\mathbf{xx}}^{-1}\boldsymbol{\sigma}_{\mathbf{x}y}}{\sigma_y^2}\right) \\ &= \sigma_y^2(1 - \varrho^2_{y\cdot\mathbf{x}}) = 1/\sigma^{yy} = \text{conditional variance}, \end{aligned}$$

(c) $\varrho^2_{y\cdot\mathbf{x}} = \dfrac{\boldsymbol{\sigma}'_{\mathbf{x}y}\boldsymbol{\Sigma}_{\mathbf{xx}}^{-1}\boldsymbol{\sigma}_{\mathbf{x}y}}{\sigma_y^2}$ = the squared population multiple correlation.

5.22 When $\binom{x}{y} \sim \mathrm{N}_2$, $\mathrm{cor}(x, y) = \varrho$, and $\beta := \sigma_{xy}/\sigma_x^2$, we have

(a) $\mathrm{E}(y \mid \underline{x}) = \mu_y + \beta(\underline{x} - \mu_x) = \mu_y + \varrho\frac{\sigma_y}{\sigma_x}(\underline{x} - \mu_x) = (\mu_y - \beta\mu_x) + \beta\underline{x}$,

(b) $\mathrm{var}(y \mid \underline{x}) = \sigma^2_{y\cdot x} = \sigma_y^2 - \frac{\sigma^2_{xy}}{\sigma_x^2} = \sigma_y^2(1 - \varrho^2) \le \sigma_y^2 = \mathrm{var}(y)$.

5.23 The random vector $\boldsymbol{\mu}_{\mathbf{y}} + \boldsymbol{\Sigma}_{\mathbf{yx}}\boldsymbol{\Sigma}_{\mathbf{xx}}^{-1}(\mathbf{x} - \boldsymbol{\mu}_{\mathbf{x}})$ appears to be the best linear predictor of $\mathbf{y}$ on the basis of $\mathbf{x}$, denoted as $\mathrm{BLP}(\mathbf{y}; \mathbf{x})$. In general, $\mathrm{BLP}(\mathbf{y}; \mathbf{x})$ is not the conditional expectation of $\mathbf{y}$ given $\mathbf{x}$.

5.24 The random vector

$$\mathbf{e}_{\mathbf{y}\cdot\mathbf{x}} = \mathbf{y} - \mathrm{BLP}(\mathbf{y}; \mathbf{x}) = \mathbf{y} - [\boldsymbol{\mu}_{\mathbf{y}} + \boldsymbol{\Sigma}_{\mathbf{yx}}\boldsymbol{\Sigma}_{\mathbf{xx}}^{-1}(\mathbf{x} - \boldsymbol{\mu}_{\mathbf{x}})]$$

is the vector of residuals of $\mathbf{y}$ from its regression on $\mathbf{x}$, i.e., prediction error between $\mathbf{y}$ and its best linear predictor $\mathrm{BLP}(\mathbf{y}; \mathbf{x})$. The matrix of partial covariances of $\mathbf{y}$ (holding $\mathbf{x}$ fixed) is

$$\mathrm{cov}(\mathbf{e}_{\mathbf{y}\cdot\mathbf{x}}) = \boldsymbol{\Sigma}_{\mathbf{yy}\cdot\mathbf{x}} = \boldsymbol{\Sigma}_{\mathbf{yy}} - \boldsymbol{\Sigma}_{\mathbf{yx}}\boldsymbol{\Sigma}_{\mathbf{xx}}^{-1}\boldsymbol{\Sigma}_{\mathbf{xy}} = \boldsymbol{\Sigma}/\boldsymbol{\Sigma}_{\mathbf{xx}}.$$

If $\mathbf{z}$ is not multinormally distributed, the matrix $\boldsymbol{\Sigma}_{\mathbf{yy}\cdot\mathbf{x}}$ is not necessarily the covariance matrix of the conditional distribution.

5.25 The population partial correlations. The ij-element of the matrix of partial correlations of $\mathbf{y}$ (eliminating $\mathbf{x}$) is

$$\varrho_{ij\cdot\mathbf{x}} = \frac{\sigma_{ij\cdot\mathbf{x}}}{\sqrt{\sigma_{ii\cdot\mathbf{x}}\sigma_{jj\cdot\mathbf{x}}}} = \mathrm{cor}(e_{y_i\cdot\mathbf{x}}, e_{y_j\cdot\mathbf{x}}),$$

$$\{\sigma_{ij\cdot\mathbf{x}}\} = \boldsymbol{\Sigma}_{\mathbf{yy}\cdot\mathbf{x}}, \quad \{\varrho_{ij\cdot\mathbf{x}}\} = \mathrm{cor}(\mathbf{e}_{\mathbf{y}\cdot\mathbf{x}}),$$

and $e_{y_i\cdot\mathbf{x}} = y_i - \mu_{y_i} - \boldsymbol{\sigma}'_{\mathbf{x}y_i}\boldsymbol{\Sigma}_{\mathbf{xx}}^{-1}(\mathbf{x} - \boldsymbol{\mu}_{\mathbf{x}})$, $\boldsymbol{\sigma}_{\mathbf{x}y_i} = \mathrm{cov}(\mathbf{x}, y_i)$. In particular,

$$\varrho_{xy\cdot z} = \frac{\varrho_{xy} - \varrho_{xz}\,\varrho_{yz}}{\sqrt{(1-\varrho_{xz}^2)(1-\varrho_{yz}^2)}}.$$

5.26 The conditional expectation $\mathrm{E}(\mathbf{y} \mid \mathbf{x})$, where $\mathbf{x}$ is now a random vector, is $\mathrm{BP}(\mathbf{y}; \mathbf{x})$, the best predictor of $\mathbf{y}$ on the basis of $\mathbf{x}$. Notice that $\mathrm{BLP}(\mathbf{y}; \mathbf{x})$ is the best *linear* predictor.

5.27 In the multinormal distribution, $\mathrm{BLP}(\mathbf{y}; \mathbf{x}) = \mathrm{E}(\mathbf{y} \mid \mathbf{x}) = \mathrm{BP}(\mathbf{y}; \mathbf{x}) =$ the best predictor of $\mathbf{y}$ on the basis of $\mathbf{x}$; here $\mathrm{E}(\mathbf{y} \mid \mathbf{x})$ is a random vector.

5.28 The conditional mean $\mathrm{E}(y \mid \underline{\mathbf{x}})$ is called (in the world of random variables) the regression function (true mean of y when $\mathbf{x}$ is held at a selected value $\underline{\mathbf{x}}$) and similarly $\mathrm{var}(y \mid \underline{\mathbf{x}})$ is called the variance function. Note that in the multinormal case $\mathrm{E}(y \mid \underline{\mathbf{x}})$ is simply a linear function of $\underline{\mathbf{x}}$ and $\mathrm{var}(y \mid \underline{\mathbf{x}})$ does not depend on $\underline{\mathbf{x}}$ at all.

5.29 Let $\binom{x}{y}$ be a random vector and let $\mathrm{E}(y \mid x) := m(x)$ be a random variable taking the value $\mathrm{E}(y \mid x = \underline{x})$ when x takes the value $\underline{x}$, and $\mathrm{var}(y \mid x) := v(x)$ is a random variable taking the value $\mathrm{var}(y \mid x = \underline{x})$ when $x = \underline{x}$. Then

$$\mathrm{E}(y) = \mathrm{E}[\mathrm{E}(y \mid x)] = \mathrm{E}[m(x)],$$
$$\mathrm{var}(y) = \mathrm{var}[\mathrm{E}(y \mid x)] + \mathrm{E}[\mathrm{var}(y \mid x)] = \mathrm{var}[m(x)] + \mathrm{E}[v(x)].$$

5.30 Let $\binom{x}{y}$ be a random vector such that $\mathrm{E}(y \mid x = \underline{x}) = \alpha + \beta\underline{x}$. Then $\beta = \sigma_{xy}/\sigma_x^2$ and $\alpha = \mu_y - \beta\mu_x$.

5.31 If $\mathbf{z} \sim (\boldsymbol{\mu}, \boldsymbol{\Sigma})$ and $\mathbf{A}$ is symmetric, then $\mathrm{E}(\mathbf{z}'\mathbf{A}\mathbf{z}) = \mathrm{tr}(\mathbf{A}\boldsymbol{\Sigma}) + \boldsymbol{\mu}'\mathbf{A}\boldsymbol{\mu}$.

5.32 Central χ^2-distribution: $\mathbf{z} \sim \mathrm{N}_n(\mathbf{0}, \mathbf{I}_n)$: $\mathbf{z}'\mathbf{z} = \chi_n^2 \sim \chi^2(n)$

5.33 Noncentral χ^2-distribution: $\mathbf{z} \sim \mathrm{N}_n(\boldsymbol{\mu}, \mathbf{I}_n)$: $\mathbf{z}'\mathbf{z} = \chi_{n,\delta}^2 \sim \chi^2(n, \delta)$, $\delta = \boldsymbol{\mu}'\boldsymbol{\mu}$

5.34 $\mathbf{z} \sim \mathrm{N}_n(\boldsymbol{\mu}, \sigma^2\mathbf{I}_n)$: $\mathbf{z}'\mathbf{z}/\sigma^2 \sim \chi^2(n, \boldsymbol{\mu}'\boldsymbol{\mu}/\sigma^2)$

5.35 Let $\mathbf{z} \sim \mathrm{N}_n(\boldsymbol{\mu}, \boldsymbol{\Sigma})$ where $\boldsymbol{\Sigma}$ is pd and let $\mathbf{A}$ and $\mathbf{B}$ be symmetric. Then

(a) $\mathbf{z}'\mathbf{A}\mathbf{z} \sim \chi^2(r, \delta)$ iff $\mathbf{A}\boldsymbol{\Sigma}\mathbf{A} = \mathbf{A}$, in which case $r = \mathrm{tr}(\mathbf{A}\boldsymbol{\Sigma}) = \mathrm{r}(\mathbf{A}\boldsymbol{\Sigma})$, $\delta = \boldsymbol{\mu}'\mathbf{A}\boldsymbol{\mu}$,

(b) $\mathbf{z}'\boldsymbol{\Sigma}^{-1}\mathbf{z} = \mathbf{z}'[\mathrm{cov}(\mathbf{z})]^{-1}\mathbf{z} \sim \chi^2(r, \delta)$, where $r = n$, $\delta = \boldsymbol{\mu}'\boldsymbol{\Sigma}^{-1}\boldsymbol{\mu}$,

(c) $(\mathbf{z} - \boldsymbol{\mu})'\boldsymbol{\Sigma}^{-1}(\mathbf{z} - \boldsymbol{\mu}) = \mathrm{MHLN}^2(\mathbf{z}, \boldsymbol{\mu}, \boldsymbol{\Sigma}) \sim \chi^2(n)$,

(d) $\mathbf{z}'\mathbf{A}\mathbf{z}$ and $\mathbf{z}'\mathbf{B}\mathbf{z}$ are independent iff $\mathbf{A}\boldsymbol{\Sigma}\mathbf{B} = \mathbf{0}$,

(e) $\mathbf{z}'\mathbf{A}\mathbf{z}$ and $\mathbf{b}'\mathbf{z}$ are independent iff $\mathbf{A}\boldsymbol{\Sigma}\mathbf{b} = \mathbf{0}$.

5.36 Let $\mathbf{z} \sim \mathrm{N}_n(\mathbf{0}, \boldsymbol{\Sigma})$ where $\boldsymbol{\Sigma}$ is nnd and let $\mathbf{A}$ and $\mathbf{B}$ be symmetric. Then

(a) $\mathbf{z}'\mathbf{A}\mathbf{z} \sim \chi^2(r)$ iff $\boldsymbol{\Sigma}\mathbf{A}\boldsymbol{\Sigma}\mathbf{A}\boldsymbol{\Sigma} = \boldsymbol{\Sigma}\mathbf{A}\boldsymbol{\Sigma}$, in which case $r = \mathrm{tr}(\mathbf{A}\boldsymbol{\Sigma}) = \mathrm{r}(\mathbf{A}\boldsymbol{\Sigma})$,

(b) $\mathbf{z}'\mathbf{A}\mathbf{z}$ and $\mathbf{x}'\mathbf{B}\mathbf{x}$ are independent $\iff \boldsymbol{\Sigma}\mathbf{A}\boldsymbol{\Sigma}\mathbf{B}\boldsymbol{\Sigma} = \mathbf{0}$,

(c) $\mathbf{z}'\boldsymbol{\Sigma}^-\mathbf{z} = \mathbf{z}'[\mathrm{cov}(\mathbf{z})]^-\mathbf{z} \sim \chi^2(r)$ for any choice of $\boldsymbol{\Sigma}^-$ and $\mathrm{r}(\boldsymbol{\Sigma}) = r$.

5.37 Let $\mathbf{z} \sim \mathrm{N}_n(\boldsymbol{\mu}, \boldsymbol{\Sigma})$ where $\boldsymbol{\Sigma}$ is pd and let $\mathbf{A}$ and $\mathbf{B}$ be symmetric. Then

(a) $\mathrm{var}(\mathbf{z}'\mathbf{A}\mathbf{z}) = 2\,\mathrm{tr}[(\mathbf{A}\boldsymbol{\Sigma})^2] + 4\boldsymbol{\mu}'\mathbf{A}\boldsymbol{\Sigma}\mathbf{A}\boldsymbol{\mu}$,

(b) $\mathrm{cov}(\mathbf{z}'\mathbf{A}\mathbf{z}, \mathbf{z}'\mathbf{B}\mathbf{z}) = 2\,\mathrm{tr}(\mathbf{A}\boldsymbol{\Sigma}\mathbf{B}\boldsymbol{\Sigma}) + 4\boldsymbol{\mu}'\mathbf{A}\boldsymbol{\Sigma}\mathbf{B}\boldsymbol{\mu}$.

5.38 Noncentral F-distribution: $F = \dfrac{\chi^2_{m,\delta}/m}{\chi^2_n/n} \sim \mathrm{F}(m, n, \delta)$, where $\chi^2_{m,\delta}$ and χ^2_n are independent

5.39 t-distribution: $\mathrm{t}^2(n) = \mathrm{F}(1, n)$

5.40 Let $\mathbf{y} \sim \mathrm{N}_n(\mathbf{X}\boldsymbol{\beta}, \sigma^2\mathbf{I})$, where $\mathbf{X} = (\mathbf{1} : \mathbf{X}_0)$, $\mathrm{r}(\mathbf{X}) = k + 1$. Then

(a) $\mathbf{y}'\mathbf{y}/\sigma^2 \sim \chi^2(n, \boldsymbol{\beta}'\mathbf{X}'\mathbf{X}\boldsymbol{\beta}/\sigma^2)$,

(b) $\mathbf{y}'(\mathbf{I}-\mathbf{J})\mathbf{y}/\sigma^2 \sim \chi^2[n-1, \boldsymbol{\beta}'\mathbf{X}'(\mathbf{I}-\mathbf{J})\mathbf{X}\boldsymbol{\beta}/\sigma^2] = \chi^2[n-1, \boldsymbol{\beta}_{\mathrm{x}}'\mathbf{T}_{\mathrm{xx}}\boldsymbol{\beta}_{\mathrm{x}}/\sigma^2]$,

(c) $\mathbf{y}'(\mathbf{H}-\mathbf{J})\mathbf{y}/\sigma^2 \sim \chi^2[k, \boldsymbol{\beta}_{\mathrm{x}}'\mathbf{T}_{\mathrm{xx}}\boldsymbol{\beta}_{\mathrm{x}}/\sigma^2]$,

(d) $\mathbf{y}'(\mathbf{I}-\mathbf{H})\mathbf{y}/\sigma^2 = \mathrm{SSE}/\sigma^2 \sim \chi^2(n-k-1)$.

5.41 Suppose $\mathbf{I}_n = \mathbf{A}_1 + \cdots + \mathbf{A}_m$. Then the following statements are equivalent:

(a) $n = \mathrm{r}(\mathbf{A}_1) + \cdots + \mathrm{r}(\mathbf{A}_m)$,

(b) $\mathbf{A}_i^2 = \mathbf{A}_i$ for $i = 1, \dots, m$,

(c) $\mathbf{A}_i\mathbf{A}_j = \mathbf{0}$ for all $i \neq j$.

5.42 Cochran's theorem. Let $\mathbf{z} \sim \mathrm{N}_n(\boldsymbol{\mu}, \mathbf{I})$ and let $\mathbf{z}'\mathbf{z} = \mathbf{z}'\mathbf{A}_1\mathbf{z} + \cdots + \mathbf{z}'\mathbf{A}_m\mathbf{z}$. Then any of 5.41a–5.41c is a necessary and sufficient condition for $\mathbf{z}'\mathbf{A}_i\mathbf{z}$ to be independently distributed as $\chi^2[\mathrm{r}(\mathbf{A}_i), \cdot]$.

5.43 Wishart-distribution. Let $\mathbf{U}' = (\mathbf{u}_{(1)} : \ldots : \mathbf{u}_{(n)})$ be a random sample from $\mathrm{N}_p(\mathbf{0}, \boldsymbol{\Sigma})$, i.e., $\mathbf{u}_{(i)}$'s are independent and each $\mathbf{u}_{(i)} \sim \mathrm{N}_p(\mathbf{0}, \boldsymbol{\Sigma})$. Then $\mathbf{W} = \mathbf{U}'\mathbf{U} = \sum_{i=1}^n \mathbf{u}_{(i)}\mathbf{u}_{(i)}'$ is said to have a Wishart-distribution with n degrees of freedom and scale matrix $\boldsymbol{\Sigma}$, and we write $\mathbf{W} \sim \mathrm{W}_p(n, \boldsymbol{\Sigma})$.

5.44 Hotelling's T^2 distribution. Suppose $\mathbf{v} \sim \mathrm{N}_p(\mathbf{0}, \boldsymbol{\Sigma})$, $\mathbf{W} \sim \mathrm{W}_p(m, \boldsymbol{\Sigma})$, $\mathbf{v}$ and $\mathbf{W}$ are independent, $\boldsymbol{\Sigma}$ is pd. Hotelling's T^2 distribution is the distribution of

$$\begin{aligned} T^2 &= m \cdot \mathbf{v}'\mathbf{W}^{-1}\mathbf{v} \\ &= \mathbf{v}'\left(\frac{\mathbf{W}}{m}\right)^{-1}\mathbf{v} = (\text{normal r.v.})'\left(\frac{\text{Wishart}}{\text{df}}\right)^{-1}(\text{normal r.v.}) \end{aligned}$$

and is denoted as $T^2 \sim \mathrm{T}^2(p, m)$.

5.45 Let $\mathbf{U}' = (\mathbf{u}_{(1)} : \mathbf{u}_{(2)} : \dots : \mathbf{u}_{(n)})$ be a random sample from $\mathrm{N}_p(\boldsymbol{\mu}, \boldsymbol{\Sigma})$. Then:

(a) The (transposed) rows of $\mathbf{U}$, i.e., $\mathbf{u}_{(1)}, \mathbf{u}_{(2)}, \dots, \mathbf{u}_{(n)}$, are independent random vectors, each $\mathbf{u}_{(i)} \sim \mathrm{N}_p(\boldsymbol{\mu}, \boldsymbol{\Sigma})$.

(b) The columns of $\mathbf{U}$, i.e., $\mathbf{u}_1, \mathbf{u}_2, \dots, \mathbf{u}_p$ are n-dimensional random vectors: $\mathbf{u}_i \sim \mathrm{N}_n(\mu_i \mathbf{1}_n, \sigma_i^2 \mathbf{I}_n)$, $\mathrm{cov}(\mathbf{u}_i, \mathbf{u}_j) = \sigma_{ij}\mathbf{I}_n$.

(c) $\mathbf{z} = \mathrm{vec}(\mathbf{U}) = \begin{pmatrix} \mathbf{u}_1 \\ \vdots \\ \mathbf{u}_p \end{pmatrix}, \quad \mathrm{E}(\mathbf{z}) = \begin{pmatrix} \mu_1 \mathbf{1}_n \\ \vdots \\ \mu_p \mathbf{1}_n \end{pmatrix} = \boldsymbol{\mu} \otimes \mathbf{1}_n.$

(d) $\mathrm{cov}(\mathbf{z}) = \begin{pmatrix} \sigma_1^2 \mathbf{I}_n & \sigma_{12}\mathbf{I}_n & \dots & \sigma_{1p}\mathbf{I}_n \\ \vdots & \vdots & & \vdots \\ \sigma_{p1}\mathbf{I}_n & \sigma_{p2}\mathbf{I}_n & \dots & \sigma_p^2 \mathbf{I}_n \end{pmatrix} = \boldsymbol{\Sigma} \otimes \mathbf{I}_n.$

(e) $\bar{\mathbf{u}} = \frac{1}{n}(\mathbf{u}_{(1)} + \mathbf{u}_{(2)} + \dots + \mathbf{u}_{(n)}) = \frac{1}{n}\mathbf{U}'\mathbf{1}_n = (\bar{u}_1, \bar{u}_2, \dots, \bar{u}_p)'.$

(f) $$\begin{aligned} \mathbf{S} &= \tfrac{1}{n-1}\mathbf{T} = \tfrac{1}{n-1}\mathbf{U}'(\mathbf{I} - \mathbf{J})\mathbf{U} = \tfrac{1}{n-1}\sum_{i=1}^{n}(\mathbf{u}_{(i)} - \bar{\mathbf{u}})(\mathbf{u}_{(i)} - \bar{\mathbf{u}})' \\ &= \tfrac{1}{n-1}\Big(\sum_{i=1}^{n}\mathbf{u}_{(i)}\mathbf{u}_{(i)}' - n\bar{\mathbf{u}}\bar{\mathbf{u}}'\Big). \end{aligned}$$

(g) $\mathrm{E}(\bar{\mathbf{u}}) = \boldsymbol{\mu}$, $\mathrm{E}(\mathbf{S}) = \boldsymbol{\Sigma}$, $\bar{\mathbf{u}} \sim \mathrm{N}_p\big(\boldsymbol{\mu}, \frac{1}{n}\boldsymbol{\Sigma}\big)$.

(h) $\bar{\mathbf{u}}$ and $\mathbf{T} = \mathbf{U}'(\mathbf{I} - \mathbf{J})\mathbf{U}$ are independent and $\mathbf{T} \sim \mathrm{W}(n-1, \boldsymbol{\Sigma})$.

(i) Hotelling's T^2: $T^2 = n(\bar{\mathbf{u}} - \boldsymbol{\mu}_0)'\mathbf{S}^{-1}(\bar{\mathbf{u}} - \boldsymbol{\mu}_0) = n \cdot \mathrm{MHLN}^2(\bar{\mathbf{u}}, \boldsymbol{\mu}_0, \mathbf{S})$,

$$\tfrac{n-p}{(n-1)p}T^2 \sim \mathrm{F}(p, n-p, \theta), \quad \theta = n(\boldsymbol{\mu} - \boldsymbol{\mu}_0)'\boldsymbol{\Sigma}^{-1}(\boldsymbol{\mu} - \boldsymbol{\mu}_0).$$

(j) Hypothesis $\boldsymbol{\mu} = \boldsymbol{\mu}_0$ is rejected at risk level α, if

$$n(\bar{\mathbf{u}} - \boldsymbol{\mu}_0)'\mathbf{S}^{-1}(\bar{\mathbf{u}} - \boldsymbol{\mu}_0) > \tfrac{p(n-1)}{n-p}F_{\alpha;p,n-p}.$$

(k) A $100(1-\alpha)\%$ confidence region for the mean of the $\mathrm{N}_p(\boldsymbol{\mu}, \boldsymbol{\Sigma})$ is the ellipsoid determined by all $\boldsymbol{\mu}$ such that

$$n(\bar{\mathbf{u}} - \boldsymbol{\mu})'\mathbf{S}^{-1}(\bar{\mathbf{u}} - \boldsymbol{\mu}) \le \tfrac{p(n-1)}{n-p}F_{\alpha;p,n-p}.$$

(l) $\displaystyle\max_{\mathbf{a} \neq \mathbf{0}} \frac{[\mathbf{a}'(\bar{\mathbf{u}} - \boldsymbol{\mu}_0)]^2}{\mathbf{a}'\mathbf{S}\mathbf{a}} = (\bar{\mathbf{u}} - \boldsymbol{\mu}_0)'\mathbf{S}^{-1}(\bar{\mathbf{u}} - \boldsymbol{\mu}_0) = \mathrm{MHLN}^2(\bar{\mathbf{u}}, \boldsymbol{\mu}_0, \mathbf{S}).$

5.46 Let $\mathbf{U}_1'$ and $\mathbf{U}_2'$ be independent random samples from $\mathrm{N}_p(\boldsymbol{\mu}_1, \boldsymbol{\Sigma})$ and $\mathrm{N}_p(\boldsymbol{\mu}_2, \boldsymbol{\Sigma})$, respectively. Denote $\mathbf{T}_i = \mathbf{U}_i'\mathbf{C}_{n_i}\mathbf{U}_i$ and

$$\mathbf{S}_* = \frac{1}{n_1 + n_2 - 2}(\mathbf{T}_1 + \mathbf{T}_2), \quad T^2 = \frac{n_1 n_2}{n_1 + n_2}(\bar{\mathbf{u}}_1 - \bar{\mathbf{u}}_2)'\mathbf{S}_*^{-1}(\bar{\mathbf{u}}_1 - \bar{\mathbf{u}}_2).$$

If $\boldsymbol{\mu}_1 = \boldsymbol{\mu}_2$, then $\frac{n_1+n_2-p-1}{(n_1+n_2-2)p}T^2 \sim \mathrm{F}(p, n_1 + n_2 - p - 1)$.

5.47 If $n_1 = 1$, then Hotelling's T^2 becomes

$$T^2 = \tfrac{n_2}{n_2+1}(\mathbf{u}_{(1)} - \bar{\mathbf{u}}_2)'\mathbf{S}_2^{-1}(\mathbf{u}_{(1)} - \bar{\mathbf{u}}_2).$$

5.48 Let $\mathbf{U}' = (\mathbf{u}_{(1)} : \ldots : \mathbf{u}_{(n)})$ be a random sample from $\mathrm{N}_p(\boldsymbol{\mu}, \boldsymbol{\Sigma})$. Then the likelihood function and its logarithm are

(a) $L = (2\pi)^{-\frac{pn}{2}}|\boldsymbol{\Sigma}|^{-\frac{n}{2}} \exp\Big[-\frac{1}{2}\sum_{i=1}^{n}(\mathbf{u}_{(i)} - \boldsymbol{\mu})'\boldsymbol{\Sigma}^{-1}(\mathbf{u}_{(i)} - \boldsymbol{\mu})\Big]$,

(b) $\log L = -\frac{1}{2}pn\log(2\pi) - \frac{1}{2}n\log|\boldsymbol{\Sigma}| - \sum_{i=1}^{n}(\mathbf{u}_{(i)} - \boldsymbol{\mu})'\boldsymbol{\Sigma}^{-1}(\mathbf{u}_{(i)} - \boldsymbol{\mu})$.

The function L is considered as a function of $\boldsymbol{\mu}$ and $\boldsymbol{\Sigma}$ while $\mathbf{U}'$ is being fixed. The maximum likelihood estimators, MLEs, of $\boldsymbol{\mu}$ and $\boldsymbol{\Sigma}$ are the vector $\boldsymbol{\mu}_*$ and the positive definite matrix $\boldsymbol{\Sigma}_*$ that maximize L:

(c) $\boldsymbol{\mu}_* = \frac{1}{n}\mathbf{U}'\mathbf{1} = (\bar{u}_1, \bar{u}_2, \ldots, \bar{u}_p)' = \bar{\mathbf{u}}, \quad \boldsymbol{\Sigma}_* = \frac{1}{n}\mathbf{U}'(\mathbf{I} - \mathbf{J})\mathbf{U} = \frac{n-1}{n}\mathbf{S}$.

(d) The maximum of L is $\max_{\boldsymbol{\mu},\boldsymbol{\Sigma}} L(\boldsymbol{\mu}, \boldsymbol{\Sigma}) = \dfrac{1}{(2\pi)^{np/2}|\boldsymbol{\Sigma}_*|^{n/2}}\mathrm{e}^{-np/2}$.

(e) Denote $\max_{\boldsymbol{\Sigma}} L(\boldsymbol{\mu}_0, \boldsymbol{\Sigma}) = \dfrac{1}{(2\pi)^{np/2}|\boldsymbol{\Sigma}_0|^{n/2}}\mathrm{e}^{-np/2}$,

where $\boldsymbol{\Sigma}_0 = \frac{1}{n}\sum_{i=1}^{n}(\mathbf{u}_{(i)} - \boldsymbol{\mu}_0)(\mathbf{u}_{(i)} - \boldsymbol{\mu}_0)'$. Then the likelihood ratio is

$$\Lambda = \frac{\max_{\boldsymbol{\Sigma}} L(\boldsymbol{\mu}_0, \boldsymbol{\Sigma})}{\max_{\boldsymbol{\mu},\boldsymbol{\Sigma}} L(\boldsymbol{\mu}, \boldsymbol{\Sigma})} = \left(\frac{|\boldsymbol{\Sigma}_*|}{|\boldsymbol{\Sigma}_0|}\right)^{n/2}.$$

(f) Wilks's lambda $= \Lambda^{2/n} = \dfrac{1}{1 + \frac{1}{n-1}T^2}$, $T^2 = n(\bar{\mathbf{u}} - \boldsymbol{\mu}_0)'\mathbf{S}^{-1}(\bar{\mathbf{u}} - \boldsymbol{\mu}_0)$.

5.49 Let $\mathbf{U}' = (\mathbf{u}_{(1)} : \ldots : \mathbf{u}_{(n)})$ be a random sample from $\mathrm{N}_{k+1}(\boldsymbol{\mu}, \boldsymbol{\Sigma})$, where

$$\mathbf{u}_{(i)} = \begin{pmatrix} \mathbf{x}_{(i)} \\ y_i \end{pmatrix}, \quad \mathrm{E}(\mathbf{u}_{(i)}) = \begin{pmatrix} \boldsymbol{\mu}_{\mathbf{x}} \\ \mu_y \end{pmatrix},$$

$$\mathrm{cov}(\mathbf{u}_{(i)}) = \begin{pmatrix} \boldsymbol{\Sigma}_{\mathbf{xx}} & \boldsymbol{\sigma}_{\mathbf{x}y} \\ \boldsymbol{\sigma}_{\mathbf{x}y}' & \sigma_y^2 \end{pmatrix}, \quad i = 1, \ldots, n.$$

Then the conditional mean and variance of y, given that $\mathbf{x} = \underline{\mathbf{x}}$, are

$$\mathrm{E}(y \mid \underline{\mathbf{x}}) = \mu_y + \boldsymbol{\sigma}'_{\mathbf{x}y}\boldsymbol{\Sigma}_{\mathbf{xx}}^{-1}(\underline{\mathbf{x}} - \boldsymbol{\mu}_{\mathbf{x}}) = \beta_0 + \boldsymbol{\beta}'_{\mathbf{x}}\underline{\mathbf{x}},$$
$$\mathrm{var}(y \mid \underline{\mathbf{x}}) = \sigma^2_{y\cdot\mathbf{x}},$$

where $\beta_0 = \mu_y - \boldsymbol{\sigma}'_{\mathbf{x}y}\boldsymbol{\Sigma}_{\mathbf{xx}}^{-1}\boldsymbol{\mu}_{\mathbf{x}}$ and $\boldsymbol{\beta}_{\mathbf{x}} = \boldsymbol{\Sigma}_{\mathbf{xx}}^{-1}\boldsymbol{\sigma}_{\mathbf{x}y}$. Then

$$\mathrm{MLE}(\beta_0) = \bar{y} - \mathbf{s}'_{\mathbf{x}y}\mathbf{S}_{\mathbf{xx}}^{-1}\bar{\mathbf{x}} = \hat{\beta}_0, \quad \mathrm{MLE}(\boldsymbol{\beta}_{\mathbf{x}}) = \mathbf{S}_{\mathbf{xx}}^{-1}\mathbf{s}_{\mathbf{x}y} = \hat{\boldsymbol{\beta}}_{\mathbf{x}},$$
$$\mathrm{MLE}(\sigma^2_{y\cdot\mathbf{x}}) = \tfrac{1}{n}(t^2_{yy} - \mathbf{t}'_{\mathbf{x}y}\mathbf{T}_{\mathbf{xx}}^{-1}\mathbf{t}_{\mathbf{x}y}) = \tfrac{1}{n}\mathrm{SSE}.$$

The squared population multiple correlation $\varrho^2_{y\cdot\mathbf{x}} = \boldsymbol{\sigma}'_{\mathbf{x}y}\boldsymbol{\Sigma}_{\mathbf{xx}}^{-1}\boldsymbol{\sigma}_{\mathbf{x}y}/\sigma^2_y$ equals 0 iff $\boldsymbol{\beta}_{\mathbf{x}} = \boldsymbol{\Sigma}_{\mathbf{xx}}^{-1}\boldsymbol{\sigma}_{\mathbf{x}y} = \mathbf{0}$. The hypothesis $\boldsymbol{\beta}_{\mathbf{x}} = \mathbf{0}$, i.e., $\varrho_{y\cdot\mathbf{x}} = 0$, can be tested by

$$F = \frac{R^2_{y\cdot\mathbf{x}}/k}{(1 - R^2_{y\cdot\mathbf{x}})/(n - k - 1)}.$$

6 Best linear predictor

6.1 Let $f(\mathbf{x})$ be a scalar valued function of the random vector $\mathbf{x}$. The mean squared error of $f(\mathbf{x})$ with respect to y (y being a random variable or a fixed constant) is

$$\mathrm{MSE}[f(\mathbf{x}); y] = \mathrm{E}[y - f(\mathbf{x})]^2.$$

Correspondingly, for the random vectors $\mathbf{y}$ and $\mathbf{f}(\mathbf{x})$, the mean squared error matrix of $\mathbf{f}(\mathbf{x})$ with respect to $\mathbf{y}$ is

$$\mathrm{MSEM}[\mathbf{f}(\mathbf{x}); \mathbf{y}] = \mathrm{E}[\mathbf{y} - \mathbf{f}(\mathbf{x})][\mathbf{y} - \mathbf{f}(\mathbf{x})]'.$$

6.2 We might be interested in predicting the random variable y on the basis of some function of the random vector $\mathbf{x}$; denote this function as $f(\mathbf{x})$. Then $f(\mathbf{x})$ is called a predictor of y on the basis of $\mathbf{x}$. Choosing $f(\mathbf{x})$ so that it minimizes the mean squared error $\mathrm{MSE}[f(\mathbf{x}); y] = \mathrm{E}[y - f(\mathbf{x})]^2$ gives the best predictor $\mathrm{BP}(y; \mathbf{x})$. Then the $\mathrm{BP}(y; \mathbf{x})$ has the property

$$\min_{f(\mathbf{x})} \mathrm{MSE}[f(\mathbf{x}); y] = \min_{f(\mathbf{x})} \mathrm{E}[y - f(\mathbf{x})]^2 = \mathrm{E}[y - \mathrm{BP}(y; \mathbf{x})]^2.$$

It appears that the conditional expectation $\mathrm{E}(y \mid \mathbf{x})$ is the best predictor of y: $\mathrm{E}(y \mid \mathbf{x}) = \mathrm{BP}(y; \mathbf{x})$. Here we have to consider $\mathrm{E}(y \mid \mathbf{x})$ as a random variable, not a real number.

6.3 $\mathrm{bias}[f(\mathbf{x}); y] = \mathrm{E}[y - f(\mathbf{x})]$

6.4 The mean squared error $\mathrm{MSE}(\mathbf{a}'\mathbf{x} + b; y)$ of the linear (inhomogeneous) predictor $\mathbf{a}'\mathbf{x} + b$ with respect to y can be expressed as

$$\begin{aligned}\mathrm{E}[y-(\mathbf{a}'\mathbf{x}+b)]^2 &= \mathrm{var}(y-\mathbf{a}'\mathbf{x})+[\mu_y-(\mathbf{a}'\boldsymbol{\mu}_{\mathbf{x}}+b)]^2\\ &= \text{variance}+\text{bias}^2,\end{aligned}$$

and in the general case, the mean squared error matrix $\mathrm{MSEM}(\mathbf{Ax}+\mathbf{b};\mathbf{y})$ is

$$\begin{aligned}&\mathrm{E}[\mathbf{y}-(\mathbf{Ax}+\mathbf{b})][\mathbf{y}-(\mathbf{Ax}+\mathbf{b})]'\\ &=\mathrm{cov}(\mathbf{y}-\mathbf{Ax})+\|\boldsymbol{\mu}_{\mathbf{y}}-(\mathbf{A}\boldsymbol{\mu}_{\mathbf{x}}+\mathbf{b})\|^2.\end{aligned}$$

6.5 BLP: Best linear predictor. Let $\mathbf{x}$ and $\mathbf{y}$ be random vectors such that

$$\mathrm{E}\begin{pmatrix}\mathbf{x}\\ \mathbf{y}\end{pmatrix}=\begin{pmatrix}\boldsymbol{\mu}_{\mathbf{x}}\\ \boldsymbol{\mu}_{\mathbf{y}}\end{pmatrix},\quad \mathrm{cov}\begin{pmatrix}\mathbf{x}\\ \mathbf{y}\end{pmatrix}=\begin{pmatrix}\boldsymbol{\Sigma}_{\mathbf{xx}} & \boldsymbol{\Sigma}_{\mathbf{xy}}\\ \boldsymbol{\Sigma}_{\mathbf{yx}} & \boldsymbol{\Sigma}_{\mathbf{yy}}\end{pmatrix}.$$

Then a linear predictor $\mathbf{Gx}+\mathbf{g}$ is said to be the best linear predictor, BLP, for $\mathbf{y}$, if the Löwner ordering

$$\mathrm{MSEM}(\mathbf{Gx}+\mathbf{g};\mathbf{y})\leq_{\mathsf{L}}\mathrm{MSEM}(\mathbf{Fx}+\mathbf{f};\mathbf{y})$$

holds for every linear predictor $\mathbf{Fx}+\mathbf{f}$ of $\mathbf{y}$.

6.6 Let $\mathrm{cov}(\mathbf{z})=\mathrm{cov}\begin{pmatrix}\mathbf{x}\\ \mathbf{y}\end{pmatrix}=\boldsymbol{\Sigma}=\begin{pmatrix}\boldsymbol{\Sigma}_{\mathbf{xx}} & \boldsymbol{\Sigma}_{\mathbf{xy}}\\ \boldsymbol{\Sigma}_{\mathbf{yx}} & \boldsymbol{\Sigma}_{\mathbf{yy}}\end{pmatrix}$, and denote

$$\mathbf{Bz}=\begin{pmatrix}\mathbf{I} & \mathbf{0}\\ -\boldsymbol{\Sigma}_{\mathbf{yx}}\boldsymbol{\Sigma}_{\mathbf{xx}}^{-} & \mathbf{I}\end{pmatrix}\begin{pmatrix}\mathbf{x}\\ \mathbf{y}\end{pmatrix}=\begin{pmatrix}\mathbf{x}\\ \mathbf{y}-\boldsymbol{\Sigma}_{\mathbf{yx}}\boldsymbol{\Sigma}_{\mathbf{xx}}^{-}\mathbf{x}\end{pmatrix}.$$

Then in view the block diagonalization theorem of a nonnegative definite matrix, see 20.26 (p. 89), we have

(a) $$\mathrm{cov}(\mathbf{Bz})=\mathbf{B}\boldsymbol{\Sigma}\mathbf{B}'=\begin{pmatrix}\mathbf{I} & \mathbf{0}\\ -\boldsymbol{\Sigma}_{\mathbf{yx}}\boldsymbol{\Sigma}_{\mathbf{xx}}^{-} & \mathbf{I}\end{pmatrix}\begin{pmatrix}\boldsymbol{\Sigma}_{\mathbf{xx}} & \boldsymbol{\Sigma}_{\mathbf{xy}}\\ \boldsymbol{\Sigma}_{\mathbf{yx}} & \boldsymbol{\Sigma}_{\mathbf{yy}}\end{pmatrix}\begin{pmatrix}\mathbf{I} & -(\boldsymbol{\Sigma}_{\mathbf{xx}}^{-})'\boldsymbol{\Sigma}_{\mathbf{xy}}\\ \mathbf{0} & \mathbf{I}\end{pmatrix}$$
$$=\begin{pmatrix}\boldsymbol{\Sigma}_{\mathbf{xx}} & \mathbf{0}\\ \mathbf{0} & \boldsymbol{\Sigma}_{\mathbf{yy}\cdot\mathbf{x}}\end{pmatrix},\quad \boldsymbol{\Sigma}_{\mathbf{yy}\cdot\mathbf{x}}=\boldsymbol{\Sigma}_{\mathbf{yy}}-\boldsymbol{\Sigma}_{\mathbf{yx}}\boldsymbol{\Sigma}_{\mathbf{xx}}^{-}\boldsymbol{\Sigma}_{\mathbf{xy}},$$

(b) $\mathrm{cov}(\mathbf{x},\mathbf{y}-\boldsymbol{\Sigma}_{\mathbf{yx}}\boldsymbol{\Sigma}_{\mathbf{xx}}^{-}\mathbf{x})=\mathbf{0}$, and

6.7 (a) $\mathrm{cov}(\mathbf{y}-\boldsymbol{\Sigma}_{\mathbf{yx}}\boldsymbol{\Sigma}_{\mathbf{xx}}^{-}\mathbf{x})=\boldsymbol{\Sigma}_{\mathbf{yy}\cdot\mathbf{x}}\leq_{\mathsf{L}}\mathrm{cov}(\mathbf{y}-\mathbf{Fx})$ for all $\mathbf{F}$,

(b) $\mathrm{BLP}(\mathbf{y};\mathbf{x})=\boldsymbol{\mu}_{\mathbf{y}}+\boldsymbol{\Sigma}_{\mathbf{yx}}\boldsymbol{\Sigma}_{\mathbf{xx}}^{-}(\mathbf{x}-\boldsymbol{\mu}_{\mathbf{x}})=(\boldsymbol{\mu}_{\mathbf{y}}-\boldsymbol{\Sigma}_{\mathbf{yx}}\boldsymbol{\Sigma}_{\mathbf{xx}}^{-}\boldsymbol{\mu}_{\mathbf{x}})+\boldsymbol{\Sigma}_{\mathbf{yx}}\boldsymbol{\Sigma}_{\mathbf{xx}}^{-}\mathbf{x}$.

6.8 Let $\mathbf{e}_{\mathbf{y}\cdot\mathbf{x}}=\mathbf{y}-\mathrm{BLP}(\mathbf{y};\mathbf{x})$ be the prediction error. Then

(a) $\mathbf{e}_{\mathbf{y}\cdot\mathbf{x}}=\mathbf{y}-\mathrm{BLP}(\mathbf{y};\mathbf{x})=\mathbf{y}-[\boldsymbol{\mu}_{\mathbf{y}}+\boldsymbol{\Sigma}_{\mathbf{yx}}\boldsymbol{\Sigma}_{\mathbf{xx}}^{-}(\mathbf{x}-\boldsymbol{\mu}_{\mathbf{x}})]$,

(b) $\mathrm{cov}(\mathbf{e}_{\mathbf{y}\cdot\mathbf{x}})=\boldsymbol{\Sigma}_{\mathbf{yy}}-\boldsymbol{\Sigma}_{\mathbf{yx}}\boldsymbol{\Sigma}_{\mathbf{xx}}^{-}\boldsymbol{\Sigma}_{\mathbf{xy}}=\boldsymbol{\Sigma}_{\mathbf{yy}\cdot\mathbf{x}}=\mathrm{MSEM}[\mathrm{BLP}(\mathbf{y};\mathbf{x});\mathbf{y}]$,

(c) $\mathrm{cov}(\mathbf{e}_{\mathbf{y}\cdot\mathbf{x}},\mathbf{x})=\mathbf{0}$.

6.9 According to 4.24 (p. 18), for the data matrix $\mathbf{U}=(\mathbf{X}:\mathbf{Y})$ we have

$$\mathbf{S}_{\mathbf{yy}} - \mathbf{S}_{\mathbf{yx}}\mathbf{S}_{\mathbf{xx}}^{-1}\mathbf{S}_{\mathbf{xy}} \leq_{\mathsf{L}} \operatorname{cov}_{\mathrm{d}}(\mathbf{Y} - \mathbf{XB}) = \operatorname{cov}_{\mathrm{s}}(\mathbf{y} - \mathbf{B}'\mathbf{x}) \quad \text{for all } \mathbf{B},$$
$$s_y^2 - \mathbf{s}_{\mathbf{x}y}'\mathbf{S}_{\mathbf{xx}}^{-1}\mathbf{s}_{\mathbf{x}y} \leq \operatorname{var}_{\mathrm{d}}(\mathbf{y} - \mathbf{Xb}) = \operatorname{var}_{\mathrm{s}}(y - \mathbf{b}'\mathbf{x}) \quad \text{for all } \mathbf{b},$$

where the minimum is attained when $\mathbf{B} = \mathbf{S}_{\mathbf{xx}}^{-1}\mathbf{S}_{\mathbf{xy}} = \mathbf{T}_{\mathbf{xx}}^{-1}\mathbf{T}_{\mathbf{xy}}$.

6.10 $\mathbf{x} \in \mathscr{C}(\mathbf{\Sigma}_{\mathbf{xx}} : \boldsymbol{\mu}_{\mathbf{x}})$ and $\mathbf{x} - \boldsymbol{\mu}_{\mathbf{x}} \in \mathscr{C}(\mathbf{\Sigma}_{\mathbf{xx}})$ with probability 1.

6.11 The following statements are equivalent:

(a) $\operatorname{cov}(\mathbf{x}, \mathbf{y} - \mathbf{Ax}) = \mathbf{0}$.

(b) $\mathbf{A}$ is a solution to $\mathbf{A\Sigma}_{\mathbf{xx}} = \mathbf{\Sigma}_{\mathbf{yx}}$.

(c) $\mathbf{A}$ is of the form $\mathbf{A} = \mathbf{\Sigma}_{\mathbf{yx}}\mathbf{\Sigma}_{\mathbf{xx}}^{-} + \mathbf{Z}(\mathbf{I}_p - \mathbf{\Sigma}_{\mathbf{xx}}\mathbf{\Sigma}_{\mathbf{xx}}^{-})$; $\mathbf{Z}$ is free to vary.

(d) $\mathbf{A}(\mathbf{x} - \boldsymbol{\mu}_{\mathbf{x}}) = \mathbf{\Sigma}_{\mathbf{yx}}\mathbf{\Sigma}_{\mathbf{xx}}^{-}(\mathbf{x} - \boldsymbol{\mu}_{\mathbf{x}})$ with probability 1.

6.12 If $\mathbf{b}_* = \mathbf{\Sigma}_{\mathbf{xx}}^{-1}\boldsymbol{\sigma}_{\mathbf{x}y}$ (no worries to use $\mathbf{\Sigma}_{\mathbf{xx}}^{-}$), then

(a)
$$\begin{aligned}\min_{\mathbf{b}} \operatorname{var}(y - \mathbf{b}'\mathbf{x}) &= \operatorname{var}(y - \mathbf{b}_*'\mathbf{x}) \\ &= \sigma_y^2 - \boldsymbol{\sigma}_{\mathbf{x}y}'\mathbf{\Sigma}_{\mathbf{xx}}^{-1}\boldsymbol{\sigma}_{\mathbf{x}y} = \sigma_{y\cdot 12\ldots p}^2 = \sigma_{y\cdot\mathbf{x}}^2,\end{aligned}$$

(b) $\operatorname{cov}(\mathbf{x}, y - \mathbf{b}_*'\mathbf{x}) = \mathbf{0}$,

(c)
$$\begin{aligned}\max_{\mathbf{b}} \operatorname{cor}^2(y, \mathbf{b}'\mathbf{x}) &= \operatorname{cor}^2(y, \mathbf{b}_*'\mathbf{x}) = \operatorname{cor}^2(y, \boldsymbol{\sigma}_{\mathbf{x}y}'\mathbf{\Sigma}_{\mathbf{xx}}^{-1}\mathbf{x}) \\ &= \frac{\boldsymbol{\sigma}_{\mathbf{x}y}'\mathbf{\Sigma}_{\mathbf{xx}}^{-1}\boldsymbol{\sigma}_{\mathbf{x}y}}{\sigma_y^2} = \varrho_{y\cdot\mathbf{x}}^2,\end{aligned}$$
squared population multiple correlation,

(d) $\sigma_{y\cdot\mathbf{x}}^2 = \sigma_y^2 - \boldsymbol{\sigma}_{\mathbf{x}y}'\mathbf{\Sigma}_{\mathbf{xx}}^{-1}\boldsymbol{\sigma}_{\mathbf{x}y} = \sigma_y^2(1 - \boldsymbol{\sigma}_{\mathbf{x}y}'\mathbf{\Sigma}_{\mathbf{xx}}^{-1}\boldsymbol{\sigma}_{\mathbf{x}y}/\sigma_y^2) = \sigma_y^2(1 - \varrho_{y\cdot\mathbf{x}}^2)$.

6.13 (a) The tasks of solving $\mathbf{b}$ from $\min_{\mathbf{b}} \operatorname{var}(y - \mathbf{b}'\mathbf{x})$ and $\max_{\mathbf{b}} \operatorname{cor}^2(y, \mathbf{b}'\mathbf{x})$ yield essentially the same solutions $\mathbf{b}_* = \mathbf{\Sigma}_{\mathbf{xx}}^{-1}\boldsymbol{\sigma}_{\mathbf{x}y}$.

(b) The tasks of solving $\mathbf{b}$ from $\min_{\mathbf{b}} \|\mathbf{y} - \mathbf{Ab}\|^2$ and $\max_{\mathbf{b}} \cos^2(\mathbf{y}, \mathbf{Ab})$ yield essentially the same solutions $\mathbf{Ab}_* = \mathbf{P}_{\mathbf{A}}\mathbf{y}$.

(c) The tasks of solving $\mathbf{b}$ from $\min_{\mathbf{b}} \operatorname{var}_{\mathrm{d}}(\mathbf{y} - \mathbf{X}_0\mathbf{b})$ and $\max_{\mathbf{b}} \operatorname{cor}_{\mathrm{d}}^2(\mathbf{y}, \mathbf{X}_0\mathbf{b})$ yield essentially the same solutions $\mathbf{b}_* = \mathbf{S}_{\mathbf{xx}}^{-1}\mathbf{s}_{\mathbf{x}y} = \hat{\boldsymbol{\beta}}_{\mathbf{x}}$.

6.14 Consider a random vector $\mathbf{z}$ with $\mathrm{E}(\mathbf{z}) = \boldsymbol{\mu}$, $\operatorname{cov}(\mathbf{z}) = \mathbf{\Sigma}$, and let the random vector $\operatorname{BLP}(\mathbf{z}; \mathbf{A}'\mathbf{z})$ denote the BLP of $\mathbf{z}$ based on $\mathbf{A}'\mathbf{z}$. Then

(a)
$$\begin{aligned}\operatorname{BLP}(\mathbf{z}; \mathbf{A}'\mathbf{z}) &= \boldsymbol{\mu} + \operatorname{cov}(\mathbf{z}, \mathbf{A}'\mathbf{z})[\operatorname{cov}(\mathbf{A}'\mathbf{z})]^{-}[\mathbf{A}'\mathbf{z} - \mathrm{E}(\mathbf{A}'\mathbf{z})] \\ &= \boldsymbol{\mu} + \mathbf{\Sigma A}(\mathbf{A}'\mathbf{\Sigma A})^{-}\mathbf{A}'(\mathbf{z} - \boldsymbol{\mu}) \\ &= \boldsymbol{\mu} + \mathbf{P}_{\mathbf{A};\mathbf{\Sigma}}'(\mathbf{z} - \boldsymbol{\mu}),\end{aligned}$$

(b) the covariance matrix of the prediction error $\mathbf{e}_{\mathbf{z}\cdot\mathbf{A}'\mathbf{z}} = \mathbf{z} - \operatorname{BLP}(\mathbf{z}; \mathbf{A}'\mathbf{z})$ is

$$\begin{aligned}\operatorname{cov}[\mathbf{z}-\mathrm{BLP}(\mathbf{z};\mathbf{A}'\mathbf{z})] &= \boldsymbol{\Sigma}-\boldsymbol{\Sigma}\mathbf{A}(\mathbf{A}'\boldsymbol{\Sigma}\mathbf{A})^{-}\mathbf{A}'\boldsymbol{\Sigma}\\ &= \boldsymbol{\Sigma}(\mathbf{I}-\mathbf{P}_{\mathbf{A};\boldsymbol{\Sigma}}) = \boldsymbol{\Sigma}^{1/2}(\mathbf{I}-\mathbf{P}_{\boldsymbol{\Sigma}^{1/2}\mathbf{A}})\boldsymbol{\Sigma}^{1/2}.\end{aligned}$$

6.15 Cook's trick. The best linear predictor of $\mathbf{z} = \binom{\mathbf{x}}{\mathbf{y}}$ on the basis of $\mathbf{x}$ is

$$\mathrm{BLP}(\mathbf{z};\mathbf{x}) = \begin{pmatrix}\mathbf{x}\\ \boldsymbol{\mu}_{\mathbf{y}}+\boldsymbol{\Sigma}_{\mathbf{yx}}\boldsymbol{\Sigma}_{\mathbf{xx}}^{-}(\mathbf{x}-\boldsymbol{\mu}_{\mathbf{x}})\end{pmatrix} = \begin{pmatrix}\mathbf{x}\\ \mathrm{BLP}(\mathbf{y};\mathbf{x})\end{pmatrix}.$$

6.16 Let $\operatorname{cov}(\mathbf{z}) = \boldsymbol{\Sigma}$ and denote $\mathbf{z}_{(i)} = (z_1,\dots,z_i)'$ and consider the following residuals (prediction errors)

$$e_{1\cdot 0} = z_1, \quad \mathbf{e}_{i\cdot 1\dots i-1} = z_i - \mathrm{BLP}(z_i;\mathbf{z}_{(i-1)}), \quad i=2,\dots,p.$$

Let $\mathbf{e}$ be a p-dimensional random vector of these residuals: $\mathbf{e} = (e_{1\cdot 0}, e_{2\cdot 1},\dots, e_{p\cdot 1\dots p-1})'$. Then $\operatorname{cov}(\mathbf{e}) := \mathbf{D}$ is a diagonal matrix and

$$\begin{aligned}\det(\boldsymbol{\Sigma}) &= \det[\operatorname{cov}(\mathbf{e})] = \sigma_1^2\sigma_{2\cdot1}^2\cdots\sigma_{p\cdot12\dots p-1}^2\\ &= (1-\varrho_{12}^2)(1-\varrho_{3\cdot12}^2)\cdots(1-\varrho_{p\cdot1\dots p-1}^2)\sigma_1^2\sigma_2^2\cdots\sigma_p^2\\ &\le \sigma_1^2\sigma_2^2\cdots\sigma_p^2.\end{aligned}$$

6.17 (Continued ...) The vector $\mathbf{e}$ can be written as $\mathbf{e} = \mathbf{F}\mathbf{z}$, where $\mathbf{F}$ is a lower triangular matrix (with ones in diagonal) and $\operatorname{cov}(\mathbf{e}) = \mathbf{D} = \mathbf{F}\boldsymbol{\Sigma}\mathbf{F}'$. Thereby $\boldsymbol{\Sigma} = \mathbf{F}^{-1}\mathbf{D}^{1/2}\mathbf{D}^{1/2}(\mathbf{F}')^{-1} := \mathbf{L}\mathbf{L}'$, where $\mathbf{L} = \mathbf{F}^{-1}\mathbf{D}^{1/2}$ is an lower triangular matrix. This gives a statistical proof of the triangular factorization of a nonnegative definite matrix.

6.18 Recursive decomposition of $1-\varrho_{y\cdot12\dots p}^2$:

$$1-\varrho_{y\cdot12\dots p}^2 = (1-\varrho_{y1}^2)(1-\varrho_{y2\cdot1}^2)(1-\varrho_{y3\cdot12}^2)\cdots(1-\varrho_{yp\cdot12\dots p-1}^2).$$

6.19 Mustonen's measure of multivariate dispersion. Let $\operatorname{cov}(\mathbf{z}) = \boldsymbol{\Sigma}_{p\times p}$. Then

$$\mathrm{Mvar}(\boldsymbol{\Sigma}) = \max\sum_{i=1}^{p}\sigma_{i\cdot12\dots i-1}^2,$$

where the maximum is sought over all permutations of $z_1,\dots,z_p$.

6.20 Consider the data matrix $\mathbf{X}_0 = (\mathbf{x}_1 : \dots : \mathbf{x}_k)$ and denote $\mathbf{e}_1 = (\mathbf{I}-\mathbf{P}_1)\mathbf{x}_1$, $\mathbf{e}_i = (\mathbf{I}-\mathbf{P}_{(1:\mathbf{x}_1:\dots:\mathbf{x}_{i-1})})\mathbf{x}_i$, $\mathbf{E} = (\mathbf{e}_1 : \mathbf{e}_2 : \dots : \mathbf{e}_k)$. Then $\mathbf{E}'\mathbf{E}$ is a diagonal matrix where $e_{ii} = \mathrm{SSE}(\mathbf{x}_i;\mathbf{1},\mathbf{x}_1,\dots,\mathbf{x}_{i-1})$, and

$$\begin{aligned}\det\left(\tfrac{1}{n-1}\mathbf{E}'\mathbf{E}\right) &= (1-r_{12}^2)(1-R_{3\cdot12}^2)\cdots(1-R_{k\cdot1\dots k-1}^2)s_1^2s_2^2\cdots s_k^2\\ &= \det(\mathbf{S}_{\mathbf{xx}}).\end{aligned}$$

6.21 AR(1)-structure. Let $y_i = \varrho y_{i-1}+u_i$, $i=1,\dots,n$, $|\varrho|<1$, where u_i's ($i=\dots,-2,-1,0,1,2,\dots$) are independent random variables, each having $\mathrm{E}(u_i) = 0$ and $\operatorname{var}(u_i) = \sigma_u^2$. Then

(a) $$\operatorname{cov}(\mathbf{y}) = \mathbf{\Sigma} = \sigma^2\mathbf{V} = \frac{\sigma_u^2}{1-\varrho^2}\begin{pmatrix} 1 & \varrho & \varrho^2 & \dots & \varrho^{n-1} \\ \varrho & 1 & \varrho & \dots & \varrho^{n-2} \\ \vdots & \vdots & \vdots & & \vdots \\ \varrho^{n-1} & \varrho^{n-2} & \varrho^{n-3} & \dots & 1 \end{pmatrix} = \frac{\sigma_u^2}{1-\varrho^2}\{\varrho^{|i-j|}\}.$$

(b) $$\operatorname{cor}(\mathbf{y}) = \mathbf{V} = \operatorname{cor}\begin{pmatrix} \mathbf{y}_{(n-1)} \\ y_n \end{pmatrix} = \begin{pmatrix} \mathbf{V}_{11} & \mathbf{v}_{12} \\ \mathbf{v}_{12}' & 1 \end{pmatrix} = \{\varrho^{|i-j|}\},$$

(c) $\mathbf{V}_{11}^{-1}\mathbf{v}_{12} = \mathbf{V}_{11}^{-1} \cdot \varrho\mathbf{V}_{11}\mathbf{i}_{n-1} = \varrho\mathbf{i}_{n-1} = (0,\dots,0,\varrho)' := \mathbf{b}_* \in \mathbb{R}^{n-1}$,

(d) $\operatorname{BLP}(y_n;\mathbf{y}_{(n-1)}) = \mathbf{b}_*'\mathbf{y}_{(n-1)}) = \mathbf{v}_{12}'\mathbf{V}_{11}^{-1}\mathbf{y}_{(n-1)} = \varrho y_{n-1}$,

(e) $e_n = y_n - \operatorname{BLP}(y_n;\mathbf{y}_{(n-1)}) = y_n - \varrho y_{n-1} = n$th prediction error,

(f) $\operatorname{cov}(\mathbf{y}_{(n-1)}, y_n - \mathbf{b}_*'\mathbf{y}_{(n-1)} = \operatorname{cov}(\mathbf{y}_{(n-1)}, y_n - \varrho y_{n-1}) = \mathbf{0} \in \mathbb{R}^{n-1}$,

(g) For each y_i, $i = 2,\dots,n$, we have

$$\operatorname{BLP}(y_i;\mathbf{y}_{(i-1)}) = \varrho y_{i-1}, \quad e_{y_i\cdot\mathbf{y}_{(i-1)}} = y_i - \varrho y_{i-1} := e_i.$$

Define $e_1 = y_1$, $e_i = y_i - \varrho y_{i-1}$, $i = 2,\dots,n$, i.e.,

$$\mathbf{e} = \begin{pmatrix} y_1 \\ y_2 - \varrho y_1 \\ y_3 - \varrho y_2 \\ \vdots \\ y_n - \varrho y_{n-1} \end{pmatrix} = \begin{pmatrix} 1 & 0 & 0 & \dots & 0 & 0 & 0 \\ -\varrho & 1 & 0 & \dots & 0 & 0 & 0 \\ \vdots & \vdots & \vdots & & \vdots & \vdots & \vdots \\ 0 & 0 & 0 & \dots & -\varrho & 1 & 0 \\ 0 & 0 & 0 & \dots & 0 & -\varrho & 1 \end{pmatrix}\mathbf{y} := \mathbf{L}\mathbf{y},$$

where $\mathbf{L} \in \mathbb{R}^{n\times n}$.

(h) $$\operatorname{cov}(\mathbf{e}) = \sigma^2\begin{pmatrix} 1 & \mathbf{0}' \\ \mathbf{0} & (1-\varrho^2)\mathbf{I}_{n-1} \end{pmatrix} := \sigma^2\mathbf{D} = \operatorname{cov}(\mathbf{L}\mathbf{y}) = \sigma^2\mathbf{L}\mathbf{V}\mathbf{L}',$$

(i) $\mathbf{L}\mathbf{V}\mathbf{L}' = \mathbf{D}$, $\quad \mathbf{V} = \mathbf{L}^{-1}\mathbf{D}(\mathbf{L}')^{-1}$, $\quad \mathbf{V}^{-1} = \mathbf{L}'\mathbf{D}^{-1}\mathbf{L}$,

(j) $$\mathbf{D}^{-1} = \tfrac{1}{1-\varrho^2}\begin{pmatrix} 1-\varrho^2 & \mathbf{0}' \\ \mathbf{0} & \mathbf{I}_{n-1} \end{pmatrix}, \quad \mathbf{D}^{-1/2} = \tfrac{1}{\sqrt{1-\varrho^2}}\begin{pmatrix} \sqrt{1-\varrho^2} & \mathbf{0}' \\ \mathbf{0} & \mathbf{I}_{n-1} \end{pmatrix},$$

(k) $$\mathbf{D}^{-1/2}\mathbf{L} = \frac{1}{\sqrt{1-\varrho^2}}\begin{pmatrix} \sqrt{1-\varrho^2} & 0 & 0 & \dots & 0 & 0 & 0 \\ -\varrho & 1 & 0 & \dots & 0 & 0 & 0 \\ \vdots & \vdots & \vdots & & \vdots & \vdots & \vdots \\ 0 & 0 & 0 & \dots & -\varrho & 1 & 0 \\ 0 & 0 & 0 & \dots & 0 & -\varrho & 1 \end{pmatrix} := \frac{1}{\sqrt{1-\varrho^2}}\mathbf{K},$$

(l) $$\mathbf{D}^{-1/2}\mathbf{L}\mathbf{V}\mathbf{L}'\mathbf{D}^{-1/2} = \frac{1}{1-\varrho^2}\mathbf{K}\mathbf{V}\mathbf{K}' = \mathbf{I}_n,$$

(m) $\mathbf{V}^{-1} = \mathbf{L}'\mathbf{D}^{-1}\mathbf{L} = \mathbf{L}'\mathbf{D}^{-1/2}\mathbf{D}^{-1/2}\mathbf{L} = \frac{1}{1-\varrho^2}\mathbf{K}'\mathbf{K}$

$$= \frac{1}{1-\varrho^2}\begin{pmatrix} 1 & -\varrho & 0 & \dots & 0 & 0 & 0 \\ -\varrho & 1+\varrho^2 & -\varrho & \dots & 0 & 0 & 0 \\ \vdots & \vdots & \vdots & & \vdots & \vdots & \vdots \\ 0 & 0 & 0 & \dots & -\varrho & 1+\varrho^2 & -\varrho \\ 0 & 0 & 0 & \dots & 0 & -\varrho & 1 \end{pmatrix},$$

(n) $\det(\mathbf{V}) = (1-\varrho^2)^{n-1}$.

(o) Consider the model $\mathscr{M} = \{\mathbf{y}, \mathbf{X}\boldsymbol{\beta}, \sigma^2\mathbf{V}\}$, where $\sigma^2 = \sigma_u^2/(1-\varrho^2)$. Premultiplying this model by $\mathbf{K}$ yields the model $\mathscr{M}_* = \{\mathbf{y}_*, \mathbf{X}_*\boldsymbol{\beta}, \sigma_u^2\mathbf{I}\}$, where $\mathbf{y}_* = \mathbf{K}\mathbf{y}$ and $\mathbf{X}_* = \mathbf{K}\mathbf{X}$.

6.22 Durbin–Watson test statistic for testing $\varrho = 0$:

$$DW = \sum_{i=2}^{n}(\hat\varepsilon_i - \hat\varepsilon_{i-1})^2 \Big/ \sum_{i=1}^{n}\hat\varepsilon_i^2 = \frac{\hat{\boldsymbol{\varepsilon}}'\mathbf{G}'\mathbf{G}\hat{\boldsymbol{\varepsilon}}}{\hat{\boldsymbol{\varepsilon}}'\hat{\boldsymbol{\varepsilon}}} \approx 2(1-\hat\varrho),$$

where $\hat\varrho = \sum_{i=2}^{n}\hat\varepsilon_{i-1}\hat\varepsilon_i \big/ \sum_{i=2}^{n}\hat\varepsilon_i^2$ and $\mathbf{G} \in \mathbb{R}^{(n-1)\times n}$ and $\mathbf{G}'\mathbf{G} \in \mathbb{R}^{n\times n}$ are

$$\mathbf{G} = \begin{pmatrix} -1 & 1 & 0 & \dots & 0 & 0 & 0 \\ 0 & -1 & 1 & \dots & 0 & 0 & 0 \\ \vdots & \vdots & \vdots & & \vdots & \vdots & \vdots \\ 0 & 0 & 0 & \dots & -1 & 1 & 0 \\ 0 & 0 & 0 & \dots & 0 & -1 & 1 \end{pmatrix},$$

$$\mathbf{G}'\mathbf{G} = \begin{pmatrix} 1 & -1 & 0 & \dots & 0 & 0 & 0 \\ -1 & 2 & -1 & \dots & 0 & 0 & 0 \\ \vdots & \vdots & \vdots & & \vdots & \vdots & \vdots \\ 0 & 0 & 0 & \dots & -1 & 2 & -1 \\ 0 & 0 & 0 & \dots & 0 & -1 & 1 \end{pmatrix}.$$

6.23 Let the sample covariance matrix of $\mathbf{x}_1, \dots, \mathbf{x}_k, \mathbf{y}$ be

$$\mathbf{S} = \{r^{|i-j|}\} = \begin{pmatrix} \mathbf{S}_{\mathbf{xx}} & \mathbf{s}_{\mathbf{x}y} \\ \mathbf{s}_{\mathbf{x}y}' & s_y^2 \end{pmatrix} \in \mathbb{R}^{(k+1)\times(k+1)}.$$

Then $\mathbf{s}_{\mathbf{x}y} = r\mathbf{S}_{\mathbf{xx}}\mathbf{i}_k$ and $\hat{\boldsymbol{\beta}}_{\mathbf{x}} = \mathbf{S}_{\mathbf{xx}}^{-1}\mathbf{s}_{\mathbf{x}y} = r\mathbf{i}_k = (0, \dots, 0, r)' \in \mathbb{R}^k$.

6.24 If $\mathbf{L} = \begin{pmatrix} 1 & 0 & 0 & \dots & 0 \\ 1 & 1 & 0 & \dots & 0 \\ 1 & 1 & 1 & \dots & 0 \\ \vdots & \vdots & \vdots & \ddots & \vdots \\ 1 & 1 & 1 & \dots & 1 \end{pmatrix} \in \mathbb{R}^{n\times n}$, then

$$(\mathbf{LL}')^{-1} = \begin{pmatrix} 1 & 1 & 1 & \dots & 1 & 1 \\ 1 & 2 & 2 & \dots & 2 & 2 \\ 1 & 2 & 3 & \dots & 3 & 3 \\ \vdots & \vdots & \vdots & \ddots & \vdots & \vdots \\ 1 & 2 & 3 & \dots & n-1 & n-1 \\ 1 & 2 & 3 & \dots & n-1 & n \end{pmatrix}^{-1} = \begin{pmatrix} 2 & -1 & 0 & \dots & 0 & 0 \\ -1 & 2 & -1 & \dots & 0 & 0 \\ 0 & -1 & 2 & \dots & 0 & 0 \\ \vdots & \vdots & \vdots & \ddots & \vdots & \vdots \\ 0 & 0 & 0 & \dots & 2 & -1 \\ 0 & 0 & 0 & \dots & -1 & 1 \end{pmatrix}.$$

7 Testing hypotheses

7.1 Consider the model $\mathscr{M} = \{\mathbf{y}, \mathbf{X}\boldsymbol{\beta}, \sigma^2\mathbf{I}\}$, where $\mathbf{y} \sim \mathrm{N}_n(\mathbf{X}\boldsymbol{\beta}, \sigma^2\mathbf{I})$, $\mathbf{X}$ is $n \times p$ $(p = k + 1)$. Most of the time we assume that $\mathbf{X} = (\mathbf{1} : \mathbf{X}_0)$ and it has full column rank. Let the hypothesis to be tested be

$$H\colon \mathbf{K}'\boldsymbol{\beta} = \mathbf{0}, \text{ where } \mathbf{K}' \in \mathbb{R}_q^{q\times p}, \text{ i.e., } \mathbf{K}_{p\times q} \text{ has a full column rank.}$$

Let $\mathscr{M}_H = \{\mathbf{y}, \mathbf{X}\boldsymbol{\beta} \mid \mathbf{K}'\boldsymbol{\beta} = \mathbf{0}, \sigma^2\mathbf{I}\}$ denote the model under H and let $\mathscr{C}(\mathbf{X}_*)$ be that subspace of $\mathscr{C}(\mathbf{X})$ where the hypothesis H holds:

$$\mathscr{C}(\mathbf{X}_*) = \{\,\mathbf{z} : \text{there exists } \mathbf{b} \text{ such that } \mathbf{z} = \mathbf{Xb} \text{ and } \mathbf{K}'\mathbf{b} = \mathbf{0}\,\}.$$

Then $\mathscr{M}_H = \{\mathbf{y}, \mathbf{X}_*\boldsymbol{\beta}, \sigma^2\mathbf{I}\}$ and the hypothesis is $H\colon \mathrm{E}(\mathbf{y}) \in \mathscr{C}(\mathbf{X}_*)$. In general we assume that $\mathbf{K}'\boldsymbol{\beta}$ is estimable, i.e., $\mathscr{C}(\mathbf{K}) \subset \mathscr{C}(\mathbf{X}')$ and the hypothesis said to be testable. If $\mathrm{r}(\mathbf{X}) = p$ then every $\mathbf{K}'\boldsymbol{\beta}$ is estimable.

7.2 The F-statistic for testing H: $F = \dfrac{Q/q}{\hat\sigma^2} = \dfrac{Q/q}{\mathrm{SSE}/(n-p)} \sim \mathrm{F}(q, n-p, \delta)$

7.3 $\mathrm{SSE}/\sigma^2 = \mathbf{y}'\mathbf{My}/\sigma^2 \sim \chi^2(n-p)$

7.4

$$\begin{aligned} Q &= \mathrm{SSE}_H - \mathrm{SSE} = \Delta\mathrm{SSE} && \text{change in SSE due to the hypothesis} \\ &= \mathrm{SSE}(\mathscr{M}_H) - \mathrm{SSE}(\mathscr{M}) && \mathrm{SSE}_H = \mathrm{SSE}(\mathscr{M}_H) = \text{SSE under } H \\ &= \mathbf{y}'(\mathbf{I} - \mathbf{P}_{\mathbf{X}_*})\mathbf{y} - \mathbf{y}'(\mathbf{I} - \mathbf{H})\mathbf{y} \\ &= \mathbf{y}'(\mathbf{H} - \mathbf{P}_{\mathbf{X}_*})\mathbf{y}, \text{ and hence} \end{aligned}$$

7.5

$$\begin{aligned} F &= \frac{[\mathrm{SSE}(\mathscr{M}_H) - \mathrm{SSE}(\mathscr{M})]/q}{\mathrm{SSE}(\mathscr{M})/(n-p)} \\ &= \frac{(R^2 - R_H^2)/q}{(1 - R^2)/(n-p)} && R_H^2 = R^2 \text{ under } H \\ &= \frac{[\mathrm{SSE}(\mathscr{M}_H) - \mathrm{SSE}(\mathscr{M})]/[\mathrm{r}(\mathbf{X}) - \mathrm{r}(\mathbf{X}_*)]}{\mathrm{SSE}(\mathscr{M})/[n - \mathrm{r}(\mathbf{X})]} \\ &= \frac{(R^2 - R_H^2)/[\mathrm{r}(\mathbf{X}) - \mathrm{r}(\mathbf{X}_*)]}{(1 - R^2)/[n - \mathrm{r}(\mathbf{X})]} \end{aligned}$$

7.6 $Q/\sigma^2 \sim \chi^2(q,\delta), \quad q = \mathrm{r}(\mathbf{X}) - \mathrm{r}(\mathbf{X}_*)$ central χ^2 if H true

7.7 $\delta = \boldsymbol{\beta}'\mathbf{X}'(\mathbf{H} - \mathbf{P}_{\mathbf{X}_*})\mathbf{X}\boldsymbol{\beta}/\sigma^2 = \boldsymbol{\beta}'\mathbf{X}'(\mathbf{I} - \mathbf{P}_{\mathbf{X}_*})\mathbf{X}\boldsymbol{\beta}/\sigma^2$

7.8 $\mathscr{C}(\mathbf{X}_*) = \mathscr{C}(\mathbf{X}\mathbf{K}^{\perp}), \quad \mathrm{r}(\mathbf{X}_*) = \mathrm{r}(\mathbf{X}) - \dim \mathscr{C}(\mathbf{X}') \cap \mathscr{C}(\mathbf{K}) = \mathrm{r}(\mathbf{X}) - \mathrm{r}(\mathbf{K})$

7.9 Consider hypothesis H: $\beta_{k-q+1} = \cdots = \beta_k = 0$, i.e., $\boldsymbol{\beta}_2 = \mathbf{0}$, when

$$\boldsymbol{\beta} = \begin{pmatrix} \boldsymbol{\beta}_1 \\ \boldsymbol{\beta}_2 \end{pmatrix}, \quad \boldsymbol{\beta}_2 \in \mathbb{R}^q, \quad \mathbf{X} = (\mathbf{X}_1 : \mathbf{X}_2), \quad \mathbf{X}_1 = (\mathbf{1} : \mathbf{x}_1 : \ldots : \mathbf{x}_{k-q}).$$

Then $\mathbf{K}' = (\mathbf{0} : \mathbf{I}_q)$, $\mathbf{X}_* = \mathbf{X}_1 = \mathbf{X}\mathbf{K}^{\perp}$, and 7.10–7.16 hold.

7.10
$$\begin{aligned} Q &= \mathbf{y}'(\mathbf{H} - \mathbf{P}_{\mathbf{X}_1})\mathbf{y} \\ &= \mathbf{y}'\mathbf{P}_{\mathbf{M}_1\mathbf{X}_2}\mathbf{y} \quad [\mathbf{M}_1 = \mathbf{I} - \mathbf{P}_{\mathbf{X}_1}] && \delta = \boldsymbol{\beta}_2'\mathbf{X}_2'\mathbf{M}_1\mathbf{X}_2\boldsymbol{\beta}_2/\sigma^2 \\ &= \mathbf{y}'\mathbf{M}_1\mathbf{X}_2(\mathbf{X}_2'\mathbf{M}_1\mathbf{X}_2)^{-1}\mathbf{X}_2'\mathbf{M}_1\mathbf{y} \\ &= \hat{\boldsymbol{\beta}}_2'\mathbf{X}_2'\mathbf{M}_1\mathbf{X}_2\hat{\boldsymbol{\beta}}_2 = \hat{\boldsymbol{\beta}}_2'\mathbf{T}_{22\cdot1}\hat{\boldsymbol{\beta}}_2 && \mathbf{T}_{22\cdot1} = \mathbf{X}_2'\mathbf{M}_1\mathbf{X}_2 \\ &= \hat{\boldsymbol{\beta}}_2'[\mathrm{cov}(\hat{\boldsymbol{\beta}}_2)]^{-1}\hat{\boldsymbol{\beta}}_2\sigma^2 && \mathrm{cov}(\hat{\boldsymbol{\beta}}_2) = \sigma^2(\mathbf{X}_2'\mathbf{M}_1\mathbf{X}_2)^{-1} \end{aligned}$$

7.11 $\dfrac{(\hat{\boldsymbol{\beta}}_2 - \boldsymbol{\beta}_2)'\mathbf{T}_{22\cdot1}(\hat{\boldsymbol{\beta}}_2 - \boldsymbol{\beta}_2)/q}{\hat{\sigma}^2} \le F_{\alpha;q,n-k-1}$ confidence ellipsoid for $\boldsymbol{\beta}_2$

7.12 The left-hand side of 7.11 can be written as

$$\frac{(\hat{\boldsymbol{\beta}}_2 - \boldsymbol{\beta}_2)'[\mathrm{cov}(\hat{\boldsymbol{\beta}}_2)]^{-1}(\hat{\boldsymbol{\beta}}_2 - \boldsymbol{\beta}_2)\sigma^2/q}{\hat{\sigma}^2} = (\hat{\boldsymbol{\beta}}_2 - \boldsymbol{\beta}_2)'[\widehat{\mathrm{cov}}(\hat{\boldsymbol{\beta}}_2)]^{-1}(\hat{\boldsymbol{\beta}}_2 - \boldsymbol{\beta}_2)/q$$

and hence the confidence region for $\boldsymbol{\beta}_2$ is

$$(\hat{\boldsymbol{\beta}}_2 - \boldsymbol{\beta}_2)'[\widehat{\mathrm{cov}}(\hat{\boldsymbol{\beta}}_2)]^{-1}(\hat{\boldsymbol{\beta}}_2 - \boldsymbol{\beta}_2)/q \le F_{\alpha;q,n-k-1}.$$

7.13 Consider the last two regressors: $\mathbf{X} = (\mathbf{X}_1 : \mathbf{x}_{k-1}, \mathbf{x}_k)$. Then 3.29 (p. 14) implies that

$$\mathrm{cor}(\hat{\boldsymbol{\beta}}_2) = \begin{pmatrix} 1 & -r_{k-1,k\cdot\mathbf{X}_1} \\ -r_{k-1,k\cdot\mathbf{X}_1} & 1 \end{pmatrix}, \quad r_{k-1,k\cdot\mathbf{X}_1} = r_{k-1,k\cdot12\ldots k-2},$$

and hence the orientation of the ellipse (the directions of the major and minor axes) is determined by the partial correlation between $\mathbf{x}_{k-1}$ and $\mathbf{x}_k$.

7.14 The volume of the ellipsoid $(\hat{\boldsymbol{\beta}}_2 - \boldsymbol{\beta}_2)'[\widehat{\mathrm{cov}}(\hat{\boldsymbol{\beta}}_2)]^{-1}(\hat{\boldsymbol{\beta}}_2 - \boldsymbol{\beta}_2) = c^2$ is proportional to $c^q \det[\widehat{\mathrm{cov}}(\hat{\boldsymbol{\beta}}_2)]$.

7.15 $\dfrac{(\hat{\boldsymbol{\beta}}_{\mathbf{x}} - \boldsymbol{\beta}_{\mathbf{x}})'\mathbf{T}_{\mathbf{xx}}(\hat{\boldsymbol{\beta}}_{\mathbf{x}} - \boldsymbol{\beta}_{\mathbf{x}})/k}{\hat{\sigma}^2} \le F_{\alpha;k,n-k-1}$ confidence region for $\boldsymbol{\beta}_{\mathbf{x}}$

7.16 $Q = \|\mathrm{res}(\mathbf{y};\mathbf{X}_1) - \mathrm{res}(\mathbf{y};\mathbf{X})\|^2 = \mathrm{SSE}(\mathbf{y};\mathbf{X}_1) - \mathrm{SSE}(\mathbf{y};\mathbf{X}_1,\mathbf{X}_2)$
$= \Delta\mathrm{SSE} =$ change in SSE when adding $\mathbf{X}_2$ to the model
$= \mathrm{SSR} - \mathrm{SSR}_H = \mathbf{y}'(\mathbf{H}-\mathbf{J})\mathbf{y} - \mathbf{y}'(\mathbf{P}_{\mathbf{X}_*} - \mathbf{J})\mathbf{y} = \Delta\mathrm{SSR}$

7.17 If the hypothesis is $H\colon \mathbf{K}'\boldsymbol{\beta} = \mathbf{d}$, where $\mathbf{K}' \in \mathbb{R}_q^{q\times p}$, then

(a) $Q = (\mathbf{K}'\hat{\boldsymbol{\beta}} - \mathbf{d})'[\mathbf{K}'(\mathbf{X}'\mathbf{X})^{-1}\mathbf{K}]^{-1}(\mathbf{K}'\hat{\boldsymbol{\beta}} - \mathbf{d})$
$= (\mathbf{K}'\hat{\boldsymbol{\beta}} - \mathbf{d})'[\mathrm{cov}(\mathbf{K}'\hat{\boldsymbol{\beta}})]^{-1}(\mathbf{K}'\hat{\boldsymbol{\beta}} - \mathbf{d})\sigma^2 := \mathbf{u}'[\mathrm{cov}(\mathbf{u})]^{-1}\mathbf{u}\sigma^2,$

(b) $Q/\sigma^2 = \chi^2(q,\delta), \qquad \delta = (\mathbf{K}'\boldsymbol{\beta} - \mathbf{d})'[\mathbf{K}'(\mathbf{X}'\mathbf{X})^{-1}\mathbf{K}]^{-1}(\mathbf{K}'\boldsymbol{\beta} - \mathbf{d})/\sigma^2$

(c) $F = \dfrac{Q/q}{\hat{\sigma}^2} = (\mathbf{K}'\hat{\boldsymbol{\beta}} - \mathbf{d})'[\widehat{\mathrm{cov}}(\mathbf{K}'\hat{\boldsymbol{\beta}})]^{-1}(\mathbf{K}'\hat{\boldsymbol{\beta}} - \mathbf{d})/q \sim \mathrm{F}(q, p-q, \cdot),$

(d) $\mathbf{K}'\hat{\boldsymbol{\beta}} - \mathbf{d} \sim \mathrm{N}_q[\mathbf{K}'\boldsymbol{\beta} - \mathbf{d}, \sigma^2\mathbf{K}'(\mathbf{X}'\mathbf{X})^{-1}\mathbf{K}],$

(e) $\mathrm{SSE}_H = \min_{\mathbf{K}'\boldsymbol{\beta}=\mathbf{d}} \|\mathbf{y} - \mathbf{X}\boldsymbol{\beta}\|^2 = \|\mathbf{y} - \mathbf{X}\hat{\boldsymbol{\beta}}_r\|^2,$

(f) $\hat{\boldsymbol{\beta}}_r = \hat{\boldsymbol{\beta}} - (\mathbf{X}'\mathbf{X})^{-1}\mathbf{K}[\mathbf{K}'(\mathbf{X}'\mathbf{X})^{-1}\mathbf{K}]^{-1}(\mathbf{K}'\hat{\boldsymbol{\beta}} - \mathbf{d}).$ restricted OLSE

7.18 The restricted OLSE $\hat{\boldsymbol{\beta}}_r$ is the solution to equation

$$\begin{pmatrix} \mathbf{X}'\mathbf{X} & \mathbf{K} \\ \mathbf{K}' & \mathbf{0} \end{pmatrix} \begin{pmatrix} \hat{\boldsymbol{\beta}}_r \\ \boldsymbol{\lambda} \end{pmatrix} = \begin{pmatrix} \mathbf{X}'\mathbf{y} \\ \mathbf{d} \end{pmatrix}, \quad \text{where } \boldsymbol{\lambda} \text{ is the Lagrangian multiplier.}$$

The equation above has a unique solution iff $\mathrm{r}(\mathbf{X}' : \mathbf{K}) = p + q$, and hence $\hat{\boldsymbol{\beta}}_r$ may be unique even though $\hat{\boldsymbol{\beta}}$ is not unique.

7.19 If the restriction is $\mathbf{K}'\boldsymbol{\beta} = \mathbf{0}$ and $\mathbf{L}$ is a matrix (of full column rank) such that $\mathbf{L} \in \{\mathbf{K}^{\perp}\}$, then $\mathbf{X}_* = \mathbf{X}\mathbf{L}$, and the following holds:

(a) $\mathbf{X}\hat{\boldsymbol{\beta}}_r = \mathbf{X}_*(\mathbf{X}_*'\mathbf{X}_*)^{-1}\mathbf{X}_*'\mathbf{y} = \mathbf{X}\mathbf{L}(\mathbf{L}'\mathbf{X}'\mathbf{X}\mathbf{L})^{-1}\mathbf{L}'\mathbf{X}'\mathbf{y},$

(b) $\hat{\boldsymbol{\beta}}_r = \big(\mathbf{I}_p - (\mathbf{X}'\mathbf{X})^{-1}\mathbf{K}[\mathbf{K}'(\mathbf{X}'\mathbf{X})^{-1}\mathbf{K}]\mathbf{K}'\big)\hat{\boldsymbol{\beta}}$
$= (\mathbf{I}_p - \mathbf{P}'_{\mathbf{K};(\mathbf{X}'\mathbf{X})^{-1}})\hat{\boldsymbol{\beta}} = \mathbf{P}_{\mathbf{L};\mathbf{X}'\mathbf{X}}\hat{\boldsymbol{\beta}}$
$= \mathbf{L}(\mathbf{L}'\mathbf{X}'\mathbf{X}\mathbf{L})^{-1}\mathbf{L}'\mathbf{X}'\mathbf{X}\hat{\boldsymbol{\beta}} = \mathbf{L}(\mathbf{L}'\mathbf{X}'\mathbf{X}\mathbf{L})^{-1}\mathbf{L}'\mathbf{X}'\mathbf{y}.$

(c) $\mathrm{cov}(\hat{\boldsymbol{\beta}}_r) = \sigma^2\mathbf{L}(\mathbf{L}'\mathbf{X}'\mathbf{X}\mathbf{L})^{-1}\mathbf{L}'$
$= \sigma^2\big[(\mathbf{X}'\mathbf{X})^{-1} - (\mathbf{X}'\mathbf{X})^{-1}\mathbf{K}[\mathbf{K}'(\mathbf{X}'\mathbf{X})^{-1}\mathbf{K}]^{-1}\mathbf{K}'(\mathbf{X}'\mathbf{X})^{-1}\big].$

7.20 If we want to find $\mathbf{K}$ so that there is only one solution to $\mathbf{X}'\mathbf{X}\boldsymbol{\beta} = \mathbf{X}'\mathbf{y}$ which satisfies the constraint $\mathbf{K}'\boldsymbol{\beta} = \mathbf{0}$, we need to consider the equation

$$\begin{pmatrix} \mathbf{X}'\mathbf{X} \\ \mathbf{K}' \end{pmatrix} \boldsymbol{\beta} = \begin{pmatrix} \mathbf{X}'\mathbf{y} \\ \mathbf{0} \end{pmatrix}, \quad \text{or equivalently,} \quad \begin{pmatrix} \mathbf{X} \\ \mathbf{K}' \end{pmatrix} \boldsymbol{\beta} = \begin{pmatrix} \mathbf{P}_{\mathbf{X}}\mathbf{y} \\ \mathbf{0} \end{pmatrix}.$$

The above equation has a unique solution for every given $\mathbf{P}_{\mathbf{X}}\mathbf{y}$ iff $\mathscr{C}(\mathbf{X}') \cap \mathscr{C}(\mathbf{K}) = \{\mathbf{0}\}$ and $\binom{\mathbf{X}}{\mathbf{K}'}$ has a full column rank in which case the solution for

$\boldsymbol{\beta}$ is $(\mathbf{X}'\mathbf{X}+\mathbf{K}\mathbf{K}')^{-1}\mathbf{X}'\mathbf{y}$. Requirement $\mathscr{C}(\mathbf{X}') \cap \mathscr{C}(\mathbf{K}) = \{\mathbf{0}\}$ means that $\mathbf{K}'\boldsymbol{\beta}$ is not estimable.

7.21 $H\colon \beta_1 = \dots = \beta_k = 0$, i.e., $H\colon \mathrm{E}(\mathbf{y}) \in \mathscr{C}(\mathbf{1})$:

$$F = F_{\text{overall}} = \frac{\text{MSR}}{\text{MSE}} = \frac{R^2/k}{(1-R^2)/(n-k-1)} \sim \mathrm{F}(k, n-k-1, \cdot)$$

7.22 $H\colon \beta_i = 0\colon F = F(\hat{\beta}_i) = \dfrac{\hat{\beta}_i^2}{\mathrm{se}^2(\hat{\beta}_i)} = \dfrac{\Delta\text{SSE}}{\text{MSE}} = \dfrac{\Delta R^2}{(1-R^2)/(n-k-1)}$,

$$= \frac{r^2_{yi\cdot\text{rest}}}{(1-r^2_{yi\cdot\text{rest}})/(n-k-1)} \sim \mathrm{F}(1, n-k-1, \cdot),$$

where $\Delta R^2 = R^2 - R^2_{(i)}$ = change in R^2 when x_i deleted, $R^2_{(i)} = R^2(\mathbf{y}; \mathbf{X}_{(-i)})$, $\mathbf{X}_{(-i)} = \{$all other regressors except the ith$\} = \{$rest$\}$.

7.23 $H\colon \beta_i = 0\colon t = t(\hat{\beta}_i) = \dfrac{\hat{\beta}_i}{\mathrm{se}(\hat{\beta}_i)} = \dfrac{\hat{\beta}_i}{\sqrt{\widehat{\mathrm{var}}(\hat{\beta}_i)}} = \sqrt{F(\hat{\beta}_i)} \sim \mathrm{t}(n-k-1)$

7.24 $H\colon \mathbf{k}'\boldsymbol{\beta} = 0\colon F = \dfrac{(\mathbf{k}'\hat{\boldsymbol{\beta}})^2}{\hat{\sigma}^2\mathbf{k}'(\mathbf{X}'\mathbf{X})^{-1}\mathbf{k}} = \dfrac{(\mathbf{k}'\hat{\boldsymbol{\beta}})^2}{\widehat{\mathrm{var}}(\mathbf{k}'\hat{\boldsymbol{\beta}})} \sim \mathrm{F}(1, n-k-1, \cdot)$

7.25 $H\colon \mathbf{K}'\boldsymbol{\beta} = \mathbf{d}\colon F = \dfrac{\Delta R^2/q}{(1-R^2)/(n-p)} \sim \mathrm{F}(q, n-p, \cdot)$,

where $\Delta R^2 = R^2 - R^2_H$ = change in R^2 due to the hypothesis.

7.26 In the no-intercept model R^2 has to be replaced with R^2_c.

7.27 Consider the model $\mathscr{M} = \{\mathbf{y}, \mathbf{X}\boldsymbol{\beta}, \sigma^2\mathbf{V}\}$, where $\mathbf{X} \in \mathbb{R}^{n\times p}_p$, $\mathbf{V}$ is pd, $\mathbf{y} \sim \mathrm{N}_n(\mathbf{X}\boldsymbol{\beta}, \sigma^2\mathbf{V})$, and $F(V)$ is the F-statistic for testing linear hypothesis $H\colon \mathbf{K}'\boldsymbol{\beta} = \mathbf{d}$ ($\mathbf{K}' \in \mathbb{R}^{q\times p}_q$). Then

(a) $F(V) = \dfrac{Q(V)/q}{\tilde{\sigma}^2} = \dfrac{Q(V)/q}{\mathrm{SSE}(V)/(n-p)} \sim \mathrm{F}(q, n-p, \delta)$,

(b) $\tilde{\sigma}^2 = \mathrm{SSE}(V)/f, \quad f = n-p$, unbiased estimator of σ^2 under $\{\mathbf{y}, \mathbf{X}\boldsymbol{\beta}, \sigma^2\mathbf{V}\}$

(c) $\mathrm{SSE}(V) = \min_{\boldsymbol{\beta}} \|\mathbf{y} - \mathbf{X}\boldsymbol{\beta}\|^2_{\mathbf{V}^{-1}}$ weighted SSE

$$= \min_{\boldsymbol{\beta}} (\mathbf{y} - \mathbf{X}\boldsymbol{\beta})'\mathbf{V}^{-1}(\mathbf{y} - \mathbf{X}\boldsymbol{\beta}) = (\mathbf{y} - \mathbf{X}\tilde{\boldsymbol{\beta}})'\mathbf{V}^{-1}(\mathbf{y} - \mathbf{X}\tilde{\boldsymbol{\beta}}),$$

(d) $\mathrm{SSE}(V)/\sigma^2 \sim \chi^2(n-p)$,

(e) $\tilde{\boldsymbol{\beta}} = (\mathbf{X}'\mathbf{V}^{-1}\mathbf{X})^{-1}\mathbf{X}'\mathbf{V}^{-1}\mathbf{y} = \mathrm{BLUE}(\boldsymbol{\beta})$ under $\{\mathbf{y}, \mathbf{X}\boldsymbol{\beta}, \sigma^2\mathbf{V}\}$,

(f) $\operatorname{cov}(\mathbf{K}'\tilde{\boldsymbol{\beta}}) = \sigma^2\mathbf{K}'(\mathbf{X}'\mathbf{V}^{-1}\mathbf{X})^{-1}\mathbf{K} = \mathbf{K}'\operatorname{cov}(\tilde{\boldsymbol{\beta}})\mathbf{K}$,

(g) $\widetilde{\operatorname{cov}}(\mathbf{K}'\tilde{\boldsymbol{\beta}}) = \tilde{\sigma}^2\mathbf{K}'(\mathbf{X}'\mathbf{V}^{-1}\mathbf{X})^{-1}\mathbf{K}$,

(h) $Q(V) = \text{SSE}_H(V) - \text{SSE}(V) = \Delta\text{SSE}(V)$ change in $\text{SSE}(V)$ due to H,

(i) $$Q(V) = (\mathbf{K}'\tilde{\boldsymbol{\beta}} - \mathbf{d})'[\mathbf{K}'(\mathbf{X}'\mathbf{V}^{-1}\mathbf{X})^{-1}\mathbf{K}]^{-1}(\mathbf{K}'\tilde{\boldsymbol{\beta}} - \mathbf{d})$$
$$= (\mathbf{K}'\tilde{\boldsymbol{\beta}} - \mathbf{d})'[\operatorname{cov}(\mathbf{K}'\tilde{\boldsymbol{\beta}})]^{-1}(\mathbf{K}'\tilde{\boldsymbol{\beta}} - \mathbf{d})\sigma^2 := \mathbf{u}'[\operatorname{cov}(\mathbf{u})]^{-1}\mathbf{u}\sigma^2,$$

(j) $Q(V)/\sigma^2 = (\mathbf{K}'\tilde{\boldsymbol{\beta}} - \mathbf{d})'[\operatorname{cov}(\mathbf{K}'\tilde{\boldsymbol{\beta}})]^{-1}(\mathbf{K}'\tilde{\boldsymbol{\beta}} - \mathbf{d}) \sim \chi^2(q, \delta)$,

(k) $\delta = (\mathbf{K}'\boldsymbol{\beta} - \mathbf{d})'[\operatorname{cov}(\mathbf{K}'\tilde{\boldsymbol{\beta}})]^{-1}(\mathbf{K}'\boldsymbol{\beta} - \mathbf{d})/\sigma^2$,

(l) $$F(V) = \frac{(\mathbf{K}'\tilde{\boldsymbol{\beta}} - \mathbf{d})'[\operatorname{cov}(\mathbf{K}'\tilde{\boldsymbol{\beta}})]^{-1}(\mathbf{K}'\tilde{\boldsymbol{\beta}} - \mathbf{d})\sigma^2/q}{\tilde{\sigma}^2}$$
$$= (\mathbf{K}'\tilde{\boldsymbol{\beta}} - \mathbf{d})'[\widetilde{\operatorname{cov}}(\mathbf{K}'\tilde{\boldsymbol{\beta}})]^{-1}(\mathbf{K}'\tilde{\boldsymbol{\beta}} - \mathbf{d})/q \sim \mathrm{F}(q, f, \delta),$$

(m) $\mathbf{K}'\tilde{\boldsymbol{\beta}} - \mathbf{d} \sim \mathrm{N}_q[\mathbf{K}'\boldsymbol{\beta} - \mathbf{d}, \sigma^2\mathbf{K}'(\mathbf{X}'\mathbf{V}^{-1}\mathbf{X})^{-1}\mathbf{K}]$,

(n) $\text{SSE}_H(V) = \min_{\mathbf{K}'\boldsymbol{\beta}=\mathbf{d}}\|\mathbf{y} - \mathbf{X}\boldsymbol{\beta}\|^2_{\mathbf{V}^{-1}} = \|\mathbf{y} - \mathbf{X}\tilde{\boldsymbol{\beta}}_r\|^2_{\mathbf{V}^{-1}}$, where

(o) $\tilde{\boldsymbol{\beta}}_r = \tilde{\boldsymbol{\beta}} - (\mathbf{X}'\mathbf{V}^{-1}\mathbf{X})^{-1}\mathbf{K}[\mathbf{K}'(\mathbf{X}'\mathbf{V}^{-1}\mathbf{X})^{-1}\mathbf{K}]^{-1}(\mathbf{K}'\tilde{\boldsymbol{\beta}} - \mathbf{d})$ restricted BLUE.

7.28 Consider general situation, when $\mathbf{V}$ is possibly singular. Denote

$$\text{SSE}(W) = (\mathbf{y} - \mathbf{X}\tilde{\boldsymbol{\beta}})'\mathbf{W}^{-}(\mathbf{y} - \mathbf{X}\tilde{\boldsymbol{\beta}}), \quad \text{where } \mathbf{W} = \mathbf{V} + \mathbf{XUX}'$$

with $\mathbf{U}$ satisfying $\mathscr{C}(\mathbf{W}) = \mathscr{C}(\mathbf{X} : \mathbf{V})$ and $\mathbf{X}\tilde{\boldsymbol{\beta}} = \text{BLUE}(\mathbf{X}\boldsymbol{\beta})$. Then

$$\tilde{\sigma}^2 = \text{SSE}(W)/f, \quad \text{where } f = \mathrm{r}(\mathbf{X} : \mathbf{V}) - \mathrm{r}(\mathbf{X}) = \mathrm{r}(\mathbf{VM}),$$

is an unbiased estimator for σ^2. Let $\mathbf{K}'\boldsymbol{\beta}$ be estimable, i.e., $\mathscr{C}(\mathbf{K}) \subset \mathscr{C}(\mathbf{X}')$, denote $\mathbf{u} = \mathbf{K}'\tilde{\boldsymbol{\beta}} - \mathbf{d}$, $\operatorname{cov}(\mathbf{K}'\tilde{\boldsymbol{\beta}}) = \sigma^2\mathbf{D}$, $m = \mathrm{r}(\mathbf{D})$, and assume that $\mathbf{DD}^{-}\mathbf{u} = \mathbf{u}$. Then $(\mathbf{u}'\mathbf{D}^{-}\mathbf{u}/m)/\sigma^2 \sim \mathrm{F}(m, f, \cdot)$, i.e.,

$$(\mathbf{K}'\tilde{\boldsymbol{\beta}} - \mathbf{d})'[\widetilde{\operatorname{cov}}(\mathbf{K}'\tilde{\boldsymbol{\beta}})]^{-}(\mathbf{K}'\tilde{\boldsymbol{\beta}} - \mathbf{d})/m \sim \mathrm{F}(m, f, \cdot).$$

7.29 If $\mathrm{r}(\mathbf{X}) < p$ but $\mathbf{K}'\boldsymbol{\beta}$ is estimable, then the hypothesis $\mathbf{K}'\boldsymbol{\beta} = \mathbf{d}$ can be tested as presented earlier by replacing appropriated inverses with generalized inverses and replacing p $(= k + 1)$ with $r = \mathrm{r}(\mathbf{X})$.

7.30 One-way ANOVA for g groups; hypothesis is $H\colon \mu_1 = \cdots = \mu_g$.

group 1: $y_{11}, y_{12}, \ldots, y_{1n_1}$, $\quad \text{SS}_1 = \sum_{j=1}^{n_1}(y_{1j} - \bar{y}_1)^2$

group g: $y_{g1}, y_{g2}, \ldots, y_{gn_g}$, $\quad \text{SS}_g = \sum_{j=1}^{n_g}(y_{gj} - \bar{y}_g)^2$

(a) $\text{SSE} = \text{SSEWithin} = \text{SS}_1 + \cdots + \text{SS}_g$, $\quad \text{SSBetween} = \displaystyle\sum_{i=1}^{g} n_i(\bar{y}_i - \bar{y})^2$,

(b) $F = \dfrac{\mathrm{SSB}/(g-1)}{\mathrm{SSE}/(n-g)} \sim \mathrm{F}(g-1, n-g)$ if H is true.

7.31 t-test for the equality of the expectations of 2 groups: $H: \mu_1 = \mu_2$.

$$F = \frac{(\bar{y}_1 - \bar{y}_2)^2}{\dfrac{\mathrm{SS}_1 + \mathrm{SS}_2}{n-2}\left(\dfrac{1}{n_1} + \dfrac{1}{n_2}\right)} = \frac{n_1(\bar{y}_1 - \bar{y})^2 + n_2(\bar{y}_2 - \bar{y})^2}{\dfrac{\mathrm{SS}_1 + \mathrm{SS}_2}{n-2}}$$

$$= \frac{n_1 n_2}{n} \cdot (\bar{y}_1 - \bar{y}_2) \cdot \left(\frac{\mathrm{SS}_1 + \mathrm{SS}_2}{n-2}\right)^{-1} \cdot (\bar{y}_1 - \bar{y}_2)$$

$$= \frac{(\bar{y}_1 - \bar{y}_2)^2}{\dfrac{\mathrm{SSE}}{n-2}\dfrac{n}{n_1 n_2}} \sim \mathrm{F}(1, n-2) = \mathrm{t}^2(n-2), \quad n_1 + n_2 = n$$

7.32 One-way ANOVA in matrix terms.

$$\mathbf{y} = \mathbf{X}\boldsymbol{\mu} + \boldsymbol{\varepsilon}, \quad \mathbf{y}_i = (y_{i1}, y_{i2}, \dots, y_{in_i})', \quad i = 1, \dots, g,$$

$$\begin{pmatrix} \mathbf{y}_1 \\ \vdots \\ \mathbf{y}_g \end{pmatrix} = \begin{pmatrix} \mathbf{1}_{n_1} & \dots & \mathbf{0} \\ \vdots & \ddots & \vdots \\ \mathbf{0} & \dots & \mathbf{1}_{n_g} \end{pmatrix} \begin{pmatrix} \mu_1 \\ \vdots \\ \mu_g \end{pmatrix} + \begin{pmatrix} \boldsymbol{\varepsilon}_1 \\ \vdots \\ \boldsymbol{\varepsilon}_g \end{pmatrix},$$

$$\mathbf{H} = \begin{pmatrix} \mathbf{J}_{n_1} & \dots & \mathbf{0} \\ \vdots & \ddots & \vdots \\ \mathbf{0} & \dots & \mathbf{J}_{n_g} \end{pmatrix}, \quad \mathbf{M} = \begin{pmatrix} \mathbf{I}_{n_1} - \mathbf{J}_{n_1} & \dots & \mathbf{0} \\ \vdots & \ddots & \vdots \\ \mathbf{0} & \dots & \mathbf{I}_{n_g} - \mathbf{J}_{n_g} \end{pmatrix},$$

$$\mathbf{Hy} = \begin{pmatrix} \bar{y}_1 \mathbf{1}_{n_1} \\ \vdots \\ \bar{y}_g \mathbf{1}_{n_g} \end{pmatrix} = \begin{pmatrix} \bar{\bar{\mathbf{y}}}_1 \\ \vdots \\ \bar{\bar{\mathbf{y}}}_g \end{pmatrix}, \quad \mathbf{My} = \begin{pmatrix} \mathbf{C}_{n_1}\mathbf{y}_1 \\ \vdots \\ \mathbf{C}_{n_g}\mathbf{y}_g \end{pmatrix},$$

$$(\mathbf{H} - \mathbf{J}_n)\mathbf{y} = \begin{pmatrix} (\bar{y}_1 - \bar{y})\mathbf{1}_{n_1} \\ \vdots \\ (\bar{y}_g - \bar{y})\mathbf{1}_{n_g} \end{pmatrix},$$

$$\mathrm{SST} = \mathbf{y}'\mathbf{C}_n\mathbf{y} = \sum_{i=1}^{g}\sum_{j=1}^{n_i}(y_{ij} - \bar{y})^2,$$

$$\mathrm{SSE} = \mathbf{y}'\mathbf{My} = \mathrm{SS}_1 + \dots + \mathrm{SS}_g,$$

$$\mathrm{SSR} = \mathbf{y}'(\mathbf{H} - \mathbf{J}_n)\mathbf{y} = \sum_{i=1}^{g} n_i(\bar{y}_i - \bar{y})^2 = \mathrm{SSB},$$

$$\mathrm{SST} = \mathrm{SSR} + \mathrm{SSE}, \quad \text{because } \mathbf{1}_n \in \mathscr{C}(\mathbf{X}),$$

$$\mathrm{SSE}_H = \mathrm{SSE}(\mu_1 = \dots = \mu_g) = \mathrm{SST} \implies \mathrm{SSE}_H - \mathrm{SSE} = \mathrm{SSR}.$$

7.33 Consider one-way ANOVA situation with g groups. Let x be a variable indicating the group where the observation belongs to so that x has values $1, \dots,$

g. Suppose that the n data points $\binom{1}{y_{11}}, \dots, \binom{1}{y_{1n_1}}, \dots, \binom{g}{y_{g1}}, \dots, \binom{g}{y_{gn_g}}$ comprise a theoretical two-dimensional discrete distribution for the random vector $\binom{x}{y}$ where each pair appears with the same probability $1/n$. Let $\mathrm{E}(y \mid x) := m(x)$ be a random variable which takes the value $\mathrm{E}(y \mid x = i)$ when x takes the value i, and $\mathrm{var}(y \mid x) := v(x)$ is a random variable taking the value $\mathrm{var}(y \mid x = i)$ when $x = i$. Then the decomposition

$$\mathrm{var}(y) = \mathrm{var}[\mathrm{E}(y \mid x)] + \mathrm{E}[\mathrm{var}(y \mid x)] = \mathrm{var}[m(x)] + \mathrm{E}[v(x)]$$

(multiplied by n) can be expressed as

$$\sum_{i=1}^{g}\sum_{j=1}^{n_i}(y_{ij} - \bar{y})^2 = \sum_{i=1}^{g} n_i(\bar{y}_i - \bar{y})^2 + \sum_{i=1}^{g}\sum_{j=1}^{n_i}(y_{ij} - \bar{y}_i)^2,$$

$$\mathrm{SST} = \mathrm{SSR} + \mathrm{SSE}, \quad \mathbf{y}'(\mathbf{I} - \mathbf{J})\mathbf{y} = \mathbf{y}'(\mathbf{H} - \mathbf{J})\mathbf{y} + \mathbf{y}'(\mathbf{I} - \mathbf{H})\mathbf{y},$$

where $\mathbf{J}$ and $\mathbf{H}$ are as in 7.32.

7.34 One-way ANOVA using another parametrization.

$$\mathbf{y} = \mathbf{X}\boldsymbol{\beta} + \boldsymbol{\varepsilon}, \quad \mathbf{y}_i = (y_{i1}, y_{i2}, \dots, y_{in_i})', \quad i = 1, \dots, g,$$

$$\begin{pmatrix} \mathbf{y}_1 \\ \vdots \\ \mathbf{y}_g \end{pmatrix} = \begin{pmatrix} \mathbf{1}_{n_1} & \mathbf{1}_{n_1} & \dots & \mathbf{0} \\ \vdots & \vdots & \ddots & \vdots \\ \mathbf{1}_{n_g} & \mathbf{0} & \dots & \mathbf{1}_{n_g} \end{pmatrix} \begin{pmatrix} \mu \\ \tau_1 \\ \vdots \\ \tau_g \end{pmatrix} + \begin{pmatrix} \boldsymbol{\varepsilon}_1 \\ \vdots \\ \boldsymbol{\varepsilon}_g \end{pmatrix},$$

$$\mathbf{X}'\mathbf{X} = \begin{pmatrix} n & n_1 & \dots & n_g \\ n_1 & n_1 & \dots & 0 \\ \vdots & \vdots & \ddots & \vdots \\ n_g & 0 & \dots & n_g \end{pmatrix},$$

$$\mathbf{G} := \begin{pmatrix} 0 & 0 & \dots & 0 \\ 0 & 1/n_1 & \dots & 0 \\ \vdots & \vdots & \ddots & \vdots \\ 0 & 0 & \dots & 1/n_g \end{pmatrix} \in \{(\mathbf{X}'\mathbf{X})^-\}, \quad \hat{\boldsymbol{\beta}} = \mathbf{G}\mathbf{X}'\mathbf{y} = \begin{pmatrix} 0 \\ \bar{y}_1 \\ \vdots \\ \bar{y}_g \end{pmatrix}.$$

8 Regression diagnostics

8.1 Some particular residuals under $\{\mathbf{y}, \mathbf{X}\boldsymbol{\beta}, \sigma^2\mathbf{I}\}$:

(a) $\hat{\varepsilon}_i = y_i - \hat{y}_i$ — ordinary least squares residual

(b) $\mathrm{RESS}_i = (y_i - \hat{y}_i)/\hat{\sigma}$ — scaled residual

(c) $\mathrm{STUDI}_i = r_i = \dfrac{y_i - \hat{y}_i}{\hat{\sigma}\sqrt{1 - h_{ii}}}$ — internally Studentized residual

(d) $\text{STUDE}_i = t_i = \dfrac{y_i - \hat{y}_i}{\hat{\sigma}_{(i)}\sqrt{1-h_{ii}}}$ externally Studentized residual

$\hat{\sigma}_{(i)}$ = the estimate of σ in the model where the ith case is excluded, $\text{STUDE}_i \sim \text{t}(n-k-2, \cdot)$.

8.2 Denote

$$\mathbf{X} = \begin{pmatrix} \mathbf{X}_{(i)} \\ \mathbf{x}'_{(i*)} \end{pmatrix}, \quad \mathbf{y} = \begin{pmatrix} \mathbf{y}_{(i)} \\ y_i \end{pmatrix}, \quad \boldsymbol{\gamma} = \begin{pmatrix} \boldsymbol{\beta} \\ \delta \end{pmatrix},$$

$$\mathbf{Z} = (\mathbf{X} : \mathbf{i}_i), \quad \mathbf{i}_i = (0, \dots, 0, 1)',$$

and let $\mathscr{M}_{(i)} = \{\mathbf{y}_{(i)}, \mathbf{X}_{(i)}\boldsymbol{\beta}, \sigma^2\mathbf{I}_{n-1}\}$ be the model where the ith (the last, for notational convenience) observation is omitted, and let $\mathscr{M}_Z = \{\mathbf{y}, \mathbf{Z}\boldsymbol{\gamma}, \sigma^2\mathbf{I}\}$ be the extended (mean-shift) model. Denote $\hat{\boldsymbol{\beta}}_{(i)} = \hat{\boldsymbol{\beta}}(\mathscr{M}_{(i)})$, $\hat{\boldsymbol{\beta}}_Z = \hat{\boldsymbol{\beta}}(\mathscr{M}_Z)$, and $\hat{\delta} = \hat{\delta}(\mathscr{M}_Z)$. Then

(a) $\hat{\boldsymbol{\beta}}_Z = [\mathbf{X}'(\mathbf{I} - \mathbf{i}_i\mathbf{i}_i')\mathbf{X}]^{-1}\mathbf{X}'(\mathbf{I} - \mathbf{i}_i\mathbf{i}_i')\mathbf{y} = (\mathbf{X}'_{(i)}\mathbf{X}_{(i)})^{-1}\mathbf{X}'_{(i)}\mathbf{y}_{(i)} = \hat{\boldsymbol{\beta}}_{(i)}$,

(b) $\hat{\delta} = \dfrac{\mathbf{i}_i'\mathbf{M}\mathbf{y}}{\mathbf{i}_i'\mathbf{M}\mathbf{i}_i} = \dfrac{\hat{\varepsilon}_i}{m_{ii}}$; we assume that $m_{ii} > 0$, i.e., δ is estimable,

(c)
$$\begin{aligned}\text{SSE}_Z &= \text{SSE}(\mathscr{M}_Z) \\ &= \mathbf{y}'(\mathbf{I} - \mathbf{P}_\mathbf{Z})\mathbf{y} = \mathbf{y}'(\mathbf{M} - \mathbf{P}_{\mathbf{Mi}_i})\mathbf{y} = \mathbf{y}'(\mathbf{I} - \mathbf{i}_i\mathbf{i}_i' - \mathbf{P}_{(\mathbf{I}-\mathbf{i}_i\mathbf{i}_i')\mathbf{X}})\mathbf{y} \\ &= \text{SSE}_{(i)} = \text{SSE} - \mathbf{y}'\mathbf{P}_{\mathbf{Mi}_i}\mathbf{y} = \text{SSE} - m_{ii}\hat{\delta}^2 = \text{SSE} - \hat{\varepsilon}_i^2/m_{ii},\end{aligned}$$

(d) $\text{SSE} - \text{SSE}_Z = \mathbf{y}'\mathbf{P}_{\mathbf{Mi}_i}\mathbf{y}$ = change in SSE due to hypothesis $\delta = 0$,

(e) $(\mathbf{I}_n - \mathbf{P}_\mathbf{Z})\mathbf{y} = \begin{pmatrix} (\mathbf{I}_{n-1} - \mathbf{P}_{\mathbf{X}_{(i)}})\mathbf{y}_{(i)} \\ 0 \end{pmatrix}$,

(f) $t_i^2 = \dfrac{\mathbf{y}'\mathbf{P}_{\mathbf{Mi}_i}\mathbf{y}}{\mathbf{y}'(\mathbf{M} - \mathbf{P}_{\mathbf{Mi}_i})\mathbf{y}/(n-k-2)} = \dfrac{\text{SSE} - \text{SSE}_Z}{\text{SSE}_Z/(n-k-2)} = \dfrac{\hat{\delta}^2}{\text{se}^2(\hat{\delta})}$,

(g) $t_i^2 \sim \text{F}(1, n-k-2, \delta^2 m_{ii}/\sigma^2)$, F-test statistic for testing $\delta = 0$ in $\mathscr{M}_Z$,

(h) $\text{se}^2(\hat{\delta}) = \hat{\sigma}_Z^2/m_{ii} = \hat{\sigma}_{(i)}^2/m_{ii}$,

(i)
$$\begin{aligned} m_{ii} > 0 &\iff \mathbf{Mi}_i \neq \mathbf{0} \iff \mathbf{i}_i \notin \mathscr{C}(\mathbf{X}) \iff \mathbf{x}_{(i*)} \in \mathscr{C}(\mathbf{X}'_{(i)}) \\ &\iff \delta \text{ is estimable under } \mathscr{M}_Z \\ &\iff \boldsymbol{\beta} \text{ is estimable under } \mathscr{M}_Z.\end{aligned}$$

8.3 If $\mathbf{Z} = (\mathbf{1} : \mathbf{i}_n)$, then $t_n = \dfrac{y_n - \bar{y}_{(n)}}{s_{(n)}/\sqrt{1-1/n}} = \dfrac{y_n - \bar{y}}{s_{(n)}\sqrt{1-1/n}}$,

where $s_{(n)}^2$ = sample variance of $y_1, \dots, y_{n-1}$ and $\bar{y}_{(n)} = (y_1 + \dots + y_{n-1})/(n-1)$.

8.4 $\text{DFBETA}_i = \hat{\boldsymbol\beta} - \hat{\boldsymbol\beta}_{(i)} = (\mathbf{X}'\mathbf{X})^{-1}\mathbf{X}'\mathbf{i}_i\hat\delta, \quad \mathbf{X}(\hat{\boldsymbol\beta} - \hat{\boldsymbol\beta}_{(i)}) = \mathbf{H}\mathbf{i}_i\hat\delta$

8.5 $\mathbf{y} - \mathbf{X}\hat{\boldsymbol\beta}_{(i)} = \mathbf{M}\mathbf{y} + \mathbf{H}\mathbf{i}_i\hat\delta \implies \hat\varepsilon_{(i)} = \mathbf{i}_i'(\mathbf{y} - \mathbf{X}\hat{\boldsymbol\beta}_{(i)}) = y_i - \mathbf{x}_{(i*)}'\hat{\boldsymbol\beta}_{(i)} = \hat\delta,$

$\hat\varepsilon_{(i)}$ = the ith predicted residual: $\hat\varepsilon_{(i)}$ is based on a fit to the data with the ith case deleted

8.6 $\hat\varepsilon_{(1)}^2 + \cdots + \hat\varepsilon_{(n)}^2 = \text{PRESS}$ the predicted residual sum of squares

8.7 $\text{COOK}_i^2 = D_i^2 = (\hat{\boldsymbol\beta} - \hat{\boldsymbol\beta}_{(i)})'\mathbf{X}'\mathbf{X}(\hat{\boldsymbol\beta} - \hat{\boldsymbol\beta}_{(i)})/(p\hat\sigma^2)$ Cook's distance

$= (\hat{\mathbf{y}} - \hat{\mathbf{y}}_{[i]})'(\hat{\mathbf{y}} - \hat{\mathbf{y}}_{[i]})/(p\hat\sigma^2) = \hat\delta^2 h_{ii}/(p\hat\sigma^2)$ $\hat{\mathbf{y}}_{[i]} = \mathbf{X}\hat{\boldsymbol\beta}_{(i)}$

8.8 $\text{COOK}_i^2 = \dfrac{1}{p}\dfrac{\hat\varepsilon_i^2}{\hat\sigma^2 m_{ii}}\dfrac{h_{ii}}{m_{ii}} = \dfrac{r_i^2}{p}\dfrac{h_{ii}}{m_{ii}}, \quad r_i = \text{STUDI}_i$

8.9 $h_{ii} = \mathbf{x}_{(i*)}'(\mathbf{X}'\mathbf{X})^{-1}\mathbf{x}_{(i*)},$ h_{ii} = leverage

$\mathbf{x}_{(i*)}' = (1, \mathbf{x}_{(i)}')$ = the ith row of $\mathbf{X}$

8.10 $h_{11} + h_{22} + \cdots + h_{nn} = p\ (= k+1) = \text{trace}(\mathbf{H}) = \text{r}(\mathbf{X})$

8.11 $\mathbf{H}\mathbf{1} = \mathbf{1}$ $\mathbf{1} \in \mathscr{C}(\mathbf{X})$

8.12 $\dfrac{1}{n} \le h_{ii} \le \dfrac{1}{c}, \quad c$ = # of rows of $\mathbf{X}$ which are identical with $\mathbf{x}_{(i*)}'$ $(c \ge 1)$

8.13 $h_{ij}^2 \le \dfrac{1}{4},$ for all $i \ne j$

8.14 $h_{ii} = \frac{1}{n} + \tilde{\mathbf{x}}_{(i)}'\mathbf{T}_{\mathbf{xx}}^{-1}\tilde{\mathbf{x}}_{(i)}$

$= \frac{1}{n} + (\mathbf{x}_{(i)} - \bar{\mathbf{x}})'\mathbf{T}_{\mathbf{xx}}^{-1}(\mathbf{x}_{(i)} - \bar{\mathbf{x}})$ $\mathbf{x}_{(i)}'$ = the ith row of $\mathbf{X}_0$

$= \frac{1}{n} + \frac{1}{n-1}(\mathbf{x}_{(i)} - \bar{\mathbf{x}})'\mathbf{S}_{\mathbf{xx}}^{-1}(\mathbf{x}_{(i)} - \bar{\mathbf{x}}) = \frac{1}{n} + \frac{1}{n-1}\text{MHLN}^2(\mathbf{x}_{(i)}, \bar{\mathbf{x}}, \mathbf{S}_{\mathbf{xx}})$

8.15 $h_{ii} = \dfrac{1}{n} + \dfrac{(x_i - \bar x)^2}{\text{SS}_x},$ when $k = 1$

8.16 $\text{MHLN}^2(\mathbf{x}_{(i)}, \bar{\mathbf{x}}, \mathbf{S}_{\mathbf{xx}}) = (\mathbf{x}_{(i)} - \bar{\mathbf{x}})'\mathbf{S}_{\mathbf{xx}}^{-1}(\mathbf{x}_{(i)} - \bar{\mathbf{x}}) = (n-1)\tilde h_{ii}$ $\tilde{\mathbf{H}} = \mathbf{P}_{\mathbf{C}\mathbf{X}_0}$

8.17 $\kappa(\mathbf{X}'\mathbf{X}) = \left[\dfrac{\text{ch}_{\max}(\mathbf{X}'\mathbf{X})}{\text{ch}_{\min}(\mathbf{X}'\mathbf{X})}\right]^{1/2} = \dfrac{\text{sg}_{\max}(\mathbf{X})}{\text{sg}_{\min}(\mathbf{X})}$ condition number of $\mathbf{X}$

8.18 $\eta_i(\mathbf{X}'\mathbf{X}) = \left[\dfrac{\text{ch}_{\max}(\mathbf{X}'\mathbf{X})}{\text{ch}_i(\mathbf{X}'\mathbf{X})}\right]^{1/2} = \dfrac{\text{sg}_{\max}(\mathbf{X})}{\text{sg}_i(\mathbf{X})}$ condition indexes of $\mathbf{X}$

8.19 Variables (columns) $\mathbf{x}_1, \dots, \mathbf{x}_p$ are exactly collinear if one of the $\mathbf{x}_i$'s is an exact linear combination of the others. This is exact collinearity, i.e., linear dependence, and by term collinearity or near dependence we mean inexact collinear relations.

8.20 Let $\mathscr{M}_Z = \{\mathbf{y}, \mathbf{Z}\boldsymbol{\gamma}, \sigma^2\mathbf{I}\}$ denote the extended model where $\mathbf{Z} = (\mathbf{X} : \mathbf{U})$, $\mathscr{M}_* = \{\mathbf{y}_1, \mathbf{X}_{(1)}\boldsymbol{\beta}, \mathbf{I}_{n-b}\}$ is the model without the last b observations and

$$\mathbf{X}_{n\times p} = \begin{pmatrix} \mathbf{X}_{(1)} \\ \mathbf{X}_{(2)} \end{pmatrix}, \quad \mathbf{X}_{(2)} \in \mathbb{R}^{b\times p}, \quad \mathbf{U}_{n\times b} = \begin{pmatrix} \mathbf{0} \\ \mathbf{I}_b \end{pmatrix}, \quad \boldsymbol{\gamma} = \begin{pmatrix} \boldsymbol{\beta} \\ \boldsymbol{\delta} \end{pmatrix}.$$

(a) $\mathbf{M} = \begin{pmatrix} \mathbf{I}_a - \mathbf{H}_{11} & -\mathbf{H}_{12} \\ -\mathbf{H}_{21} & \mathbf{I}_b - \mathbf{H}_{22} \end{pmatrix}, \quad a + b = n,$

$\mathbf{M}_{22} = \mathbf{I}_b - \mathbf{H}_{22} = \mathbf{U}'(\mathbf{I}_n - \mathbf{H})\mathbf{U},$

(b) $\mathrm{r}(\mathbf{M}_{22}) = \mathrm{r}(\mathbf{U}'\mathbf{M}) = b - \dim \mathscr{C}(\mathbf{U}) \cap \mathscr{C}(\mathbf{X}),$

(c) $\mathrm{r}[\mathbf{X}'(\mathbf{I}_n - \mathbf{U}\mathbf{U}')] = \mathrm{r}(\mathbf{X}_{(1)}) = \mathrm{r}(\mathbf{X}) - \dim \mathscr{C}(\mathbf{X}) \cap \mathscr{C}(\mathbf{U}),$

(d) $\dim \mathscr{C}(\mathbf{X}) \cap \mathscr{C}(\mathbf{U}) = \mathrm{r}(\mathbf{X}) - \mathrm{r}(\mathbf{X}_{(1)}) = \mathrm{r}(\mathbf{X}_{(2)}) - \dim \mathscr{C}(\mathbf{X}'_{(1)}) \cap \mathscr{C}(\mathbf{X}'_{(2)}),$

(e) $\mathrm{r}(\mathbf{M}_{22}) = b - \mathrm{r}(\mathbf{X}_{(2)}) + \dim \mathscr{C}(\mathbf{X}'_{(1)}) \cap \mathscr{C}(\mathbf{X}'_{(2)}),$

(f) $\mathbf{M}_{22}$ is pd $\iff \mathscr{C}(\mathbf{X}'_{(2)}) \subset \mathscr{C}(\mathbf{X}'_{(1)}) \iff \mathrm{r}(\mathbf{X}_{(1)}) = \mathrm{r}(\mathbf{X})$

$\iff \mathrm{ch}_1(\mathbf{U}'\mathbf{H}\mathbf{U}) = \mathrm{ch}_1(\mathbf{H}_{22}) < 1$

$\iff \mathscr{C}(\mathbf{X}) \cap \mathscr{C}(\mathbf{U}) = \{\mathbf{0}\}$

$\iff \boldsymbol{\delta}$ (and then $\boldsymbol{\beta}$) is estimable under $\mathscr{M}_Z$.

8.21 If $\mathbf{X}_{(1)}$ has full column rank, then $\hat{\boldsymbol{\beta}}(\mathscr{M}_*) = \hat{\boldsymbol{\beta}}(\mathscr{M}_Z) := \hat{\boldsymbol{\beta}}_*$ and

(a) $\hat{\boldsymbol{\delta}} = (\mathbf{U}'\mathbf{M}\mathbf{U})^{-1}\mathbf{U}'\mathbf{M}\mathbf{y} = \mathbf{M}_{22}^{-1}\hat{\boldsymbol{\varepsilon}}_2$ $\qquad$ $\hat{\boldsymbol{\varepsilon}}_2 =$ lower part of $\hat{\boldsymbol{\varepsilon}} = \mathbf{M}\mathbf{y}$

$= (\mathbf{M}_{22}^{-1}\mathbf{M}_{21} : \mathbf{I}_b)\mathbf{y} = \mathbf{M}_{22}^{-1}\mathbf{M}_{21}\mathbf{y}_1 + \mathbf{y}_2$ $\qquad$ $\mathbf{y}_2 =$ lower part of $\mathbf{y}$

(b) $\mathrm{SSE}_Z = \mathbf{y}'(\mathbf{I}_n - \mathbf{P}_\mathbf{Z})\mathbf{y} = \mathbf{y}'(\mathbf{M} - \mathbf{P}_{\mathbf{MU}})\mathbf{y} = \mathbf{y}'[\mathbf{I}_n - \mathbf{U}\mathbf{U}' - \mathbf{P}_{(\mathbf{I}_n - \mathbf{U}\mathbf{U}')\mathbf{X}}]\mathbf{y}$

$= \mathrm{SSE}_* = \mathrm{SSE} - \mathbf{y}'\mathbf{P}_{\mathbf{MU}}\mathbf{y}$

$= \mathrm{SSE} - \hat{\boldsymbol{\varepsilon}}_2'\mathbf{M}_{22}^{-1}\hat{\boldsymbol{\varepsilon}}_2 = \mathrm{SSE} - \hat{\boldsymbol{\delta}}'\mathbf{M}_{22}\hat{\boldsymbol{\delta}}$

(c) $\hat{\boldsymbol{\beta}} - \hat{\boldsymbol{\beta}}_* = (\mathbf{X}'\mathbf{X})^{-1}\mathbf{X}'\mathbf{U}\mathbf{M}_{22}^{-1}\mathbf{U}'\mathbf{M}\mathbf{y}$

$= (\mathbf{X}'\mathbf{X})^{-1}\mathbf{X}'_{(2)}\mathbf{M}_{22}^{-1}\hat{\boldsymbol{\varepsilon}}_2 = (\mathbf{X}'\mathbf{X})^{-1}\mathbf{X}'\mathbf{U}\hat{\boldsymbol{\delta}}$

(d) $t_*^2 = \dfrac{\mathbf{y}'\mathbf{P}_{\mathbf{MU}}\mathbf{y}/b}{\mathbf{y}'(\mathbf{M} - \mathbf{P}_{\mathbf{MU}})\mathbf{y}/(n-p-b)} = \dfrac{\mathbf{y}'\mathbf{M}\mathbf{U}(\mathbf{U}'\mathbf{M}\mathbf{U})^{-1}\mathbf{U}'\mathbf{M}\mathbf{y}/b}{\mathrm{SSE}_*/(n-p-b)}$

$= \dfrac{\hat{\boldsymbol{\varepsilon}}_2'\mathbf{M}_{22}^{-1}\hat{\boldsymbol{\varepsilon}}_2/b}{\mathrm{SSE}_*/(n-p-b)} \sim \mathrm{F}(b, n-p-b, \boldsymbol{\delta}'\mathbf{M}_{22}\boldsymbol{\delta}/\sigma^2)$

$=$ F-test statistic for testing hypothesis $\boldsymbol{\delta} = \mathbf{0}$ in $\mathscr{M}_Z$

$=$ the multiple case analogue of externally Studentized residual

(e) $\text{COOK}_* = (\hat{\boldsymbol\beta} - \hat{\boldsymbol\beta}_*)'\mathbf{X}'\mathbf{X}(\hat{\boldsymbol\beta} - \hat{\boldsymbol\beta}_*)/(p\hat\sigma^2) = \hat{\boldsymbol\delta}'\mathbf{H}_{22}\hat{\boldsymbol\delta}/(p\hat\sigma^2)$
$= \hat{\boldsymbol\varepsilon}_2'\mathbf{M}_{22}^{-1}\mathbf{H}_{22}\mathbf{M}_{22}^{-1}\hat{\boldsymbol\varepsilon}_2/(p\hat\sigma^2)$

8.22 $|\mathbf{Z}'\mathbf{Z}| = |\mathbf{X}'(\mathbf{I}_n - \mathbf{U}\mathbf{U}')\mathbf{X}| = |\mathbf{X}_{(1)}'\mathbf{X}_{(1)}| = |\mathbf{M}_{22}||\mathbf{X}'\mathbf{X}|$

8.23 $|\mathbf{X}'\mathbf{X}|(1 - \mathbf{i}_i'\mathbf{H}\mathbf{i}_i) = |\mathbf{X}'(\mathbf{I}_n - \mathbf{i}_i\mathbf{i}_i')\mathbf{X}|$, i.e., $m_{ii} = |\mathbf{X}_{(i)}'\mathbf{X}_{(i)}|/|\mathbf{X}'\mathbf{X}|$

8.24 $\mathscr{C}(\mathbf{X}_{(1)}') \cap \mathscr{C}(\mathbf{X}_{(2)}') = \{\mathbf{0}\} \iff \mathbf{H} = \begin{pmatrix} \mathbf{P}_{\mathbf{X}_{(1)}} & \mathbf{0} \\ \mathbf{0} & \mathbf{P}_{\mathbf{X}_{(2)}} \end{pmatrix}$

8.25 $\mathbf{P}_\mathbf{Z} = \begin{pmatrix} \mathbf{P}_{\mathbf{X}_{(1)}} & \mathbf{0} \\ \mathbf{0} & \mathbf{I}_b \end{pmatrix}, \quad \mathbf{Q}_\mathbf{Z} = \mathbf{I}_n - \mathbf{P}_\mathbf{Z} = \begin{pmatrix} \mathbf{I}_a - \mathbf{P}_{\mathbf{X}_{(1)}} & \mathbf{0} \\ \mathbf{0} & \mathbf{0} \end{pmatrix}$

8.26 Corresponding to 8.2 (p. 40), consider the models $\mathscr{M} = \{\mathbf{y}, \mathbf{X}\boldsymbol\beta, \sigma^2\mathbf{V}\}$, $\mathscr{M}_{(i)} = \{\mathbf{y}_{(i)}, \mathbf{X}_{(i)}\boldsymbol\beta, \sigma^2\mathbf{V}_{(n-1)}\}$ and $\mathscr{M}_Z = \{\mathbf{y}, \mathbf{Z}\boldsymbol\gamma, \sigma^2\mathbf{V}\}$, where $\mathrm{r}(\mathbf{Z}) = p + 1$, and let $\tilde{\boldsymbol\gamma} = (\tilde{\boldsymbol\beta}_Z' : \tilde\delta)'$ denote the BLUE of $\boldsymbol\gamma$ under $\mathscr{M}_Z$. Then:

(a) $\tilde{\boldsymbol\beta}_Z = \tilde{\boldsymbol\beta}_{(i)}, \quad \tilde\delta = \dot e_i/\dot m_{ii}$,

(b) $\text{DFBETA}_i(V) = \tilde{\boldsymbol\beta} - \tilde{\boldsymbol\beta}_{(i)} = (\mathbf{X}'\mathbf{V}^{-1}\mathbf{X})^{-1}\mathbf{X}'\mathbf{V}^{-1}\mathbf{i}_i\dfrac{\dot e_i}{\dot m_{ii}}$,

where $\dot m_{ii}$ is the ith diagonal element of the matrix $\dot{\mathbf{M}}$ defined as

(c) $\dot{\mathbf{M}} = \mathbf{V}^{-1} - \mathbf{V}^{-1}\mathbf{X}(\mathbf{X}'\mathbf{V}^{-1}\mathbf{X})^{-1}\mathbf{X}'\mathbf{V}^{-1} = \mathbf{M}(\mathbf{MVM})^{-}\mathbf{M} = (\mathbf{MVM})^{+}$,

and $\dot e_i$ is the ith element of the vector $\dot{\mathbf{e}} = \dot{\mathbf{M}}\mathbf{y}$. We denote

(d) $\dot{\mathbf{H}} = \mathbf{V}^{-1}\mathbf{X}(\mathbf{X}'\mathbf{V}^{-1}\mathbf{X})^{-1}\mathbf{X}'\mathbf{V}^{-1}$ and hence $\dot{\mathbf{H}} + \dot{\mathbf{M}} = \mathbf{V}^{-1}$.

(e) $t_i^2(V) = \dfrac{\dot e_i^2}{\dot m_{ii}\tilde\sigma_{(i)}^2} = \dfrac{\dot e_i^2}{\widetilde{\mathrm{var}}(\dot e_i)} = \dfrac{\tilde\delta^2}{\widetilde{\mathrm{var}}(\tilde\delta)}$ F-test statistic for testing $\delta = 0$

(f) $\tilde\sigma^2 = \dfrac{\text{SSE}(V)}{n-p}$, where $\text{SSE}(V) = \mathbf{y}'\dot{\mathbf{M}}\mathbf{y}$, $\tilde\sigma_{(i)}^2 = \dfrac{\text{SSE}_{(i)}(V)}{n-p-1}$,

(g) $\text{SSE}_{(i)}(V) = \text{SSE}(V) - \dot e_i^2/\dot m_{ii} = \text{SSE}(V) - \tilde\delta^2\dot m_{ii} = \text{SSE}_Z(V)$,

(h) $\mathrm{var}(\dot e_i) = \sigma^2\dot m_{ii}, \quad \mathrm{var}(\tilde\delta) = \sigma^2/\dot m_{ii}$,

(i) $\text{COOK}_i^2(V) = \dfrac{(\tilde{\boldsymbol\beta} - \tilde{\boldsymbol\beta}_{(i)})'\mathbf{X}'\mathbf{V}^{-1}\mathbf{X}(\tilde{\boldsymbol\beta} - \tilde{\boldsymbol\beta}_{(i)})}{p\tilde\sigma^2} = \dfrac{r_i^2(V)}{p}\dfrac{\dot h_{ii}}{\dot m_{ii}}$.

8.27 Consider the deletion of the last b observations in model $\mathscr{M} = \{\mathbf{y}, \mathbf{X}\boldsymbol\beta, \sigma^2\mathbf{V}\}$ and let $\mathbf{Z}$ be partitioned as in 8.20 (p. 42). Then

(a) $\tilde{\boldsymbol\beta} - \tilde{\boldsymbol\beta}_* = (\mathbf{X}'\mathbf{V}^{-1}\mathbf{X})^{-1}\mathbf{X}'\mathbf{V}^{-1}\mathbf{U}\tilde{\boldsymbol\delta} = (\mathbf{X}'\mathbf{V}^{-1}\mathbf{X})^{-1}\mathbf{X}'\mathbf{V}^{-1}\mathbf{U}\dot{\mathbf{M}}_{22}^{-1}\dot{\mathbf{e}}_2$

$\tilde{\boldsymbol\beta}_*$ is calculated without the last b observations,

(b) $\tilde{\boldsymbol\delta} = (\mathbf{U}'\dot{\mathbf{M}}\mathbf{U})^{-1}\mathbf{U}'\dot{\mathbf{M}}\mathbf{y} = \dot{\mathbf{M}}_{22}^{-1}\dot{\mathbf{e}}_2,$ $\quad\dot{\mathbf{e}}_2 =$ lower part of $\dot{\mathbf{e}} = \dot{\mathbf{M}}\mathbf{y}$

(c) $$\begin{aligned}\mathrm{SSE}_Z(V) &= \mathbf{y}'\dot{\mathbf{M}}\mathbf{y} - \mathbf{y}'\dot{\mathbf{M}}\mathbf{U}(\mathbf{U}'\dot{\mathbf{M}}\mathbf{U})^{-1}\mathbf{U}'\dot{\mathbf{M}}\mathbf{y} \\ &= \mathbf{y}'\dot{\mathbf{M}}\mathbf{y} - \dot{\mathbf{e}}_2'\dot{\mathbf{M}}_{22}^{-1}\dot{\mathbf{e}}_2 \\ &= \mathbf{y}'\dot{\mathbf{M}}_Z\mathbf{y} = \mathbf{y}'\mathbf{M}_Z(\mathbf{M}_Z\mathbf{V}\mathbf{M}_Z)^-\mathbf{M}_Z\mathbf{y} \qquad \mathbf{M}_Z = \mathbf{I} - \mathbf{P}_Z \\ &= \mathrm{SSE}_*(V),\end{aligned}$$

(d) $$t_*^2(V) = \frac{\dot{\mathbf{e}}_2'\dot{\mathbf{M}}_{22}^{-1}\dot{\mathbf{e}}_2/b}{\mathrm{SSE}_*(V)/(n-p-b)} \sim \mathrm{F}(b, n-p-b, \boldsymbol\delta'\dot{\mathbf{M}}_{22}\boldsymbol\delta/\sigma^2),$$

(e) $\mathrm{COOK}_*^2(V) = (\tilde{\boldsymbol\beta} - \tilde{\boldsymbol\beta}_*)'\mathbf{X}'\mathbf{V}^{-1}\mathbf{X}(\tilde{\boldsymbol\beta} - \tilde{\boldsymbol\beta}_*)/(p\sigma^2) = \tilde{\boldsymbol\delta}'\dot{\mathbf{H}}_{22}\tilde{\boldsymbol\delta}/(p\tilde\sigma^2).$

9 BLUE: Some preliminaries

9.1 Let $\mathbf{Z}$ be any matrix such that $\mathscr{C}(\mathbf{Z}) = \mathscr{C}(\mathbf{X}^\perp) = \mathscr{N}(\mathbf{X}') = \mathscr{C}(\mathbf{M})$ and denote $\mathbf{F} = \mathbf{V}^{+1/2}\mathbf{X}$, $\mathbf{L} = \mathbf{V}^{1/2}\mathbf{M}$. Then $\mathbf{F}'\mathbf{L} = \mathbf{X}'\mathbf{P}_V\mathbf{M}$, and

$$\mathbf{X}'\mathbf{P}_V\mathbf{M} = \mathbf{0} \implies \mathbf{P}_\mathbf{F} + \mathbf{P}_\mathbf{L} = \mathbf{P}_{(\mathbf{F}:\mathbf{L})} = \mathbf{P}_V.$$

9.2 Matrix $\dot{\mathbf{M}}$. Consider the linear model $\mathscr{M} = \{\mathbf{y}, \mathbf{X}\boldsymbol\beta, \sigma^2\mathbf{V}\}$, where $\mathbf{X}$ and $\mathbf{V}$ may not have full column ranks. Let the matrices $\dot{\mathbf{M}}$ and $\ddot{\mathbf{M}}$ be defined as

$$\dot{\mathbf{M}} = \mathbf{M}(\mathbf{M}\mathbf{V}\mathbf{M})^-\mathbf{M}, \quad \ddot{\mathbf{M}} = \mathbf{P}_V\dot{\mathbf{M}}\mathbf{P}_V.$$

The matrix $\dot{\mathbf{M}}$ is unique iff $\mathscr{C}(\mathbf{X} : \mathbf{V}) = \mathbb{R}^n$. However, $\mathbf{P}_V\dot{\mathbf{M}}\mathbf{P}_V$ is always unique. Suppose that the condition $\mathbf{H}\mathbf{P}_V\mathbf{M} = \mathbf{0}$ holds. Then

(a) $\ddot{\mathbf{M}} = \mathbf{P}_V\mathbf{M}(\mathbf{M}\mathbf{V}\mathbf{M})^-\mathbf{M}\mathbf{P}_V = \mathbf{V}^+ - \mathbf{V}^+\mathbf{X}(\mathbf{X}'\mathbf{V}^+\mathbf{X})^-\mathbf{X}'\mathbf{V}^+,$

(b) $\ddot{\mathbf{M}} = \mathbf{M}\mathbf{V}^+\mathbf{M} - \mathbf{M}\mathbf{V}^+\mathbf{X}(\mathbf{X}'\mathbf{V}^+\mathbf{X})^-\mathbf{X}'\mathbf{V}^+\mathbf{M},$

(c) $\ddot{\mathbf{M}} = \mathbf{M}\ddot{\mathbf{M}} = \ddot{\mathbf{M}}\mathbf{M} = \mathbf{M}\ddot{\mathbf{M}}\mathbf{M},$

(d) $\mathbf{M}(\mathbf{M}\mathbf{V}\mathbf{M})^+\mathbf{M} = (\mathbf{M}\mathbf{V}\mathbf{M})^+\mathbf{M} = \mathbf{M}(\mathbf{M}\mathbf{V}\mathbf{M})^+ = (\mathbf{M}\mathbf{V}\mathbf{M})^+,$

(e) $\ddot{\mathbf{M}}\mathbf{V}\ddot{\mathbf{M}} = \ddot{\mathbf{M}}$, i.e., $\mathbf{V} \in \{(\ddot{\mathbf{M}})^-\}$,

(f) $\mathrm{r}(\ddot{\mathbf{M}}) = \mathrm{r}(\mathbf{V}\mathbf{M}) = \mathrm{r}(\mathbf{V}) - \dim\mathscr{C}(\mathbf{X}) \cap \mathscr{C}(\mathbf{V}) = \mathrm{r}(\mathbf{X} : \mathbf{V}) - \mathrm{r}(\mathbf{X}),$

(g) If $\mathbf{Z}$ is a matrix with property $\mathscr{C}(\mathbf{Z}) = \mathscr{C}(\mathbf{M})$, then

$$\ddot{\mathbf{M}} = \mathbf{P}_V\mathbf{Z}(\mathbf{Z}'\mathbf{V}\mathbf{Z})^-\mathbf{Z}'\mathbf{P}_V, \quad \mathbf{V}\dot{\mathbf{M}}\mathbf{V} = \mathbf{V}\mathbf{Z}(\mathbf{Z}'\mathbf{V}\mathbf{Z})^-\mathbf{Z}'\mathbf{V}.$$

(h) Let $(\mathbf{X} : \mathbf{Z})$ be orthogonal. Then always (even if $\mathbf{H}\mathbf{P}_V\mathbf{M} \neq \mathbf{0}$)

$$\left[(\mathbf{X} : \mathbf{Z})'\mathbf{V}(\mathbf{X} : \mathbf{Z})\right]^+ = (\mathbf{X} : \mathbf{Z})'\mathbf{V}^+(\mathbf{X} : \mathbf{Z}).$$

Moreover, if in addition we have $\mathbf{H}\mathbf{P}_V\mathbf{M} = \mathbf{0}$, then

$$[(\mathbf{X}:\mathbf{Z})'\mathbf{V}(\mathbf{X}:\mathbf{Z})]^+ = \begin{pmatrix} \mathbf{X}'\mathbf{V}^+\mathbf{X} & \mathbf{X}'\mathbf{V}^+\mathbf{Z} \\ \mathbf{Z}'\mathbf{V}^+\mathbf{X} & \mathbf{Z}'\mathbf{V}^+\mathbf{Z} \end{pmatrix} =$$

$$\begin{pmatrix} [\mathbf{X}'\mathbf{V}\mathbf{X} - \mathbf{X}'\mathbf{V}\mathbf{Z}(\mathbf{Z}'\mathbf{V}\mathbf{Z})^-\mathbf{Z}'\mathbf{V}\mathbf{X}]^+ & -\mathbf{X}'\mathbf{V}^+\mathbf{X}\mathbf{X}'\mathbf{V}\mathbf{Z}(\mathbf{Z}'\mathbf{V}\mathbf{Z})^+ \\ -(\mathbf{Z}'\mathbf{V}\mathbf{Z})^+\mathbf{Z}'\mathbf{V}\mathbf{X}\mathbf{X}'\mathbf{V}^+\mathbf{X} & [\mathbf{Z}'\mathbf{V}\mathbf{Z} - \mathbf{Z}'\mathbf{V}\mathbf{X}(\mathbf{X}'\mathbf{V}\mathbf{X})^-\mathbf{X}'\mathbf{V}\mathbf{Z}]^+ \end{pmatrix}.$$

(i) If $\mathbf{V}$ is positive definite and $\mathscr{C}(\mathbf{Z}) = \mathscr{C}(\mathbf{M})$, then

(i) $$\dot{\mathbf{M}} = \ddot{\mathbf{M}} = \mathbf{M}(\mathbf{M}\mathbf{V}\mathbf{M})^-\mathbf{M} = (\mathbf{M}\mathbf{V}\mathbf{M})^+ = \mathbf{Z}(\mathbf{Z}'\mathbf{V}\mathbf{Z})^-\mathbf{Z}'$$
$$= \mathbf{V}^{-1} - \mathbf{V}^{-1}\mathbf{X}(\mathbf{X}'\mathbf{V}^{-1}\mathbf{X})^-\mathbf{X}'\mathbf{V}^{-1} = \mathbf{V}^{-1}(\mathbf{I} - \mathbf{P}_{\mathbf{X};\mathbf{V}^{-1}}),$$

(ii) $$\mathbf{X}(\mathbf{X}'\mathbf{V}^{-1}\mathbf{X})^-\mathbf{X}' = \mathbf{V} - \mathbf{V}\mathbf{Z}(\mathbf{Z}'\mathbf{V}\mathbf{Z})^-\mathbf{Z}'\mathbf{V} = \mathbf{V} - \mathbf{V}\dot{\mathbf{M}}\mathbf{V},$$

(iii) $$\mathbf{X}(\mathbf{X}'\mathbf{V}^{-1}\mathbf{X})^-\mathbf{X}'\mathbf{V}^{-1} = \mathbf{I} - \mathbf{V}\mathbf{Z}(\mathbf{Z}'\mathbf{V}\mathbf{Z})^-\mathbf{Z}'$$
$$= \mathbf{I} - \mathbf{V}\dot{\mathbf{M}} = \mathbf{I} - \mathbf{P}'_{\mathbf{Z};\mathbf{V}} = \mathbf{I} - \mathbf{P}_{\mathbf{V}\mathbf{Z};\mathbf{V}^{-1}}.$$

(j) If $\mathbf{V}$ is positive definite and $(\mathbf{X}:\mathbf{Z})$ is orthogonal, then

$$(\mathbf{X}'\mathbf{V}^{-1}\mathbf{X})^{-1} = \mathbf{X}'\mathbf{V}\mathbf{X} - \mathbf{X}'\mathbf{V}\mathbf{Z}(\mathbf{Z}'\mathbf{V}\mathbf{Z})^{-1}\mathbf{Z}'\mathbf{V}\mathbf{X}.$$

9.3 If $\mathbf{V}$ is positive definite, then

(a) $$\dot{\mathbf{M}} = \mathbf{V}^{-1} - \dot{\mathbf{H}}$$
$$= \mathbf{V}^{-1} - \mathbf{V}^{-1}\mathbf{X}(\mathbf{X}'\mathbf{V}^{-1}\mathbf{X})^-\mathbf{X}'\mathbf{V}^{-1} = \mathbf{V}^{-1}(\mathbf{I} - \mathbf{P}_{\mathbf{X};\mathbf{V}^{-1}}),$$

(b) $$\dot{\mathbf{M}} = \mathbf{M}(\mathbf{M}\mathbf{V}\mathbf{M})^-\mathbf{M} = \mathbf{M}\dot{\mathbf{M}}\mathbf{M} = \mathbf{M}\dot{\mathbf{M}} = \dot{\mathbf{M}}\mathbf{M}$$
$$= \mathbf{M}(\mathbf{M}\mathbf{V}\mathbf{M})^+\mathbf{M} = (\mathbf{M}\mathbf{V}\mathbf{M})^+\mathbf{M} = \mathbf{M}(\mathbf{M}\mathbf{V}\mathbf{M})^+ = (\mathbf{M}\mathbf{V}\mathbf{M})^+.$$

9.4 $\mathbf{M}(\mathbf{M}\mathbf{V}\mathbf{M})^-\mathbf{M} = \mathbf{M}(\mathbf{M}\mathbf{V}\mathbf{M})^+\mathbf{M}$ iff $\mathscr{C}(\mathbf{M}) \subset \mathscr{C}(\mathbf{M}\mathbf{V})$ iff $\mathrm{r}(\mathbf{X}:\mathbf{V}) = n$

9.5 Matrix $\dot{\mathbf{M}}$: general case. Consider the model $\{\mathbf{y}, \mathbf{X}\boldsymbol{\beta}, \mathbf{V}\}$. Let $\mathbf{U}$ be any matrix such that $\mathbf{W} = \mathbf{V} + \mathbf{X}\mathbf{U}\mathbf{X}'$ has the property $\mathscr{C}(\mathbf{W}) = \mathscr{C}(\mathbf{X}:\mathbf{V})$. Then

$$\mathbf{P}_\mathbf{W}\mathbf{M}(\mathbf{M}\mathbf{V}\mathbf{M})^-\mathbf{M}\mathbf{P}_\mathbf{W} = (\mathbf{M}\mathbf{V}\mathbf{M})^+$$
$$= \mathbf{W}^+ - \mathbf{W}^+\mathbf{X}(\mathbf{X}'\mathbf{W}^-\mathbf{X})^-\mathbf{X}'\mathbf{W}^+,$$

i.e.,

$$\mathbf{P}_\mathbf{W}\dot{\mathbf{M}}\mathbf{P}_\mathbf{W} := \ddot{\mathbf{M}}_\mathbf{W} = \mathbf{W}^+ - \mathbf{W}^+\mathbf{X}(\mathbf{X}'\mathbf{W}^-\mathbf{X})^-\mathbf{X}'\mathbf{W}^+.$$

The matrix $\ddot{\mathbf{M}}_\mathbf{W}$ has the corresponding properties as $\ddot{\mathbf{M}}$ in 9.2 (p. 44).

9.6 $\mathbf{W} = \mathbf{V}\mathbf{M}(\mathbf{M}\mathbf{V}\mathbf{M})^-\mathbf{M}\mathbf{V} + \mathbf{X}(\mathbf{X}'\mathbf{W}^-\mathbf{X})^-\mathbf{X}'$

9.7 $$\mathbf{V} - \mathbf{V}\mathbf{M}(\mathbf{M}\mathbf{V}\mathbf{M})^-\mathbf{M}\mathbf{V} = \mathbf{X}(\mathbf{X}'\mathbf{W}^-\mathbf{X})^-\mathbf{X}' - \mathbf{X}'\mathbf{U}\mathbf{X}$$
$$= \mathbf{H}\mathbf{V}\mathbf{H} - \mathbf{H}\mathbf{V}\mathbf{M}(\mathbf{M}\mathbf{V}\mathbf{M})^-\mathbf{M}\mathbf{V}\mathbf{H}$$

9.8 $\mathbf{X}(\mathbf{X}'\mathbf{W}^{+}\mathbf{X})^{-}\mathbf{X}'\mathbf{W}^{+} = \mathbf{X}(\mathbf{X}'\mathbf{W}^{-}\mathbf{X})^{-}\mathbf{X}'\mathbf{W}^{+} = [\mathbf{I} - \mathbf{W}\mathbf{M}(\mathbf{M}\mathbf{W}\mathbf{M})^{-}\mathbf{M}]\mathbf{P}_{\mathbf{W}}$
$= [\mathbf{I} - \mathbf{V}\mathbf{M}(\mathbf{M}\mathbf{V}\mathbf{M})^{-}\mathbf{M}]\mathbf{P}_{\mathbf{W}} = \mathbf{H} - \mathbf{H}\mathbf{V}\mathbf{M}(\mathbf{M}\mathbf{V}\mathbf{M})^{-}\mathbf{M}\mathbf{P}_{\mathbf{W}}$

9.9 Let $\mathbf{W} = \mathbf{V} + \mathbf{X}\mathbf{U}\mathbf{U}'\mathbf{X} \in \mathrm{NND}_n$ have property $\mathscr{C}(\mathbf{X} : \mathbf{V}) = \mathscr{C}(\mathbf{W})$. Then $\mathbf{W}^{-}$ is a generalized inverse of $\mathbf{V}$ iff $\mathscr{C}(\mathbf{V}) \cap \mathscr{C}(\mathbf{X}\mathbf{U}\mathbf{U}'\mathbf{X}') = \{\mathbf{0}\}$.

9.10 Properties of $\mathbf{X}'\mathbf{W}^{-}\mathbf{X}$. Suppose that $\mathbf{V} \in \mathrm{NND}_{n\times n}$, $\mathbf{X} \in \mathbb{R}^{n\times p}$, and $\mathbf{W} = \mathbf{V} + \mathbf{X}\mathbf{U}\mathbf{X}'$, where $\mathbf{U} \in \mathbb{R}^{p\times p}$. Then the following statements are equivalent:

(a) $\mathscr{C}(\mathbf{X}) \subset \mathscr{C}(\mathbf{W})$,

(b) $\mathscr{C}(\mathbf{X} : \mathbf{V}) = \mathscr{C}(\mathbf{W})$,

(c) $\mathrm{r}(\mathbf{X} : \mathbf{V}) = \mathrm{r}(\mathbf{W})$,

(d) $\mathbf{X}'\mathbf{W}^{-}\mathbf{X}$ is invariant for any choice of $\mathbf{W}^{-}$,

(e) $\mathscr{C}(\mathbf{X}'\mathbf{W}^{-}\mathbf{X})$ is invariant for any choice of $\mathbf{W}^{-}$,

(f) $\mathscr{C}(\mathbf{X}'\mathbf{W}^{-}\mathbf{X}) = \mathscr{C}(\mathbf{X}')$ for any choice of $\mathbf{W}^{-}$,

(g) $\mathrm{r}(\mathbf{X}'\mathbf{W}^{-}\mathbf{X}) = \mathrm{r}(\mathbf{X})$ irrespective of the choice of $\mathbf{W}^{-}$,

(h) $\mathrm{r}(\mathbf{X}'\mathbf{W}^{-}\mathbf{X})$ is invariant with respect to the choice of $\mathbf{W}^{-}$,

(i) $\mathbf{X}(\mathbf{X}'\mathbf{W}^{-}\mathbf{X})^{-}\mathbf{X}'\mathbf{W}^{-}\mathbf{X} = \mathbf{X}$ for any choices of the g-inverses involved.

Moreover, each of these statements is equivalent to (a') $\mathscr{C}(\mathbf{X}) \subset \mathscr{C}(\mathbf{W}')$, and hence to the statements (b')–(i') obtained from (b)–(i), by setting $\mathbf{W}'$ in place of $\mathbf{W}$. We will denote

(w) $\mathcal{W} = \{\, \mathbf{W}_{n\times n} : \mathbf{W} = \mathbf{V} + \mathbf{X}\mathbf{U}\mathbf{X}',\ \mathscr{C}(\mathbf{W}) = \mathscr{C}(\mathbf{X} : \mathbf{V}) \,\}$.

9.11 $\mathscr{C}(\mathbf{V}\mathbf{X}^{\perp})^{\perp} = \mathscr{C}(\mathbf{W}^{-}\mathbf{X} : \mathbf{I} - \mathbf{W}^{-}\mathbf{W})$ $\qquad \mathbf{W} = \mathbf{V} + \mathbf{X}\mathbf{U}\mathbf{X}' \in \mathcal{W}$

9.12 Consider the linear model $\{\mathbf{y}, \mathbf{X}\boldsymbol{\beta}, \mathbf{V}\}$ and denote $\mathbf{W} = \mathbf{V} + \mathbf{X}\mathbf{U}\mathbf{X}' \in \mathcal{W}$, and let $\mathbf{W}^{-}$ be an arbitrary g-inverse of $\mathbf{W}$. Then

$$\mathscr{C}(\mathbf{W}^{-}\mathbf{X}) \oplus \mathscr{C}(\mathbf{X})^{\perp} = \mathbb{R}^n, \qquad \mathscr{C}(\mathbf{W}^{-}\mathbf{X})^{\perp} \oplus \mathscr{C}(\mathbf{X}) = \mathbb{R}^n,$$
$$\mathscr{C}[(\mathbf{W}^{-})'\mathbf{X}] \oplus \mathscr{C}(\mathbf{X})^{\perp} = \mathbb{R}^n, \quad \mathscr{C}[(\mathbf{W}^{-})'\mathbf{X}]^{\perp} \oplus \mathscr{C}(\mathbf{X}) = \mathbb{R}^n.$$

9.13 $\mathbf{P}_{\mathbf{A}}(\mathbf{P}_{\mathbf{A}}\mathbf{N}\mathbf{P}_{\mathbf{A}})^{+}\mathbf{P}_{\mathbf{A}} = (\mathbf{P}_{\mathbf{A}}\mathbf{N}\mathbf{P}_{\mathbf{A}})^{+}\mathbf{P}_{\mathbf{A}}$
$= \mathbf{P}_{\mathbf{A}}(\mathbf{P}_{\mathbf{A}}\mathbf{N}\mathbf{P}_{\mathbf{A}})^{+} = (\mathbf{P}_{\mathbf{A}}\mathbf{N}\mathbf{P}_{\mathbf{A}})^{+}$ for any $\mathbf{N}$

9.14 Let $\mathbf{X}_*$ be any matrix such that $\mathscr{C}(\mathbf{X}) = \mathscr{C}(\mathbf{X}_*)$. Then

(a) $\mathbf{P}_{\mathbf{V}}\mathbf{X}_*(\mathbf{X}_*'\mathbf{V}\mathbf{X}_*)^{-}\mathbf{X}_*'\mathbf{P}_{\mathbf{V}} = \mathbf{P}_{\mathbf{V}}\mathbf{X}(\mathbf{X}'\mathbf{V}\mathbf{X})^{-}\mathbf{X}'\mathbf{P}_{\mathbf{V}}$,

(b) $\mathbf{P}_{\mathbf{V}}\mathbf{X}_*(\mathbf{X}_*'\mathbf{V}^{+}\mathbf{X}_*)^{-}\mathbf{X}_*'\mathbf{P}_{\mathbf{V}} = \mathbf{P}_{\mathbf{V}}\mathbf{X}(\mathbf{X}'\mathbf{V}^{+}\mathbf{X})^{-}\mathbf{X}'\mathbf{P}_{\mathbf{V}}$,

(c) $\mathscr{C}(\mathbf{X}) \subset \mathscr{C}(\mathbf{V}) \implies$

$$\mathbf{X}_*(\mathbf{X}_*'\mathbf{V}^-\mathbf{X}_*)^-\mathbf{X}_*' = \mathbf{X}(\mathbf{X}'\mathbf{V}^-\mathbf{X})^-\mathbf{X}' = \mathbf{H}(\mathbf{H}'\mathbf{V}^-\mathbf{H})^-\mathbf{H} = (\mathbf{H}'\mathbf{V}^-\mathbf{H})^+.$$

9.15 Let $\mathbf{X}_o$ be such that $\mathscr{C}(\mathbf{X}) = \mathscr{C}(\mathbf{X}_o)$ and $\mathbf{X}_o'\mathbf{X}_o = \mathbf{I}_r$, where $r = \mathrm{r}(\mathbf{X})$. Then

$$\mathbf{X}_o(\mathbf{X}_o'\mathbf{V}^+\mathbf{X}_o)^+\mathbf{X}_o' = \mathbf{H}(\mathbf{H}'\mathbf{V}^+\mathbf{H})^+\mathbf{H} = (\mathbf{H}'\mathbf{V}^+\mathbf{H})^+,$$

whereas $\mathbf{X}_o(\mathbf{X}_o'\mathbf{V}^+\mathbf{X}_o)^+\mathbf{X}_o' = \mathbf{X}(\mathbf{X}'\mathbf{V}^+\mathbf{X})^+\mathbf{X}'$ iff $\mathscr{C}(\mathbf{X}'\mathbf{X}\mathbf{X}'\mathbf{V}) = \mathscr{C}(\mathbf{X}'\mathbf{V})$.

9.16 Let $\mathbf{V} \in \mathrm{NND}_{n\times n}$, $\mathbf{X} \in \mathbb{R}^{n\times p}$, and $\mathbf{U} \in \mathbb{R}^{p\times p}$ be such such that $\mathbf{W} = \mathbf{V} + \mathbf{X}\mathbf{U}\mathbf{X}'$ satifies the condition $\mathscr{C}(\mathbf{W}) = \mathscr{C}(\mathbf{X} : \mathbf{V})$. Then the equality

$$\mathbf{W} = \mathbf{V}\mathbf{B}(\mathbf{B}'\mathbf{V}\mathbf{B})^-\mathbf{B}'\mathbf{V} + \mathbf{X}(\mathbf{X}'\mathbf{W}^-\mathbf{X})^-\mathbf{X}'$$

holds for an $n \times q$ matrix $\mathbf{B}$ iff

$$\mathscr{C}(\mathbf{V}\mathbf{W}^-\mathbf{X}) \subset \mathscr{C}(\mathbf{B})^\perp \quad \text{and} \quad \mathscr{C}(\mathbf{V}\mathbf{X}^\perp) \subset \mathscr{C}(\mathbf{V}\mathbf{B}),$$

or, equivalently, $\mathscr{C}(\mathbf{V}\mathbf{W}^-\mathbf{X}) = \mathscr{C}(\mathbf{B})^\perp \cap \mathscr{C}(\mathbf{V})$, the subspace $\mathscr{C}(\mathbf{V}\mathbf{W}^-\mathbf{X})$ being independent of the choice of $\mathbf{W}^-$.

9.17 Let $\mathbf{V} \in \mathrm{NND}_{n\times n}$, $\mathbf{X} \in \mathbb{R}^{n\times p}$ and $\mathscr{C}(\mathbf{X}) \subset \mathscr{C}(\mathbf{V})$. Then the equality

$$\mathbf{V} = \mathbf{V}\mathbf{B}(\mathbf{B}'\mathbf{V}\mathbf{B})^-\mathbf{B}'\mathbf{V} + \mathbf{X}(\mathbf{X}'\mathbf{V}^-\mathbf{X})^-\mathbf{X}'$$

holds for an $n \times q$ matrix $\mathbf{B}$ iff $\mathscr{C}(\mathbf{X}) = \mathscr{C}(\mathbf{B})^\perp \cap \mathscr{C}(\mathbf{V})$.

9.18 Definition of estimability: Let $\mathbf{K}' \in \mathbb{R}^{q\times p}$. Then $\mathbf{K}'\boldsymbol{\beta}$ is estimable if there exists a linear estimator $\mathbf{F}\mathbf{y}$ which is unbiased for $\mathbf{K}'\boldsymbol{\beta}$.

In what follows we consider the model $\{\mathbf{y}, \mathbf{X}_1\boldsymbol{\beta}_1 + \mathbf{X}_2\boldsymbol{\beta}_2, \mathbf{V}\}$, where $\mathbf{X}_i \in \mathbb{R}^{n\times p_i}$, $i = 1, 2$, $p_1 + p_2 = p$.

9.19 $\mathbf{K}'\boldsymbol{\beta}$ is estimable iff $\mathscr{C}(\mathbf{K}) \subset \mathscr{C}(\mathbf{X}')$ iff $\mathbf{K} = \mathbf{X}'\mathbf{A}$ for some $\mathbf{A}$ iff $\mathbf{K}'\hat{\boldsymbol{\beta}} = \mathbf{K}'(\mathbf{X}'\mathbf{X})^-\mathbf{X}'\mathbf{y}$ is unique for all $(\mathbf{X}'\mathbf{X})^-$.

9.20 $\mathbf{L}\boldsymbol{\beta}_2$ is estimable iff $\mathscr{C}(\mathbf{L}') \subset \mathscr{C}(\mathbf{X}_2'\mathbf{M}_1)$, i.e., $\mathbf{L} = \mathbf{B}\mathbf{M}_1\mathbf{X}_2$ for some $\mathbf{B}$.

9.21 The following statements are equivalent:

(a) $\boldsymbol{\beta}_2$ is estimable,

(b) $\mathscr{C}\begin{pmatrix} \mathbf{0} \\ \mathbf{I}_{p_2} \end{pmatrix} \subset \mathscr{C}(\mathbf{X}')$,

(c) $\mathscr{C}(\mathbf{X}_1) \cap \mathscr{C}(\mathbf{X}_2) = \{\mathbf{0}\}$ and $\mathrm{r}(\mathbf{X}_2) = p_2$,

(d) $\mathrm{r}(\mathbf{M}_1\mathbf{X}_2) = \mathrm{r}(\mathbf{X}_2) = p_2$.

9.22 Denoting $\mathbf{P}_{\mathbf{X}_2\cdot\mathbf{X}_1} = \mathbf{X}_2(\mathbf{X}_2'\mathbf{M}_1\mathbf{X}_2)^-\mathbf{X}_2'\mathbf{M}_1$, the following statements are equivalent:

(a) $\mathbf{X}_2\boldsymbol{\beta}_2$ is estimable, (b) $\mathrm{r}(\mathbf{X}_2') = \mathrm{r}(\mathbf{X}_2'\mathbf{M}_1)$,

(c) $\mathscr{C}(\mathbf{X}_1) \cap \mathscr{C}(\mathbf{X}_2) = \{\mathbf{0}\}$, (d) $\mathbf{P}_{\mathbf{X}_2 \cdot \mathbf{X}_1} \mathbf{X}_2 = \mathbf{X}_2$,

(e) $\mathbf{P}_{\mathbf{X}_2 \cdot \mathbf{X}_1}$ is invariant with respect to the choice of $(\mathbf{X}_2' \mathbf{M}_1 \mathbf{X}_2)^-$,

(f) $\mathbf{P}_{\mathbf{X}_2 \cdot \mathbf{X}_1}$ is a projector onto $\mathscr{C}(\mathbf{X}_2)$ along $\mathscr{C}(\mathbf{X}_1) \oplus \mathscr{C}(\mathbf{X})^\perp$,

(g) $\mathbf{H} = \mathbf{P}_{\mathbf{X}_2 \cdot \mathbf{X}_1} + \mathbf{P}_{\mathbf{X}_1 \cdot \mathbf{X}_2}$.

9.23 β_k is estimable $\iff \mathbf{x}_k \notin \mathscr{C}(\mathbf{X}_1) \iff \mathbf{M}_1 \mathbf{x}_k \neq \mathbf{0}$

9.24 $\mathbf{k}'\boldsymbol{\beta}$ is estimable $\iff \mathbf{k} \in \mathscr{C}(\mathbf{X}')$

9.25 $\boldsymbol{\beta}$ is estimable $\iff \mathrm{r}(\mathbf{X}_{n \times p}) = p$

9.26 Consider the one-way ANOVA using the following parametrization:

$$\mathbf{y} = \mathbf{X}\boldsymbol{\beta} + \boldsymbol{\varepsilon}, \quad \boldsymbol{\beta} = \begin{pmatrix} \mu \\ \boldsymbol{\tau} \end{pmatrix},$$

$$\begin{pmatrix} \mathbf{y}_1 \\ \vdots \\ \mathbf{y}_g \end{pmatrix} = \begin{pmatrix} \mathbf{1}_{n_1} & \mathbf{1}_{n_1} & \dots & \mathbf{0} \\ \vdots & \vdots & \ddots & \vdots \\ \mathbf{1}_{n_g} & \mathbf{0} & \dots & \mathbf{1}_{n_g} \end{pmatrix} \begin{pmatrix} \mu \\ \tau_1 \\ \vdots \\ \tau_g \end{pmatrix} + \begin{pmatrix} \boldsymbol{\varepsilon}_1 \\ \vdots \\ \boldsymbol{\varepsilon}_g \end{pmatrix}.$$

Then the parametric function $\mathbf{a}'\boldsymbol{\beta} = (b, \mathbf{c}')\boldsymbol{\beta} = b\mu + \mathbf{c}'\boldsymbol{\tau}$ is estimable iff

$$\mathbf{a}'\mathbf{u} = 0, \text{ where } \mathbf{u}' = (-1, \mathbf{1}_g'), \quad \text{i.e.,} \quad b - (c_1 + \dots + c_g) = 0,$$

and $\mathbf{c}'\boldsymbol{\tau}$ is estimable iff $\mathbf{c}'\mathbf{1}_g = 0$. Such a function $\mathbf{c}'\boldsymbol{\tau}$ is called a contrast, and the contrasts of the form $\tau_i - \tau_j$, $i \neq j$, are called elementary contrasts.

9.27 (Continued ...) Denote the model matrix above as $\mathbf{X} = (\mathbf{1}_n : \mathbf{X}_0) \in \mathbb{R}^{n \times (g+1)}$, and let $\mathbf{C} \in \mathbb{R}^{n \times n}$ be the centering matrix. Then

(a) $\mathscr{N}(\mathbf{X}) = \mathscr{C}(\mathbf{u})$, where $\mathbf{u}' = (-1, \mathbf{1}_g')$,

(b) $\mathscr{N}(\mathbf{C}\mathbf{X}_0) = \mathscr{C}(\mathbf{X}_0'\mathbf{C})^\perp = \mathscr{C}(\mathbf{1}_g)$, and hence $\mathscr{C}(\mathbf{X}_0'\mathbf{C}) = \mathscr{C}(\mathbf{1}_g)^\perp$.

10 Best linear unbiased estimator

10.1 Definition 1: Let $\mathbf{k}'\boldsymbol{\beta}$ be an estimable parametric function. Then $\mathbf{g}'\mathbf{y}$ is $\mathrm{BLUE}(\mathbf{k}'\boldsymbol{\beta})$ under $\{\mathbf{y}, \mathbf{X}\boldsymbol{\beta}, \sigma^2\mathbf{V}\}$ if $\mathbf{g}'\mathbf{y}$ is an unbiased estimator of $\mathbf{k}'\boldsymbol{\beta}$ and it has the minimum variance among all linear unbiased estimators of $\mathbf{k}'\boldsymbol{\beta}$:

$$\mathrm{E}(\mathbf{g}'\mathbf{y}) = \mathbf{k}'\boldsymbol{\beta} \text{ and } \mathrm{var}(\mathbf{g}'\mathbf{y}) \leq \mathrm{var}(\mathbf{f}'\mathbf{y}) \text{ for any } \mathbf{f}\colon \mathrm{E}(\mathbf{f}'\mathbf{y}) = \mathbf{k}'\boldsymbol{\beta}.$$

10.2 Definition 2: $\mathbf{G}\mathbf{y}$ is $\mathrm{BLUE}(\mathbf{X}\boldsymbol{\beta})$ under $\{\mathbf{y}, \mathbf{X}\boldsymbol{\beta}, \sigma^2\mathbf{V}\}$ if

$$\mathrm{E}(\mathbf{G}\mathbf{y}) = \mathbf{X}\boldsymbol{\beta} \text{ and } \mathrm{cov}(\mathbf{G}\mathbf{y}) \leq_{\mathrm{L}} \mathrm{cov}(\mathbf{F}\mathbf{y}) \text{ for any } \mathbf{F}\colon \mathrm{E}(\mathbf{F}\mathbf{y}) = \mathbf{X}\boldsymbol{\beta}.$$

10.3 $\mathbf{Gy} = \text{BLUE}(\mathbf{X}\boldsymbol{\beta})$ under $\{\mathbf{y}, \mathbf{X}\boldsymbol{\beta}, \sigma^2\mathbf{V}\} \iff \mathbf{G}(\mathbf{X} : \mathbf{VM}) = (\mathbf{X} : \mathbf{0})$.

10.4 Gauss–Markov theorem: $\text{OLSE}(\mathbf{X}\boldsymbol{\beta}) = \text{BLUE}(\mathbf{X}\boldsymbol{\beta})$ under $\{\mathbf{y}, \mathbf{X}\boldsymbol{\beta}, \sigma^2\mathbf{I}\}$.

10.5 The notation $\mathbf{Gy} = \text{BLUE}(\mathbf{X}\boldsymbol{\beta}) = \widetilde{\mathbf{X}\boldsymbol{\beta}} = \mathbf{X}\tilde{\boldsymbol{\beta}}$ should be understood as

$$\mathbf{Gy} \in \{\text{BLUE}(\mathbf{X}\boldsymbol{\beta} \mid \mathscr{M})\}, \quad \text{i.e.,} \quad \mathbf{G} \in \{\mathbf{P}_{\mathbf{X}|\mathbf{VM}}\},$$

where $\{\text{BLUE}(\mathbf{X}\boldsymbol{\beta} \mid \mathscr{M})\}$ refers to the set of all representations of the BLUE. The matrix $\mathbf{G}$ is unique iff $\mathscr{C}(\mathbf{X} : \mathbf{V}) = \mathbb{R}^n$, but the value of $\mathbf{Gy}$ is always unique after observing $\mathbf{y}$.

10.6 A gentle warning regarding notation 10.5 may be worth giving. Namely in 10.5 $\tilde{\boldsymbol{\beta}}$ refers now to any vector $\tilde{\boldsymbol{\beta}} = \mathbf{Ay}$ such that $\mathbf{X}\tilde{\boldsymbol{\beta}}$ is the $\text{BLUE}(\mathbf{X}\boldsymbol{\beta})$. The vector $\tilde{\boldsymbol{\beta}}$ in 10.5 need *not* be the $\text{BLUE}(\boldsymbol{\beta})$—the parameter vector $\boldsymbol{\beta}$ may not even be estimable.

10.7 Let $\mathbf{K}'\boldsymbol{\beta}$ be an estimable parametric function under $\{\mathbf{y}, \mathbf{X}\boldsymbol{\beta}, \sigma^2\mathbf{V}\}$. Then

$$\mathbf{Gy} = \text{BLUE}(\mathbf{K}'\boldsymbol{\beta}) \iff \mathbf{G}(\mathbf{X} : \mathbf{VM}) = (\mathbf{K}' : \mathbf{0}).$$

10.8 $\mathbf{Gy} = \text{BLUE}(\mathbf{X}\boldsymbol{\beta})$ under $\{\mathbf{y}, \mathbf{X}\boldsymbol{\beta}, \sigma^2\mathbf{V}\}$ iff there exists a matrix $\mathbf{L}$ so that $\mathbf{G}$ is a solution to (Pandora's Box)

$$\begin{pmatrix} \mathbf{V} & \mathbf{X} \\ \mathbf{X}' & \mathbf{0} \end{pmatrix} \begin{pmatrix} \mathbf{G}' \\ \mathbf{L} \end{pmatrix} = \begin{pmatrix} \mathbf{0} \\ \mathbf{X}' \end{pmatrix}.$$

10.9 The general solution for $\mathbf{G}$ satisfying $\mathbf{G}(\mathbf{X} : \mathbf{VM}) = (\mathbf{X} : \mathbf{0})$ can be expressed, for example, in the following ways:

(a) $\mathbf{G}_1 = (\mathbf{X} : \mathbf{0})(\mathbf{X} : \mathbf{VM})^- + \mathbf{F}_1[\mathbf{I}_n - (\mathbf{X} : \mathbf{VM})(\mathbf{X} : \mathbf{VM})^-]$,

(b) $\mathbf{G}_2 = \mathbf{X}(\mathbf{X}'\mathbf{W}^-\mathbf{X})^-\mathbf{X}'\mathbf{W}^- + \mathbf{F}_2(\mathbf{I}_n - \mathbf{WW}^-)$,

(c) $\mathbf{G}_3 = \mathbf{I}_n - \mathbf{VM}(\mathbf{MVM})^-\mathbf{M} + \mathbf{F}_3[\mathbf{I}_n - \mathbf{MVM}(\mathbf{MVM})^-]\mathbf{M}$,

(d) $\mathbf{G}_4 = \mathbf{H} - \mathbf{HVM}(\mathbf{MVM})^-\mathbf{M} + \mathbf{F}_4[\mathbf{I}_n - \mathbf{MVM}(\mathbf{MVM})^-]\mathbf{M}$,

where $\mathbf{F}_1$, $\mathbf{F}_2$, $\mathbf{F}_3$ and $\mathbf{F}_4$ are arbitrary matrices, $\mathbf{W} = \mathbf{V} + \mathbf{XUX}'$ and $\mathbf{U}$ is any matrix such that $\mathscr{C}(\mathbf{W}) = \mathscr{C}(\mathbf{X} : \mathbf{V})$, i.e., $\mathbf{W} \in \mathcal{W}$, see 9.10w (p. 46).

10.10 If $\mathscr{C}(\mathbf{X}) \subset \mathscr{C}(\mathbf{V})$, $\{\mathbf{y}, \mathbf{X}\boldsymbol{\beta}, \sigma^2\mathbf{V}\}$ is called a weakly singular linear model.

10.11 Consistency condition for $\mathscr{M} = \{\mathbf{y}, \mathbf{X}\boldsymbol{\beta}, \sigma^2\mathbf{V}\}$. Under $\mathscr{M}$, we have

$$\mathbf{y} \in \mathscr{C}(\mathbf{X} : \mathbf{V}) \text{ with probability 1.}$$

The above statement holds if the model $\mathscr{M}$ is *indeed correct*, so to say (as all our models are supposed to be). Notice that $\mathscr{C}(\mathbf{X} : \mathbf{V})$ can be written as

$$\mathscr{C}(\mathbf{X} : \mathbf{V}) = \mathscr{C}(\mathbf{X} : \mathbf{VM}) = \mathscr{C}(\mathbf{X}) \oplus \mathscr{C}(\mathbf{VM}).$$

10.12 Under the (consistent model) $\{\mathbf{y}, \mathbf{X}\boldsymbol{\beta}, \sigma^2\mathbf{V}\}$, the estimators $\mathbf{Ay}$ and $\mathbf{By}$ are said to be equal if their realized values are equal for all $\mathbf{y} \in \mathscr{C}(\mathbf{X} : \mathbf{V})$:

$$\mathbf{A}_1\mathbf{y} \text{ equals } \mathbf{A}_2\mathbf{y} \iff \mathbf{A}_1\mathbf{y} = \mathbf{A}_2\mathbf{y} \text{ for all } \mathbf{y} \in \mathscr{C}(\mathbf{X} : \mathbf{V}) = \mathscr{C}(\mathbf{X} : \mathbf{VM}).$$

10.13 If $\mathbf{A}_1\mathbf{y}$ and $\mathbf{A}_2\mathbf{y}$ are two BLUEs under $\{\mathbf{y}, \mathbf{X}\boldsymbol{\beta}, \sigma^2\mathbf{V}\}$, then $\mathbf{A}_1\mathbf{y} = \mathbf{A}_2\mathbf{y}$ for all $\mathbf{y} \in \mathscr{C}(\mathbf{X} : \mathbf{V})$.

10.14 A linear estimator $\boldsymbol{\ell}'\mathbf{y}$ which is unbiased for zero, is called a linear zero function. Every linear zero function can be written as $\mathbf{b}'\mathbf{My}$ for some $\mathbf{b}$. Hence an unbiased estimator $\mathbf{Gy}$ is BLUE($\mathbf{X}\boldsymbol{\beta}$) iff $\mathbf{Gy}$ is uncorrelated with every linear zero function.

10.15 If $\mathbf{Gy}$ is the BLUE for an estimable $\mathbf{K}'\boldsymbol{\beta}$ under $\{\mathbf{y}, \mathbf{X}\boldsymbol{\beta}, \sigma^2\mathbf{V}\}$, then $\mathbf{LGy}$ is the BLUE for $\mathbf{LK}'\boldsymbol{\beta}$; shortly, $\{\mathbf{L}[\text{BLUE}(\mathbf{K}'\boldsymbol{\beta})]\} \subset \{\text{BLUE}(\mathbf{LK}'\boldsymbol{\beta})\}$ for any $\mathbf{L}$.

10.16 $\mathbf{Gy} = \text{BLUE}(\mathbf{GX}\boldsymbol{\beta})$ under $\{\mathbf{y}, \mathbf{X}\boldsymbol{\beta}, \sigma^2\mathbf{V}\} \iff \mathbf{GVM} = \mathbf{0}$

10.17 If $\mathbf{Gy} = \text{BLUE}(\mathbf{X}\boldsymbol{\beta})$ then $\mathbf{HGy} = \text{BLUE}(\mathbf{X}\boldsymbol{\beta})$ and thereby there exists $\mathbf{L}$ such that $\mathbf{XLy} = \text{BLUE}(\mathbf{X}\boldsymbol{\beta})$.

10.18 Consider the models $\mathscr{M} = \{\mathbf{y}, \mathbf{X}\boldsymbol{\beta}, \mathbf{V}\}$ and $\mathscr{M}_W = \{\mathbf{y}, \mathbf{X}\boldsymbol{\beta}, \mathbf{W}\}$, where $\mathbf{W} = \mathbf{V} + \mathbf{XUU}'\mathbf{X}'$, and $\mathscr{C}(\mathbf{W}) = \mathscr{C}(\mathbf{X} : \mathbf{V})$. Then

(a) $\text{cov}(\mathbf{X}\hat{\boldsymbol{\beta}} \mid \mathscr{M}_W) = \mathbf{HWH}$,
$\text{cov}(\mathbf{X}\tilde{\boldsymbol{\beta}} \mid \mathscr{M}_W) = \mathbf{X}(\mathbf{X}'\mathbf{W}^-\mathbf{X})^-\mathbf{X}' = (\mathbf{HW}^-\mathbf{H})^+$,

(b) $\text{cov}(\mathbf{X}\hat{\boldsymbol{\beta}} \mid \mathscr{M}) = \mathbf{HVH}$, $\quad \text{cov}(\mathbf{X}\tilde{\boldsymbol{\beta}} \mid \mathscr{M}) = \mathbf{X}(\mathbf{X}'\mathbf{W}^-\mathbf{X})^-\mathbf{X}' - \mathbf{XUU}'\mathbf{X}'$,

(c) $\text{cov}(\mathbf{X}\hat{\boldsymbol{\beta}} \mid \mathscr{M}_W) - \text{cov}(\mathbf{X}\tilde{\boldsymbol{\beta}} \mid \mathscr{M}_W) = \text{cov}(\mathbf{X}\hat{\boldsymbol{\beta}} \mid \mathscr{M}) - \text{cov}(\mathbf{X}\tilde{\boldsymbol{\beta}} \mid \mathscr{M})$,

(d) $\{\text{BLUE}(\mathbf{X}\boldsymbol{\beta} \mid \mathscr{M}_W)\} = \{\text{BLUE}(\mathbf{X}\boldsymbol{\beta} \mid \mathscr{M})\}$.

10.19 Under $\{\mathbf{y}, \mathbf{X}\boldsymbol{\beta}, \sigma^2\mathbf{V}\}$, the BLUE($\mathbf{X}\boldsymbol{\beta}$) has the representations

$$\begin{aligned}
\text{BLUE}(\mathbf{X}\boldsymbol{\beta}) &= \widetilde{\mathbf{X}\boldsymbol{\beta}} = \mathbf{X}\tilde{\boldsymbol{\beta}} = \tilde{\boldsymbol{\mu}} \\
&= \mathbf{P}_{\mathbf{X};\mathbf{V}^{-1}}\mathbf{y} = \mathbf{X}(\mathbf{X}'\mathbf{V}^{-1}\mathbf{X})^-\mathbf{X}'\mathbf{V}^{-1}\mathbf{y} && \mathbf{V} \text{ pd} \\
&= \mathbf{X}(\mathbf{X}_\#'\mathbf{X}_\#)^-\mathbf{X}_\#'\mathbf{y}_\# && \mathbf{y}_\# = \mathbf{V}^{-1/2}\mathbf{y},\ \mathbf{X}_\# = \mathbf{V}^{-1/2}\mathbf{X} \\
&= \mathbf{X}(\mathbf{X}'\mathbf{V}^-\mathbf{X})^-\mathbf{X}'\mathbf{V}^-\mathbf{y} && \mathscr{C}(\mathbf{X}) \subset \mathscr{C}(\mathbf{V}) \\
&= [\mathbf{H} - \mathbf{HVM}(\mathbf{MVM})^-\mathbf{M}]\mathbf{y} \\
&= \text{OLSE}(\mathbf{X}\boldsymbol{\beta}) - \mathbf{HVM}(\mathbf{MVM})^-\mathbf{My} \\
&= [\mathbf{I} - \mathbf{VM}(\mathbf{MVM})^-\mathbf{M}]\mathbf{y} \\
&= \mathbf{X}(\mathbf{X}'\mathbf{W}^-\mathbf{X})^-\mathbf{X}'\mathbf{W}^-\mathbf{y} && \mathbf{W} = \mathbf{V} + \mathbf{XUX}' \in \mathcal{W}
\end{aligned}$$

10.20 $\text{BLUE}(\boldsymbol{\beta}) = \tilde{\boldsymbol{\beta}} = (\mathbf{X}'\mathbf{X})^{-1}\mathbf{X}'\mathbf{y} - (\mathbf{X}'\mathbf{X})^{-1}\mathbf{X}'\mathbf{V}\mathbf{M}(\mathbf{M}\mathbf{V}\mathbf{M})^{-}\mathbf{M}\mathbf{y}$

$= \hat{\boldsymbol{\beta}} - (\mathbf{X}'\mathbf{X})^{-1}\mathbf{X}'\mathbf{V}\dot{\mathbf{M}}\mathbf{y}$ only $\text{r}(\mathbf{X}) = p$ requested

$= (\mathbf{X}'\mathbf{V}^{+}\mathbf{X})^{-1}\mathbf{X}'\mathbf{V}^{+}\mathbf{y}$ $\mathscr{C}(\mathbf{X}) \subset \mathscr{C}(\mathbf{V})$, any $\mathbf{V}^{-}$ will do

10.21 (a) $\text{cov}(\mathbf{X}\tilde{\boldsymbol{\beta}}) = \mathbf{X}(\mathbf{X}'\mathbf{V}^{-1}\mathbf{X})^{-}\mathbf{X}'$ $\mathbf{V}$ pd

$= \mathbf{X}(\mathbf{X}'\mathbf{V}^{-}\mathbf{X})^{-}\mathbf{X}' = (\mathbf{H}\mathbf{V}^{-}\mathbf{H})^{+}$ $\mathscr{C}(\mathbf{X}) \subset \mathscr{C}(\mathbf{V})$

$= \mathbf{H}\mathbf{V}\mathbf{H} - \mathbf{H}\mathbf{V}\mathbf{M}(\mathbf{M}\mathbf{V}\mathbf{M})^{-}\mathbf{M}\mathbf{V}\mathbf{H}$

$= \mathbf{K}_*'\mathbf{K}_* - \mathbf{K}_*'\mathbf{P}_{\mathbf{L}}\mathbf{K}_*$ $\mathbf{K}_* = \mathbf{V}^{1/2}\mathbf{H},\ \mathbf{L} = \mathbf{V}^{1/2}\mathbf{M}$

$= \text{cov}(\mathbf{X}\hat{\boldsymbol{\beta}}) - \mathbf{H}\mathbf{V}\mathbf{M}(\mathbf{M}\mathbf{V}\mathbf{M})^{-}\mathbf{M}\mathbf{V}\mathbf{H}$

$= \mathbf{V} - \mathbf{V}\mathbf{M}(\mathbf{M}\mathbf{V}\mathbf{M})^{-}\mathbf{M}\mathbf{V} = \mathbf{V}^{1/2}(\mathbf{I} - \mathbf{P}_{\mathbf{V}^{1/2}\mathbf{M}})\mathbf{V}^{1/2}$

$= \mathbf{X}(\mathbf{X}'\mathbf{W}^{-}\mathbf{X})^{-}\mathbf{X}' - \mathbf{X}\mathbf{U}\mathbf{X}'$,

(b) $\text{cov}(\tilde{\boldsymbol{\beta}})$

$= (\mathbf{X}'\mathbf{V}^{-1}\mathbf{X})^{-1}$ $\mathbf{V}$ pd, $\text{r}(\mathbf{X}) = p$

$= (\mathbf{X}'\mathbf{X})^{-1}\mathbf{X}'\mathbf{V}\mathbf{X}(\mathbf{X}'\mathbf{X})^{-1} - (\mathbf{X}'\mathbf{X})^{-1}\mathbf{X}'\mathbf{V}\mathbf{M}(\mathbf{M}\mathbf{V}\mathbf{M})^{-}\mathbf{M}\mathbf{V}\mathbf{X}(\mathbf{X}'\mathbf{X})^{-1}$

$= \text{cov}(\hat{\boldsymbol{\beta}}) - (\mathbf{X}'\mathbf{X})^{-1}\mathbf{K}'\mathbf{P}_{\mathbf{L}}\mathbf{K}(\mathbf{X}'\mathbf{X})^{-1}$ only $\text{r}(\mathbf{X}) = p$ requested

$= (\mathbf{X}'\mathbf{X})^{-1}(\mathbf{K}'\mathbf{K} - \mathbf{K}'\mathbf{P}_{\mathbf{L}}\mathbf{K})(\mathbf{X}'\mathbf{X})^{-1}$ $\mathbf{K} = \mathbf{V}^{1/2}\mathbf{X},\ \mathbf{L} = \mathbf{V}^{1/2}\mathbf{M}$

$= (\mathbf{X}'\mathbf{V}^{+}\mathbf{X})^{-1}$ $\mathscr{C}(\mathbf{X}) \subset \mathscr{C}(\mathbf{V})$, $\text{r}(\mathbf{X}) = p$

10.22 Denoting $\boldsymbol{\Sigma} = (\mathbf{H} : \mathbf{M})'\mathbf{V}(\mathbf{H} : \mathbf{M})$, we have $\text{cov}(\mathbf{X}\tilde{\boldsymbol{\beta}}) = \mathbf{M}\mathbf{V}\mathbf{M}/\boldsymbol{\Sigma}$ and hence $\text{r}(\boldsymbol{\Sigma}) = \text{r}(\mathbf{V}) = \text{r}(\mathbf{M}\mathbf{V}\mathbf{M}) + \text{r}[\text{cov}(\mathbf{X}\tilde{\boldsymbol{\beta}})]$, which yields $\text{r}[\text{cov}(\mathbf{X}\tilde{\boldsymbol{\beta}})] = \text{r}(\mathbf{V}) - \text{r}(\mathbf{V}\mathbf{M}) = \dim \mathscr{C}(\mathbf{X}) \cap \mathscr{C}(\mathbf{V})$. Moreover, $\mathscr{C}[\text{cov}(\mathbf{X}\tilde{\boldsymbol{\beta}})] = \mathscr{C}(\mathbf{X}) \cap \mathscr{C}(\mathbf{V})$ and if $\text{r}(\mathbf{X}) = p$, we have $\text{r}[\text{cov}(\tilde{\boldsymbol{\beta}})] = \dim \mathscr{C}(\mathbf{X}) \cap \mathscr{C}(\mathbf{V})$.

10.23 The BLUE of $\mathbf{X}\boldsymbol{\beta}$ and its covariance matrix remain invariant for any choice of $\mathbf{X}$ as long as $\mathscr{C}(\mathbf{X})$ remains the same.

10.24 (a) $\text{cov}(\mathbf{X}\hat{\boldsymbol{\beta}} - \mathbf{X}\tilde{\boldsymbol{\beta}}) = \text{cov}(\mathbf{X}\hat{\boldsymbol{\beta}}) - \text{cov}(\mathbf{X}\tilde{\boldsymbol{\beta}}) = \mathbf{H}\mathbf{V}\mathbf{M}(\mathbf{M}\mathbf{V}\mathbf{M})^{-}\mathbf{M}\mathbf{V}\mathbf{H}$

(b) $\text{cov}(\hat{\boldsymbol{\beta}} - \tilde{\boldsymbol{\beta}}) = \text{cov}(\hat{\boldsymbol{\beta}}) - \text{cov}(\tilde{\boldsymbol{\beta}})$

$= (\mathbf{X}'\mathbf{X})^{-1}\mathbf{X}'\mathbf{V}\mathbf{M}(\mathbf{M}\mathbf{V}\mathbf{M})^{-}\mathbf{M}\mathbf{V}\mathbf{X}(\mathbf{X}'\mathbf{X})^{-1}$

(c) $\text{cov}(\hat{\boldsymbol{\beta}}, \tilde{\boldsymbol{\beta}}) = \text{cov}(\tilde{\boldsymbol{\beta}})$

10.25 $\tilde{\boldsymbol{\varepsilon}} = \mathbf{y} - \mathbf{X}\tilde{\boldsymbol{\beta}} = \mathbf{V}\mathbf{M}(\mathbf{M}\mathbf{V}\mathbf{M})^{-}\mathbf{M}\mathbf{y} = \mathbf{V}\dot{\mathbf{M}}\mathbf{y}$ residual of the BLUE

10.26 $\text{cov}(\tilde{\boldsymbol{\varepsilon}}) = \text{cov}(\mathbf{y} - \mathbf{X}\tilde{\boldsymbol{\beta}}) = \text{cov}(\mathbf{y}) - \text{cov}(\mathbf{X}\tilde{\boldsymbol{\beta}}) = \text{cov}[\mathbf{V}\mathbf{M}(\mathbf{M}\mathbf{V}\mathbf{M})^{-}\mathbf{M}\mathbf{y}]$

$= \text{cov}(\mathbf{V}\dot{\mathbf{M}}\mathbf{y}) = \mathbf{V}\mathbf{M}(\mathbf{M}\mathbf{V}\mathbf{M})^{-}\mathbf{M}\mathbf{V}$, $\mathscr{C}[\text{cov}(\tilde{\boldsymbol{\varepsilon}})] = \mathscr{C}(\mathbf{V}\mathbf{M})$

10.27 Pandora's Box. Consider the model $\{\mathbf{y}, \mathbf{X}\boldsymbol{\beta}, \sigma^2\mathbf{V}\}$ and denote

$$\boldsymbol{\Gamma} = \begin{pmatrix} \mathbf{V} & \mathbf{X} \\ \mathbf{X}' & \mathbf{0} \end{pmatrix}, \quad \text{and} \quad \mathbf{B} = \begin{pmatrix} \mathbf{B}_1 & \mathbf{B}_2 \\ \mathbf{B}_3 & -\mathbf{B}_4 \end{pmatrix} = \begin{pmatrix} \mathbf{V} & \mathbf{X} \\ \mathbf{X}' & \mathbf{0} \end{pmatrix}^{-} \in \{\boldsymbol{\Gamma}^{-}\}.$$

Then

(a) $\mathscr{C}\begin{pmatrix} \mathbf{V} \\ \mathbf{X}' \end{pmatrix} \cap \mathscr{C}\begin{pmatrix} \mathbf{X} \\ \mathbf{0} \end{pmatrix} = \{\mathbf{0}\}$,

(b) $\operatorname{r}\begin{pmatrix} \mathbf{V} & \mathbf{X} \\ \mathbf{X}' & \mathbf{0} \end{pmatrix} = \operatorname{r}(\mathbf{V} : \mathbf{X}) + \operatorname{r}(\mathbf{X})$,

(c) $\mathbf{X}\mathbf{B}_2'\mathbf{X} = \mathbf{X}, \quad \mathbf{X}\mathbf{B}_3\mathbf{X} = \mathbf{X}$,

(d) $\mathbf{X}\mathbf{B}_4\mathbf{X}' = \mathbf{X}\mathbf{B}_4'\mathbf{X}' = \mathbf{V}\mathbf{B}_3'\mathbf{X}' = \mathbf{X}\mathbf{B}_3\mathbf{V} = \mathbf{V}\mathbf{B}_2\mathbf{X}' = \mathbf{X}\mathbf{B}_2'\mathbf{V}$,

(e) $\mathbf{X}'\mathbf{B}_1\mathbf{X}$, $\mathbf{X}'\mathbf{B}_1\mathbf{V}$ and $\mathbf{V}\mathbf{B}_1\mathbf{X}$ are all zero matrices,

(f) $\mathbf{V}\mathbf{B}_1\mathbf{V}\mathbf{B}_1\mathbf{V} = \mathbf{V}\mathbf{B}_1\mathbf{V} = \mathbf{V}\mathbf{B}_1'\mathbf{V}\mathbf{B}_1\mathbf{V} = \mathbf{V}\mathbf{B}_1'\mathbf{V}$,

(g) $\operatorname{tr}(\mathbf{V}\mathbf{B}_1) = \operatorname{r}(\mathbf{V} : \mathbf{X}) - \operatorname{r}(\mathbf{X}) = \operatorname{tr}(\mathbf{V}\mathbf{B}_1')$,

(h) $\mathbf{V}\mathbf{B}_1\mathbf{V}$ and $\mathbf{X}\mathbf{B}_4\mathbf{X}'$ are invariant for any choice of $\mathbf{B}_1$ and $\mathbf{B}_4$,

(i) $\mathbf{X}\tilde{\boldsymbol{\beta}} = \mathbf{X}\mathbf{B}_2'\mathbf{y}, \quad \operatorname{cov}(\mathbf{X}\tilde{\boldsymbol{\beta}}) = \mathbf{X}\mathbf{B}_4\mathbf{X}' = \mathbf{V} - \mathbf{V}\mathbf{B}_1\mathbf{V}$,
$\tilde{\boldsymbol{\varepsilon}} = \mathbf{y} - \mathbf{X}\tilde{\boldsymbol{\beta}} = \mathbf{V}\mathbf{B}_1\mathbf{y}$,

(j) for estimable $\mathbf{k}'\boldsymbol{\beta}$, $\mathbf{k}'\tilde{\boldsymbol{\beta}} = \mathbf{k}'\mathbf{B}_2'\mathbf{y} = \mathbf{k}'\mathbf{B}_3\mathbf{y}, \quad \operatorname{var}(\mathbf{k}'\tilde{\boldsymbol{\beta}}) = \sigma^2\mathbf{k}'\mathbf{B}_4\mathbf{k}$,

(k) $\tilde{\sigma}^2 = \mathbf{y}'\mathbf{B}_1\mathbf{y}/f$ is an unbiased estimator of σ^2; $f = \operatorname{r}(\mathbf{V}\mathbf{M})$.

10.28 $\tilde{\boldsymbol{\varepsilon}} = \mathbf{y} - \mathbf{X}\tilde{\boldsymbol{\beta}} = (\mathbf{I} - \mathbf{P}_{\mathbf{X};\mathbf{V}^{-1}})\mathbf{y}$ — residual of the BLUE, $\mathbf{V}$ pd
$= \mathbf{V}\mathbf{M}(\mathbf{M}\mathbf{V}\mathbf{M})^{-}\mathbf{M}\mathbf{y} = \mathbf{V}\dot{\mathbf{M}}\mathbf{y}$ — $\mathbf{V}$ can be singular

10.29 $\tilde{\boldsymbol{\varepsilon}}_{\#} = \mathbf{V}^{-1/2}\tilde{\boldsymbol{\varepsilon}}$ — residual in $\mathscr{M}_{\#} = \{\mathbf{y}_{\#}, \mathbf{X}_{\#}\boldsymbol{\beta}, \sigma^2\mathbf{I}\}$, $\mathbf{V}$ pd
$= \mathbf{V}^{-1/2}(\mathbf{I} - \mathbf{P}_{\mathbf{X};\mathbf{V}^{-1}})\mathbf{y}$

10.30
$$\begin{aligned}
\operatorname{SSE}(V) &= \min_{\boldsymbol{\beta}} \|\mathbf{y} - \mathbf{X}\boldsymbol{\beta}\|^2_{\mathbf{V}^{-1}} = \min_{\boldsymbol{\beta}} (\mathbf{y} - \mathbf{X}\boldsymbol{\beta})'\mathbf{V}^{-1}(\mathbf{y} - \mathbf{X}\boldsymbol{\beta}) \\
&= \|\mathbf{y} - \mathbf{P}_{\mathbf{X};\mathbf{V}^{-1}}\mathbf{y}\|^2_{\mathbf{V}^{-1}} \\
&= \|\mathbf{y} - \mathbf{X}\tilde{\boldsymbol{\beta}}\|^2_{\mathbf{V}^{-1}} = (\mathbf{y} - \mathbf{X}\tilde{\boldsymbol{\beta}})'\mathbf{V}^{-1}(\mathbf{y} - \mathbf{X}\tilde{\boldsymbol{\beta}}) = \tilde{\boldsymbol{\varepsilon}}'\mathbf{V}^{-1}\tilde{\boldsymbol{\varepsilon}} \\
&= \tilde{\boldsymbol{\varepsilon}}_{\#}'\tilde{\boldsymbol{\varepsilon}}_{\#} = \mathbf{y}_{\#}'(\mathbf{I} - \mathbf{P}_{\mathbf{X}_{\#}})\mathbf{y}_{\#} \qquad \mathbf{y}_{\#} = \mathbf{V}^{-1/2}\mathbf{y},\ \mathbf{X}_{\#} = \mathbf{V}^{-1/2}\mathbf{X} \\
&= \mathbf{y}'[\mathbf{V}^{-1} - \mathbf{V}^{-1}\mathbf{X}(\mathbf{X}'\mathbf{V}^{-1}\mathbf{X})^{-}\mathbf{X}'\mathbf{V}^{-1}]\mathbf{y} \\
&= \mathbf{y}'\mathbf{M}(\mathbf{M}\mathbf{V}\mathbf{M})^{-}\mathbf{M}\mathbf{y} = \mathbf{y}'\dot{\mathbf{M}}\mathbf{y} \qquad \text{general presentation}
\end{aligned}$$

10.31 $\tilde{\sigma}^2 = \operatorname{SSE}(V)/[n - \operatorname{r}(\mathbf{X})]$ — unbiased estimator of σ^2, $\mathbf{V}$ pd

10.32 Let $\mathbf{W}$ be defined as $\mathbf{W} = \mathbf{V} + \mathbf{X}\mathbf{U}\mathbf{X}'$, with $\mathscr{C}(\mathbf{W}) = \mathscr{C}(\mathbf{X} : \mathbf{V})$. Then an unbiased estimator for σ^2 is

(a) $\tilde{\sigma}^2 = \text{SSE}(W)/f$, where $f = \text{r}(\mathbf{X} : \mathbf{V}) - \text{r}(\mathbf{X}) = \text{r}(\mathbf{VM})$, and

(b) $$\begin{aligned}\text{SSE}(W) &= (\mathbf{y} - \mathbf{X}\tilde{\boldsymbol{\beta}})'\mathbf{W}^-(\mathbf{y} - \mathbf{X}\tilde{\boldsymbol{\beta}}) = (\mathbf{y} - \mathbf{X}\tilde{\boldsymbol{\beta}})'\mathbf{V}^-(\mathbf{y} - \mathbf{X}\tilde{\boldsymbol{\beta}}) \\ &= \tilde{\boldsymbol{\varepsilon}}'\mathbf{W}^-\tilde{\boldsymbol{\varepsilon}} = \mathbf{y}'[\mathbf{W}^- - \mathbf{W}^-\mathbf{X}(\mathbf{X}'\mathbf{W}^-\mathbf{X})^-\mathbf{X}'\mathbf{W}^-]\mathbf{y} \\ &= \mathbf{y}'\mathbf{M}(\mathbf{MVM})^-\mathbf{My} = \mathbf{y}'\dot{\mathbf{M}}\mathbf{y} = \text{SSE}(V).\end{aligned}$$ NOTE: $\mathbf{y} \in \mathscr{C}(\mathbf{W})$

10.33 $\text{SSE}(V) = \mathbf{y}'\dot{\mathbf{M}}\mathbf{y} = \mathbf{y}'\mathbf{My} = \text{SSE}(I)\ \forall\, \mathbf{y} \in \mathscr{C}(\mathbf{X} : \mathbf{V}) \iff (\mathbf{VM})^2 = \mathbf{VM}$

10.34 Let $\mathbf{V}$ be pd and let $\mathbf{X}$ be partitioned as $\mathbf{X} = (\mathbf{X}_1 : \mathbf{X}_2)$, $\text{r}(\mathbf{X}) = p$. Then, in view of 21.23–21.24 (p. 95):

(a) $$\begin{aligned}\mathbf{P}_{(\mathbf{X}_1:\mathbf{X}_2);\mathbf{V}^{-1}} &= \mathbf{X}_1(\mathbf{X}_1'\mathbf{V}^{-1}\mathbf{X}_1)^{-1}\mathbf{X}_1'\mathbf{V}^{-1} + \mathbf{V}\dot{\mathbf{M}}_1\mathbf{X}_2(\mathbf{X}_2'\dot{\mathbf{M}}_1\mathbf{X}_2)^{-1}\mathbf{X}_2'\dot{\mathbf{M}}_1 \\ &= \mathbf{X}_1(\mathbf{X}_1'\dot{\mathbf{M}}_2\mathbf{X}_1)^{-1}\mathbf{X}_1'\dot{\mathbf{M}}_2 + \mathbf{X}_2(\mathbf{X}_2'\dot{\mathbf{M}}_1\mathbf{X}_2)^{-1}\mathbf{X}_2'\dot{\mathbf{M}}_1,\end{aligned}$$

(b) $\dot{\mathbf{M}}_1 = \mathbf{V}^{-1} - \mathbf{V}^{-1}\mathbf{X}_1(\mathbf{X}_1'\mathbf{V}^{-1}\mathbf{X}_1)^{-1}\mathbf{X}_1'\mathbf{V}^{-1} = \mathbf{M}_1(\mathbf{M}_1\mathbf{V}\mathbf{M}_1)^-\mathbf{M}_1$,

(c) $\tilde{\boldsymbol{\beta}}_1 = (\mathbf{X}_1'\dot{\mathbf{M}}_2\mathbf{X}_1)^{-1}\mathbf{X}_1'\dot{\mathbf{M}}_2\mathbf{y}, \quad \tilde{\boldsymbol{\beta}}_2 = (\mathbf{X}_2'\dot{\mathbf{M}}_1\mathbf{X}_2)^{-1}\mathbf{X}_2'\dot{\mathbf{M}}_1\mathbf{y}$,

(d) $\text{cov}(\tilde{\boldsymbol{\beta}}_2) = \sigma^2(\mathbf{X}_2'\dot{\mathbf{M}}_1\mathbf{X}_2)^{-1}$,

(e) $\text{cov}(\hat{\boldsymbol{\beta}}_2) = \sigma^2(\mathbf{X}_2'\mathbf{M}_1\mathbf{X}_2)^{-1}\mathbf{X}_2'\mathbf{M}_1\mathbf{V}\mathbf{M}_1\mathbf{X}_2(\mathbf{X}_2'\mathbf{M}_1\mathbf{X}_2)^{-1}$,

(f) $$\begin{aligned}&\text{BLUE}(\mathbf{X}\boldsymbol{\beta}) \\ &= \mathbf{P}_{(\mathbf{X}_1:\mathbf{X}_2);\mathbf{V}^{-1}}\mathbf{y} = \mathbf{X}_1\tilde{\boldsymbol{\beta}}_1 + \mathbf{X}_2\tilde{\boldsymbol{\beta}}_2 \\ &= \mathbf{X}_1(\mathbf{X}_1'\dot{\mathbf{M}}_2\mathbf{X}_1)^{-1}\mathbf{X}_1'\dot{\mathbf{M}}_2\mathbf{y} + \mathbf{X}_2(\mathbf{X}_2'\dot{\mathbf{M}}_1\mathbf{X}_2)^{-1}\mathbf{X}_2'\dot{\mathbf{M}}_1\mathbf{y} \\ &= \mathbf{X}_1(\mathbf{X}_1'\mathbf{V}^{-1}\mathbf{X}_1)^{-1}\mathbf{X}_1'\mathbf{V}^{-1}\mathbf{y} + \mathbf{V}\dot{\mathbf{M}}_1\mathbf{X}_2(\mathbf{X}_2'\dot{\mathbf{M}}_1\mathbf{X}_2)^{-1}\mathbf{X}_2'\dot{\mathbf{M}}_1\mathbf{y} \\ &= \mathbf{P}_{\mathbf{X}_1;\mathbf{V}^{-1}}\mathbf{y} + \mathbf{V}\dot{\mathbf{M}}_1\mathbf{X}_2\tilde{\boldsymbol{\beta}}_2 \\ &= \mathbf{X}_1\tilde{\boldsymbol{\beta}}_1(\mathscr{M}_1) + \mathbf{V}\dot{\mathbf{M}}_1\mathbf{X}_2\tilde{\boldsymbol{\beta}}_2(\mathscr{M}_{12}), \qquad \mathscr{M}_1 = \{\mathbf{y}, \mathbf{X}_1\boldsymbol{\beta}_1, \sigma^2\mathbf{V}\}\end{aligned}$$

(g) $\tilde{\boldsymbol{\beta}}_1(\mathscr{M}_{12}) = \tilde{\boldsymbol{\beta}}_1(\mathscr{M}_1) - (\mathbf{X}_1'\mathbf{V}^{-1}\mathbf{X}_1)^{-1}\mathbf{X}_1'\mathbf{V}^{-1}\mathbf{X}_2\tilde{\boldsymbol{\beta}}_2(\mathscr{M}_{12})$.

(h) Replacing $\mathbf{V}^{-1}$ with $\mathbf{V}^+$ and denoting $\mathbf{P}_{\mathbf{A};\mathbf{V}^+} = \mathbf{A}(\mathbf{A}'\mathbf{V}^+\mathbf{A})^{-1}\mathbf{A}'\mathbf{V}^+$, the results in (c)–(g) hold under a weakly singular model.

10.35 Assume that $\mathscr{C}(\mathbf{X}_1) \cap \mathscr{C}(\mathbf{X}_2) = \{\mathbf{0}\}$, $\text{r}(\mathbf{M}_1\mathbf{X}_2) = p_2$ and denote

$$\begin{aligned}&\mathbf{W} = \mathbf{V} + \mathbf{X}_1\mathbf{U}_1\mathbf{X}_1' + \mathbf{X}_2\mathbf{U}_2\mathbf{X}_2', \quad \mathbf{W}_i = \mathbf{V} + \mathbf{X}_i\mathbf{U}_i\mathbf{X}_i', \quad i = 1, 2, \\ &\dot{\mathbf{M}}_{1\mathrm{W}} = \mathbf{M}_1(\mathbf{M}_1\mathbf{W}\mathbf{M}_1)^-\mathbf{M}_1 = \mathbf{M}_1(\mathbf{M}_1\mathbf{W}_2\mathbf{M}_1)^-\mathbf{M}_1,\end{aligned}$$

where $\mathscr{C}(\mathbf{W}_i) = \mathscr{C}(\mathbf{X}_i : \mathbf{V})$, $\mathscr{C}(\mathbf{W}) = \mathscr{C}(\mathbf{X} : \mathbf{V})$. Then

$$\text{BLUE}(\boldsymbol{\beta}_2) = \tilde{\boldsymbol{\beta}}_2(\mathscr{M}_{12}) = (\mathbf{X}_2'\dot{\mathbf{M}}_{1\mathrm{W}}\mathbf{X}_2)^{-1}\mathbf{X}_2'\dot{\mathbf{M}}_{1\mathrm{W}}\mathbf{y}.$$

10.36 (Continued ...) If the disjointness holds, and $\mathscr{C}(\mathbf{X}_2) \subset \mathscr{C}(\mathbf{X}_1 : \mathbf{V})$, then

(a) $\mathbf{X}_2\tilde{\boldsymbol{\beta}}_2(\mathscr{M}_{12}) = \mathbf{X}_2(\mathbf{X}_2'\dot{\mathbf{M}}_1\mathbf{X}_2)^-\mathbf{X}_2'\dot{\mathbf{M}}_1\mathbf{y}$,

(b) $\mathbf{X}_1\tilde{\boldsymbol{\beta}}_1(\mathscr{M}_{12}) = \mathbf{X}_1\tilde{\boldsymbol{\beta}}_1(\mathscr{M}_1) - \mathbf{X}_1(\mathbf{X}_1'\mathbf{W}^+\mathbf{X}_1)^-\mathbf{X}_1'\mathbf{W}^+\mathbf{X}_2\tilde{\boldsymbol{\beta}}_2(\mathscr{M}_{12})$,

(c) $\mathrm{BLUE}(\mathbf{X}\boldsymbol{\beta} \mid \mathscr{M}_{12})$
$$= \mathrm{BLUE}(\mathbf{X}_1\boldsymbol{\beta}_1 \mid \mathscr{M}_1) + \mathbf{W}_1\dot{\mathbf{M}}_1 \cdot \mathrm{BLUE}(\mathbf{X}_2\boldsymbol{\beta}_2 \mid \mathscr{M}_{12})$$
$$= \mathrm{BLUE}(\mathbf{X}_1\boldsymbol{\beta}_1 \mid \mathscr{M}_1) + \mathbf{W}_1 \cdot \mathrm{BLUE}(\dot{\mathbf{M}}_1\mathbf{X}_2\boldsymbol{\beta}_2 \mid \mathscr{M}_{12}).$$

10.37 $\mathrm{SSE}(V) = \min_{\boldsymbol{\beta}} \|\mathbf{y} - \mathbf{X}\boldsymbol{\beta}\|^2_{\mathbf{V}^{-1}} = \mathbf{y}'\dot{\mathbf{M}}_{12}\mathbf{y}$ $\qquad \dot{\mathbf{M}}_{12} = \dot{\mathbf{M}}$
$$= \mathbf{y}'\mathbf{V}^{-1}(\mathbf{I} - \mathbf{P}_{(\mathbf{X}_1:\mathbf{X}_2);\mathbf{V}^{-1}})\mathbf{y}$$
$$= \mathbf{y}'\mathbf{V}^{-1}(\mathbf{I} - \mathbf{P}_{\mathbf{X}_1;\mathbf{V}^{-1}})\mathbf{y} - \mathbf{y}'\dot{\mathbf{M}}_1\mathbf{X}_2(\mathbf{X}_2'\dot{\mathbf{M}}_1\mathbf{X}_2)^{-1}\mathbf{X}_2'\dot{\mathbf{M}}_1\mathbf{y}$$
$$= \mathrm{SSE}_H(V) - \Delta\mathrm{SSE}(V) \qquad H\colon \boldsymbol{\beta}_2 = \mathbf{0}$$

10.38 $\Delta\mathrm{SSE}(V) = \mathbf{y}'\dot{\mathbf{M}}_1\mathbf{X}_2(\mathbf{X}_2'\dot{\mathbf{M}}_1\mathbf{X}_2)^{-1}\mathbf{X}_2'\dot{\mathbf{M}}_1\mathbf{y}$
$$= \mathbf{y}'\dot{\mathbf{M}}_1\mathbf{y} - \mathbf{y}'\dot{\mathbf{M}}_{12}\mathbf{y} \qquad \text{change in SSE}(V) \text{ due to the hypothesis}$$
$$= \min_H \|\mathbf{y} - \mathbf{X}\boldsymbol{\beta}\|^2_{\mathbf{V}^{-1}} - \min_{\boldsymbol{\beta}} \|\mathbf{y} - \mathbf{X}\boldsymbol{\beta}\|^2_{\mathbf{V}^{-1}} \qquad H : \boldsymbol{\beta}_2 = \mathbf{0}$$
$$= \tilde{\boldsymbol{\beta}}_2'[\mathrm{cov}(\tilde{\boldsymbol{\beta}}_2)]^{-1}\tilde{\boldsymbol{\beta}}_2\sigma^2 = Q(V)$$

10.39 $\mathrm{OLSE}(\mathbf{X}\boldsymbol{\beta}) = \mathrm{BLUE}(\mathbf{X}\boldsymbol{\beta})$ iff any of the following equivalent conditions holds. (NOTE: $\mathbf{V}$ is replaceable with $\mathbf{V}^+$ and $\mathbf{H}$ and $\mathbf{M}$ can be interchanged.)

(a) $\mathbf{HV} = \mathbf{VH}$, (b) $\mathbf{HV} = \mathbf{HVH}$,

(c) $\mathbf{HVM} = \mathbf{0}$, (d) $\mathbf{X}'\mathbf{VZ} = \mathbf{0}$, where $\mathscr{C}(\mathbf{Z}) = \mathscr{C}(\mathbf{M})$,

(e) $\mathscr{C}(\mathbf{VX}) \subset \mathscr{C}(\mathbf{X})$, (f) $\mathscr{C}(\mathbf{VX}) = \mathscr{C}(\mathbf{X}) \cap \mathscr{C}(\mathbf{V})$,

(g) $\mathbf{HVH} \leq_{\mathsf{L}} \mathbf{V}$, i.e., $\mathbf{V} - \mathbf{HVH}$ is nonnegative definite,

(h) $\mathbf{HVH} \leq_{\mathsf{rs}} \mathbf{V}$, i.e., $\mathrm{r}(\mathbf{V} - \mathbf{HVH}) = \mathrm{r}(\mathbf{V}) - \mathrm{r}(\mathbf{HVH})$, i.e., $\mathbf{HVH}$ and $\mathbf{V}$ are rank-subtractive, i.e., $\mathbf{HVH}$ is below $\mathbf{V}$ w.r.t. the minus ordering,

(i) $\mathscr{C}(\mathbf{X})$ has a basis consisting of $r = \mathrm{r}(\mathbf{X})$ ortonormal eigenvectors of $\mathbf{V}$,

(j) $\mathrm{r}(\mathbf{T}_{\{1\}}'\mathbf{X}) + \cdots + \mathrm{r}(\mathbf{T}_{\{s\}}'\mathbf{X}) = \mathrm{r}(\mathbf{X})$, where $\mathbf{T}_{\{i\}}$ is a matrix consisting of the orthonormal eigenvectors corresponding to the ith largest eigenvalue $\lambda_{\{i\}}$ of $\mathbf{V}$; $\lambda_{\{1\}} > \lambda_{\{2\}} > \cdots > \lambda_{\{s\}}$, $\lambda_{\{i\}}$'s are the distinct eigenvalues of $\mathbf{V}$,

(k) $\mathbf{T}_{\{i\}}'\mathbf{HT}_{\{i\}} = (\mathbf{T}_{\{i\}}'\mathbf{HT}_{\{i\}})^2$ for all $i = 1, 2, \ldots, s$,

(l) $\mathbf{T}_{\{i\}}'\mathbf{HT}_{\{j\}} = \mathbf{0}$ for all $i, j = 1, 2, \ldots, s$, $i \neq j$,

(m) the squared nonzero canonical correlations between $\mathbf{y}$ and $\mathbf{Hy}$ are the nonzero eigenvalues of $\mathbf{V}^-\mathbf{HVH}$ for all $\mathbf{V}^-$, i.e.,
$$\mathrm{cc}_+^2(\mathbf{y}, \mathbf{Hy}) = \mathrm{nzch}(\mathbf{V}^-\mathbf{HVH}) \quad \text{for all } \mathbf{V}^-,$$

(n) $\mathbf{V}$ can be expressed as $\mathbf{V} = \mathbf{HAH} + \mathbf{MBM}$, where $\mathbf{A} \geq_{\mathsf{L}} \mathbf{0}$, $\mathbf{B} \geq_{\mathsf{L}} \mathbf{0}$, i.e.,

$$\mathbf{V} \in \mathcal{V}_1 = \{\, \mathbf{V} \geq_{\mathsf{L}} \mathbf{0} : \mathbf{V} = \mathbf{HAH} + \mathbf{MBM},\ \mathbf{A} \geq_{\mathsf{L}} \mathbf{0},\ \mathbf{B} \geq_{\mathsf{L}} \mathbf{0} \,\},$$

(o) $\mathbf{V}$ can be expressed as $\mathbf{V} = \mathbf{XCX}' + \mathbf{ZDZ}'$, where $\mathbf{C} \geq_{\mathsf{L}} \mathbf{0}$, $\mathbf{D} \geq_{\mathsf{L}} \mathbf{0}$, i.e.,

$$\mathbf{V} \in \mathcal{V}_2 = \{\, \mathbf{V} \geq_{\mathsf{L}} \mathbf{0} : \mathbf{V} = \mathbf{XCX}' + \mathbf{ZDZ}',\ \mathbf{C} \geq_{\mathsf{L}} \mathbf{0},\ \mathbf{D} \geq_{\mathsf{L}} \mathbf{0} \,\},$$

(p) $\mathbf{V}$ can be expressed as $\mathbf{V} = \alpha\mathbf{I} + \mathbf{XKX}' + \mathbf{ZLZ}'$, where $\alpha \in \mathbb{R}$, and $\mathbf{K}$ and $\mathbf{L}$ are symmetric, such that $\mathbf{V}$ is nonnegative definite, i.e.,

$$\mathbf{V} \in \mathcal{V}_3 = \{\, \mathbf{V} \geq_{\mathsf{L}} \mathbf{0} : \mathbf{V} = \alpha\mathbf{I} + \mathbf{XKX}' + \mathbf{ZLZ}',\ \mathbf{K} = \mathbf{K}',\ \mathbf{L} = \mathbf{L}' \,\}.$$

10.40 Intraclass correlation structure. Consider the model $\mathscr{M} = \{\mathbf{y}, \mathbf{X}\boldsymbol{\beta}, \sigma^2\mathbf{V}\}$, where $\mathbf{1} \in \mathscr{C}(\mathbf{X})$. If $\mathbf{V} = (1-\varrho)\mathbf{I} + \varrho\mathbf{1}\mathbf{1}'$, then $\mathrm{OLSE}(\mathbf{X}\boldsymbol{\beta}) = \mathrm{BLUE}(\mathbf{X}\boldsymbol{\beta})$.

10.41 Consider models $\mathscr{M}_\varrho = \{\mathbf{y}, \mathbf{X}\boldsymbol{\beta}, \sigma^2\mathbf{V}\}$ and $\mathscr{M}_0 = \{\mathbf{y}, \mathbf{X}\boldsymbol{\beta}, \sigma^2\mathbf{I}\}$, where $\mathbf{V}$ has intraclass correlation structure. Let the hypothesis to be tested be H: $\mathrm{E}(\mathbf{y}) \in \mathscr{C}(\mathbf{X}_*)$, where it is assumed that $\mathbf{1} \in \mathscr{C}(\mathbf{X}_*) \subset \mathscr{C}(\mathbf{X})$. Then the F-test statistics for testing H are the same under $\mathscr{M}_\varrho$ and $\mathscr{M}_0$.

10.42 Consider the models $\mathscr{M}_1 = \{\mathbf{y}, \mathbf{X}\boldsymbol{\beta}, \mathbf{V}_1\}$ and $\mathscr{M}_2 = \{\mathbf{y}, \mathbf{X}\boldsymbol{\beta}, \mathbf{V}_2\}$ where $\mathbf{V}_1$ and $\mathbf{V}_2$ are pd and $\mathbf{X} \in \mathbb{R}^{n\times p}_r$. Then the following statements are equivalent:

(a) $\mathrm{BLUE}(\mathbf{X}\boldsymbol{\beta} \mid \mathscr{M}_1) = \mathrm{BLUE}(\mathbf{X}\boldsymbol{\beta} \mid \mathscr{M}_2)$,

(b) $\mathbf{P}_{\mathbf{X};\mathbf{V}_1^{-1}} = \mathbf{P}_{\mathbf{X};\mathbf{V}_2^{-1}}$,

(c) $\mathbf{X}'\mathbf{V}_2^{-1}\mathbf{P}_{\mathbf{X};\mathbf{V}_1^{-1}} = \mathbf{X}'\mathbf{V}_2^{-1}$,

(d) $\mathbf{P}'_{\mathbf{X};\mathbf{V}_1^{-1}}\mathbf{V}_2^{-1}\mathbf{P}_{\mathbf{X};\mathbf{V}_1^{-1}} = \mathbf{V}_2^{-1}\mathbf{P}_{\mathbf{X};\mathbf{V}_1^{-1}}$,

(e) $\mathbf{V}_2^{-1}\mathbf{P}_{\mathbf{X};\mathbf{V}_1^{-1}}$ is symmetric,

(f) $\mathscr{C}(\mathbf{V}_1^{-1}\mathbf{X}) = \mathscr{C}(\mathbf{V}_2^{-1}\mathbf{X})$,

(g) $\mathscr{C}(\mathbf{V}_1\mathbf{X}^\perp) = \mathscr{C}(\mathbf{V}_2\mathbf{X}^\perp)$,

(h) $\mathscr{C}(\mathbf{V}_2\mathbf{V}_1^{-1}\mathbf{X}) = \mathscr{C}(\mathbf{X})$,

(i) $\mathbf{X}'\mathbf{V}_1^{-1}\mathbf{V}_2\mathbf{M} = \mathbf{0}$,

(j) $\mathscr{C}(\mathbf{V}_1^{-1/2}\mathbf{V}_2\mathbf{V}_1^{-1/2} \cdot \mathbf{V}_1^{-1/2}\mathbf{X}) = \mathscr{C}(\mathbf{V}_1^{-1/2}\mathbf{X})$,

(k) $\mathscr{C}(\mathbf{V}_1^{-1/2}\mathbf{X})$ has a basis $\mathbf{U} = (\mathbf{u}_1 : \ldots : \mathbf{u}_r)$ comprising a set of r eigenvectors of $\mathbf{V}_1^{-1/2}\mathbf{V}_2\mathbf{V}_1^{-1/2}$,

(l) $\mathbf{V}_1^{-1/2}\mathbf{X} = \mathbf{U}\mathbf{A}$ for some $\mathbf{A}_{r\times p}$, $\mathrm{r}(\mathbf{A}) = r$,

(m) $\mathbf{X} = \mathbf{V}_1^{1/2}\mathbf{U}\mathbf{A}$; the columns of $\mathbf{V}_1^{1/2}\mathbf{U}$ are r eigenvectors of $\mathbf{V}_2\mathbf{V}_1^{-1}$,

(n) $\mathscr{C}(\mathbf{X})$ has a basis comprising a set of r eigenvectors of $\mathbf{V}_2\mathbf{V}_1^{-1}$.

10.43 Consider the models $\mathscr{M}_1 = \{\mathbf{y}, \mathbf{X}\boldsymbol{\beta}, \mathbf{V}_1\}$ and $\mathscr{M}_2 = \{\mathbf{y}, \mathbf{X}\boldsymbol{\beta}, \mathbf{V}_2\}$, where $\mathrm{r}(\mathbf{X}) = r$. Denote $\mathbf{P}_{\mathbf{X};\mathbf{W}_1^+} = \mathbf{X}(\mathbf{X}'\mathbf{W}_1^+\mathbf{X})^-\mathbf{X}'\mathbf{W}_1^+$, where $\mathbf{W}_1 = \mathbf{V}_1 + \mathbf{X}\mathbf{U}\mathbf{U}'\mathbf{X}'$, and $\mathscr{C}(\mathbf{W}_1) = \mathscr{C}(\mathbf{X} : \mathbf{V}_1)$ and so $\mathbf{P}_{\mathbf{X};\mathbf{W}_1^+}\mathbf{y}$ is the BLUE for $\mathbf{X}\boldsymbol{\beta}$ under $\mathscr{M}_1$. Then $\mathbf{P}_{\mathbf{X};\mathbf{W}_1^+}\mathbf{y}$ is the BLUE for $\mathbf{X}\boldsymbol{\beta}$ also under $\mathscr{M}_2$ iff any of the following equivalent conditions holds:

(a) $\mathbf{X}'\mathbf{W}_1^+\mathbf{V}_2\mathbf{M} = \mathbf{0}$, (b) $\mathscr{C}(\mathbf{V}_2\mathbf{W}_1^+\mathbf{X}) \subset \mathscr{C}(\mathbf{X})$,

(c) $\mathscr{C}(\mathbf{V}_2\mathbf{M}) \subset \mathscr{N}(\mathbf{X}'\mathbf{W}_1^+) = \mathscr{C}(\mathbf{W}_1^+\mathbf{X})^\perp = \mathscr{C}(\mathbf{G})$; $\mathbf{G} \in \{(\mathbf{W}_1^+\mathbf{X})^\perp\}$,

(d) $\mathscr{C}(\mathbf{W}_1^+\mathbf{X})$ is spanned by a set of r proper eigenvectors of $\mathbf{V}_2$ w.r.t. $\mathbf{W}_1$,

(e) $\mathscr{C}(\mathbf{X})$ is spanned by a set of r eigenvectors of $\mathbf{V}_2\mathbf{W}_1^+$,

(f) $\mathbf{P}_{\mathbf{X};\mathbf{W}_1^+}\mathbf{V}_2$ is symmetric,

(g) $\mathbf{V}_2 \in \{\, \mathbf{V}_2 \in \mathrm{NND}_n : \mathbf{V}_2 = \mathbf{X}\mathbf{N}_1\mathbf{X}' + \mathbf{G}\mathbf{N}_2\mathbf{G}'$ for some $\mathbf{N}_1$ and $\mathbf{N}_2 \,\}$.

10.44 Consider the models $\mathscr{M}_1 = \{\mathbf{y}, \mathbf{X}\boldsymbol{\beta}, \mathbf{V}_1\}$ and $\mathscr{M}_2 = \{\mathbf{y}, \mathbf{X}\boldsymbol{\beta}, \mathbf{V}_2\}$, and let the notation $\{\mathrm{BLUE}(\mathbf{X}\boldsymbol{\beta} \mid \mathscr{M}_1)\} \subset \{\mathrm{BLUE}(\mathbf{X}\boldsymbol{\beta} \mid \mathscr{M}_2)\}$ mean that every representation of the BLUE for $\mathbf{X}\boldsymbol{\beta}$ under $\mathscr{M}_1$ remains the BLUE for $\mathbf{X}\boldsymbol{\beta}$ under $\mathscr{M}_2$. Then the following statements are equivalent:

(a) $\{\mathrm{BLUE}(\mathbf{X}\boldsymbol{\beta} \mid \mathscr{M}_1)\} \subset \{\mathrm{BLUE}(\mathbf{X}\boldsymbol{\beta} \mid \mathscr{M}_2)\}$,

(b) $\{\mathrm{BLUE}(\mathbf{K}'\boldsymbol{\beta} \mid \mathscr{M}_1)\} \subset \{\mathrm{BLUE}(\mathbf{K}'\boldsymbol{\beta}) \mid \mathscr{M}_2)\}$ for every estimable $\mathbf{K}'\boldsymbol{\beta}$,

(c) $\mathscr{C}(\mathbf{V}_2\mathbf{X}^\perp) \subset \mathscr{C}(\mathbf{V}_1\mathbf{X}^\perp)$,

(d) $\mathbf{V}_2 = a\mathbf{V}_1 + \mathbf{X}\mathbf{N}_1\mathbf{X}' + \mathbf{V}_1\mathbf{M}\mathbf{N}_2\mathbf{M}\mathbf{V}_1$ for some $a \in \mathbb{R}$, $\mathbf{N}_1$ and $\mathbf{N}_2$,

(e) $\mathbf{V}_2 = \mathbf{X}\mathbf{N}_3\mathbf{X}' + \mathbf{V}_1\mathbf{M}\mathbf{N}_4\mathbf{M}\mathbf{V}_1$ for some $\mathbf{N}_3$ and $\mathbf{N}_4$.

10.45 Consider the models $\mathscr{M}_1 = \{\mathbf{y}, \mathbf{X}\boldsymbol{\beta}, \mathbf{V}_1\}$ and $\mathscr{M}_2 = \{\mathbf{y}, \mathbf{X}\boldsymbol{\beta}, \mathbf{V}_2\}$. For $\mathbf{X}\boldsymbol{\beta}$ to have a common BLUE under $\mathscr{M}_1$ and $\mathscr{M}_2$ it is necessary and sufficient that

$$\mathscr{C}(\mathbf{V}_1\mathbf{X}^\perp : \mathbf{V}_2\mathbf{X}^\perp) \cap \mathscr{C}(\mathbf{X}) = \{\mathbf{0}\}.$$

10.46 Consider the linear models $\mathscr{M}_1 = \{\mathbf{y}, \mathbf{X}\boldsymbol{\beta}, \mathbf{V}_1\}$ and $\mathscr{M}_2 = \{\mathbf{y}, \mathbf{X}\boldsymbol{\beta}, \mathbf{V}_2\}$. Then the following statements are equivalent:

(a) $\mathbf{G}(\mathbf{X} : \mathbf{V}_1\mathbf{M} : \mathbf{V}_2\mathbf{M}) = (\mathbf{X} : \mathbf{0} : \mathbf{0})$ has a solution for $\mathbf{G}$,

(b) $\mathscr{C}(\mathbf{V}_1\mathbf{M} : \mathbf{V}_2\mathbf{M}) \cap \mathscr{C}(\mathbf{X}) = \{\mathbf{0}\}$,

(c) $\mathscr{C}\begin{pmatrix}\mathbf{M}\mathbf{V}_1\\ \mathbf{M}\mathbf{V}_2\end{pmatrix} \subset \mathscr{C}\begin{pmatrix}\mathbf{M}\mathbf{V}_1\mathbf{M}\\ \mathbf{M}\mathbf{V}_2\mathbf{M}\end{pmatrix}$.

10.47 Let $\mathbf{U} \in \mathbb{R}^{n\times k}$ be given such that $\mathrm{r}(\mathbf{U}) \leq n - 1$. Then for every $\mathbf{X}$ satisfying $\mathscr{C}(\mathbf{U}) \subset \mathscr{C}(\mathbf{X})$ the equality $\mathrm{OLSE}(\mathbf{X}\boldsymbol{\beta}) = \mathrm{BLUE}(\mathbf{X}\boldsymbol{\beta})$ holds under $\{\mathbf{y}, \mathbf{X}\boldsymbol{\beta}, \sigma^2\mathbf{V}\}$ iff any of the following equivalent conditions holds:

(a) $\mathbf{V}(\mathbf{I}-\mathbf{P_U}) = \dfrac{\operatorname{tr}[\mathbf{V}(\mathbf{I}-\mathbf{P_U})]}{n-\operatorname{r}(\mathbf{U})}(\mathbf{I}-\mathbf{P_U})$,

(b) $\mathbf{V}$ can be expressed as $\mathbf{V} = a\mathbf{I}+\mathbf{UAU}'$, where $a \in \mathbb{R}$, and $\mathbf{A}$ is symmetric, such that $\mathbf{V}$ is nonnegative definite.

10.48 (Continued ...) If $\mathbf{U} = \mathbf{1}_n$ then $\mathbf{V}$ in (b) above becomes $\mathbf{V} = a\mathbf{I} + b\mathbf{11}'$, i.e. $\mathbf{V}$ is a completely symmetric matrix.

10.49 Consider the model $\mathscr{M}_{12} = \{\mathbf{y}, \mathbf{X}_1\boldsymbol{\beta}_1 + \mathbf{X}_2\boldsymbol{\beta}_2, \mathbf{V}\}$ and the reduced model $\mathscr{M}_{12\cdot1} = \{\mathbf{M}_1\mathbf{y}, \mathbf{M}_1\mathbf{X}_2\boldsymbol{\beta}_2, \mathbf{M}_1\mathbf{VM}_1\}$. Then

(a) every estimable function of $\boldsymbol{\beta}_2$ is of the form $\mathbf{LM}_1\mathbf{X}_2\boldsymbol{\beta}_2$ for some $\mathbf{L}$,

(b) $\mathbf{K}'\boldsymbol{\beta}_2$ is estimable under $\mathscr{M}_{12}$ iff it is estimable under $\mathscr{M}_{12\cdot1}$.

10.50 Generalized Frisch–Waugh–Lovell theorem. Let us denote

$$\{\text{BLUE}(\mathbf{M}_1\mathbf{X}_2\boldsymbol{\beta}_2 \mid \mathscr{M}_{12})\} = \{\,\mathbf{Ay} : \mathbf{Ay} \text{ is BLUE for } \mathbf{M}_1\mathbf{X}_2\boldsymbol{\beta}_2\,\}.$$

Then every representation of the BLUE of $\mathbf{M}_1\mathbf{X}_2\boldsymbol{\beta}_2$ under $\mathscr{M}_{12}$ remains the BLUE under $\mathscr{M}_{12\cdot1}$ and vice versa, i.e., the sets of the BLUEs coincide:

$$\{\text{BLUE}(\mathbf{M}_1\mathbf{X}_2\boldsymbol{\beta}_2 \mid \mathscr{M}_{12})\} = \{\text{BLUE}(\mathbf{M}_1\mathbf{X}_2\boldsymbol{\beta}_2 \mid \mathscr{M}_{12\cdot1})\}.$$

In other words: Let $\mathbf{K}'\boldsymbol{\beta}_2$ be an arbitrary estimable parametric function under $\mathscr{M}_{12}$. Then every representation of the BLUE of $\mathbf{K}'\boldsymbol{\beta}_2$ under $\mathscr{M}_{12}$ remains the BLUE under $\mathscr{M}_{12\cdot1}$ and vice versa.

10.51 Let $\boldsymbol{\beta}_2$ be estimable under $\mathscr{M}_{12}$. Then

$$\hat{\boldsymbol{\beta}}_2(\mathscr{M}_{12}) = \hat{\boldsymbol{\beta}}_2(\mathscr{M}_{12\cdot1}), \quad \tilde{\boldsymbol{\beta}}_2(\mathscr{M}_{12}) = \tilde{\boldsymbol{\beta}}_2(\mathscr{M}_{12\cdot1}).$$

10.52 Equality of the OLSE and BLUE of the subvectors. Consider a partitioned linear model $\mathscr{M}_{12}$, where $\mathbf{X}_2$ has full column rank and $\mathscr{C}(\mathbf{X}_1) \cap \mathscr{C}(\mathbf{X}_2) = \{\mathbf{0}\}$ holds. Then the following statements are equivalent:

(a) $\hat{\boldsymbol{\beta}}_2(\mathscr{M}_{12}) = \tilde{\boldsymbol{\beta}}_2(\mathscr{M}_{12})$,

(b) $\hat{\boldsymbol{\beta}}_2(\mathscr{M}_{12\cdot1}) = \tilde{\boldsymbol{\beta}}_2(\mathscr{M}_{12\cdot1})$,

(c) $\mathscr{C}(\mathbf{M}_1\mathbf{VM}_1\mathbf{X}_2) \subset \mathscr{C}(\mathbf{M}_1\mathbf{X}_2)$,

(d) $\mathscr{C}[\mathbf{M}_1\mathbf{VM}_1(\mathbf{M}_1\mathbf{X}_2)^\perp] \subset \mathscr{C}(\mathbf{M}_1\mathbf{X}_2)^\perp$,

(e) $\mathbf{P}_{\mathbf{M}_1\mathbf{X}_2}\mathbf{M}_1\mathbf{VM}_1 = \mathbf{M}_1\mathbf{VM}_1\mathbf{P}_{\mathbf{M}_1\mathbf{X}_2}$,

(f) $\mathbf{P}_{\mathbf{M}_1\mathbf{X}_2}\mathbf{M}_1\mathbf{VM}_1\mathbf{Q}_{\mathbf{M}_1\mathbf{X}_2} = \mathbf{0}$, where $\mathbf{Q}_{\mathbf{M}_1\mathbf{X}_2} = \mathbf{I} - \mathbf{P}_{\mathbf{M}_1\mathbf{X}_2}$,

(g) $\mathbf{P}_{\mathbf{M}_1\mathbf{X}_2}\mathbf{VM}_1 = \mathbf{M}_1\mathbf{VP}_{\mathbf{M}_1\mathbf{X}_2}$,

(h) $\mathbf{P}_{\mathbf{M}_1\mathbf{X}_2}\mathbf{VM} = \mathbf{0}$,

(i) $\mathscr{C}(\mathbf{M}_1\mathbf{X}_2)$ has a basis comprising p_2 eigenvectors of $\mathbf{M}_1\mathbf{VM}_1$.

11 The relative efficiency of OLSE

11.1 The Watson efficiency ϕ under model $\mathscr{M} = \{\mathbf{y}, \mathbf{X}\boldsymbol{\beta}, \mathbf{V}\}$. The relative efficiency of OLSE vs. BLUE is defined as the ratio

$$\operatorname{eff}(\hat{\boldsymbol{\beta}}) = \phi = \frac{\det[\operatorname{cov}(\tilde{\boldsymbol{\beta}})]}{\det[\operatorname{cov}(\hat{\boldsymbol{\beta}})]} = \frac{|\mathbf{X}'\mathbf{X}|^2}{|\mathbf{X}'\mathbf{V}\mathbf{X}| \cdot |\mathbf{X}'\mathbf{V}^{-1}\mathbf{X}|}.$$

We have $0 < \phi \leq 1$, with $\phi = 1$ iff $\operatorname{OLSE}(\boldsymbol{\beta}) = \operatorname{BLUE}(\boldsymbol{\beta})$.

BWK The Bloomfield–Watson–Knott inequality. The lower bound for ϕ is

$$\frac{4\lambda_1\lambda_n}{(\lambda_1 + \lambda_n)^2} \cdot \frac{4\lambda_2\lambda_{n-1}}{(\lambda_2 + \lambda_{n-1})^2} \cdots \frac{4\lambda_p\lambda_{n-p+1}}{(\lambda_p + \lambda_{n-p+1})^2} = \tau_1^2\tau_2^2 \cdots \tau_p^2 \leq \phi,$$

i.e.,

$$\min_{\mathbf{X}} \phi = \min_{\mathbf{X}'\mathbf{X}=\mathbf{I}_p} \frac{1}{|\mathbf{X}'\mathbf{V}\mathbf{X}| \cdot |\mathbf{X}'\mathbf{V}^{-1}\mathbf{X}|} = \prod_{i=1}^{p} \frac{4\lambda_i\lambda_{n-i+1}}{(\lambda_i + \lambda_{n-i+1})^2} = \prod_{i=1}^{p} \tau_i^2,$$

where $\lambda_i = \operatorname{ch}_i(\mathbf{V})$, and $\tau_i = i$th antieigenvalue of $\mathbf{V}$. The lower bound is attained when $\mathbf{X}$ is

$$\mathbf{X}_{\text{bad}} = (\mathbf{t}_1 \pm \mathbf{t}_n : \ldots : \mathbf{t}_p \pm \mathbf{t}_{n-p+1}) = \mathbf{T}(\mathbf{i}_1 \pm \mathbf{i}_n : \ldots : \mathbf{i}_p \pm \mathbf{i}_{n-p+1}),$$

where $\mathbf{t}_i$'s are the orthonormal eigenvectors of $\mathbf{V}$ and $p \leq n/2$.

11.2 Suppose that $\operatorname{r}(\mathbf{X}) = p$ and denote $\mathbf{K} = \mathbf{V}^{1/2}\mathbf{X}$, $\mathbf{L} = \mathbf{V}^{1/2}\mathbf{M}$. Then

$$\operatorname{cov}\begin{pmatrix}\mathbf{X}'\mathbf{y}\\ \mathbf{M}\mathbf{y}\end{pmatrix} = \begin{pmatrix}\mathbf{X}'\mathbf{V}\mathbf{X} & \mathbf{X}'\mathbf{V}\mathbf{M}\\ \mathbf{M}\mathbf{V}\mathbf{X} & \mathbf{M}\mathbf{V}\mathbf{M}\end{pmatrix} = \begin{pmatrix}\mathbf{K}'\mathbf{K} & \mathbf{K}'\mathbf{L}\\ \mathbf{L}'\mathbf{K} & \mathbf{L}'\mathbf{L}\end{pmatrix},$$

$$\operatorname{cov}\begin{pmatrix}\hat{\boldsymbol{\beta}}\\ \tilde{\boldsymbol{\beta}}\end{pmatrix} = \begin{pmatrix}\operatorname{cov}(\hat{\boldsymbol{\beta}}) & \operatorname{cov}(\tilde{\boldsymbol{\beta}})\\ \operatorname{cov}(\tilde{\boldsymbol{\beta}}) & \operatorname{cov}(\tilde{\boldsymbol{\beta}})\end{pmatrix} = \begin{pmatrix}(\mathbf{X}'\mathbf{X})^{-1}\mathbf{X}'\mathbf{V}\mathbf{X}(\mathbf{X}'\mathbf{X})^{-1} & \cdot\\ \cdot & \cdot\end{pmatrix}$$

$$= \begin{pmatrix}\mathbf{F}\mathbf{K}'\mathbf{K}\mathbf{F} & \mathbf{F}\mathbf{K}'(\mathbf{I}_n - \mathbf{P}_{\mathbf{L}})\mathbf{K}\mathbf{F}\\ \mathbf{F}\mathbf{K}'(\mathbf{I}_n - \mathbf{P}_{\mathbf{L}})\mathbf{K}\mathbf{F} & \mathbf{F}\mathbf{K}'(\mathbf{I}_n - \mathbf{P}_{\mathbf{L}})\mathbf{K}\mathbf{F}\end{pmatrix}, \quad \mathbf{F} = (\mathbf{X}'\mathbf{X})^{-1},$$

$$\begin{aligned}\operatorname{cov}(\tilde{\boldsymbol{\beta}}) &= \operatorname{cov}(\hat{\boldsymbol{\beta}}) - (\mathbf{X}'\mathbf{X})^{-1}\mathbf{X}'\mathbf{V}\mathbf{M}(\mathbf{M}\mathbf{V}\mathbf{M})^{-}\mathbf{M}\mathbf{V}\mathbf{X}(\mathbf{X}'\mathbf{X})^{-1}\\ &= (\mathbf{X}'\mathbf{X})^{-1}\mathbf{K}'(\mathbf{I}_n - \mathbf{P}_{\mathbf{L}})\mathbf{K}(\mathbf{X}'\mathbf{X})^{-1} := \operatorname{cov}(\hat{\boldsymbol{\beta}}) - \mathbf{D}.\end{aligned}$$

11.3
$$\begin{aligned}\phi = \frac{|\operatorname{cov}(\tilde{\boldsymbol{\beta}})|}{|\operatorname{cov}(\hat{\boldsymbol{\beta}})|} &= \frac{|\mathbf{X}'\mathbf{V}\mathbf{X} - \mathbf{X}'\mathbf{V}\mathbf{M}(\mathbf{M}\mathbf{V}\mathbf{M})^{-}\mathbf{M}\mathbf{V}\mathbf{X}| \cdot |\mathbf{X}'\mathbf{X}|^{-2}}{|\mathbf{X}'\mathbf{V}\mathbf{X}| \cdot |\mathbf{X}'\mathbf{X}|^{-2}}\\ &= \frac{|\mathbf{X}'\mathbf{V}\mathbf{X} - \mathbf{X}'\mathbf{V}\mathbf{M}(\mathbf{M}\mathbf{V}\mathbf{M})^{-}\mathbf{M}\mathbf{V}\mathbf{X}|}{|\mathbf{X}'\mathbf{V}\mathbf{X}|}\\ &= |\mathbf{I}_p - (\mathbf{X}'\mathbf{V}\mathbf{X})^{-1}\mathbf{X}'\mathbf{V}\mathbf{M}(\mathbf{M}\mathbf{V}\mathbf{M})^{-}\mathbf{M}\mathbf{V}\mathbf{X}|\\ &= |\mathbf{I}_p - (\mathbf{K}'\mathbf{K})^{-1}\mathbf{K}'\mathbf{P}_{\mathbf{L}}\mathbf{K}|,\end{aligned}$$

where we must have $\mathrm{r}(\mathbf{VX}) = \mathrm{r}(\mathbf{X}) = p$ so that $\dim \mathscr{C}(\mathbf{X}) \cap \mathscr{C}(\mathbf{V})^{\perp} = \{\mathbf{0}\}$. In a weakly singular model with $\mathrm{r}(\mathbf{X}) = p$ the above representation is valid.

11.4 Consider a weakly singular linear model where $\mathrm{r}(\mathbf{X}) = p$ and let and $\kappa_1 \geq \kappa_2 \geq \cdots \geq \kappa_p \geq 0$ and $\theta_1 \geq \theta_2 \geq \cdots \geq \theta_p > 0$ denote the canonical correlations between $\mathbf{X}'\mathbf{y}$ and $\mathbf{My}$, and $\hat{\boldsymbol{\beta}}$ and $\tilde{\boldsymbol{\beta}}$, respectively, i.e.,

$$\kappa_i = \mathrm{cc}_i(\mathbf{X}'\mathbf{y}, \mathbf{My}), \quad \theta_i = \mathrm{cc}_i(\hat{\boldsymbol{\beta}}, \tilde{\boldsymbol{\beta}}), \quad i = 1, \dots, p.$$

Suppose that $p \leq n/2$ in which case the number of the canonical correlations, i.e., the number of pairs of canonical variables based on $\mathbf{X}'\mathbf{y}$ and $\mathbf{My}$, is p. Then

(a) $m = \mathrm{r}(\mathbf{X}'\mathbf{VM}) =$ number of nonzero κ_i's,

$p =$ number of nonzero θ_i's,

(b)
$$\begin{aligned}\{\kappa_1^2, \dots, \kappa_m^2\} &= \mathrm{nzch}[(\mathbf{X}'\mathbf{VX})^{-1}\mathbf{X}'\mathbf{VM}(\mathbf{MVM})^{-}\mathbf{MVX}] \\ &= \mathrm{nzch}(\mathbf{P}_{\mathbf{V}^{1/2}\mathbf{X}}\mathbf{P}_{\mathbf{V}^{1/2}\mathbf{M}}) = \mathrm{nzch}(\mathbf{P}_{\mathbf{V}^{1/2}\mathbf{H}}\mathbf{P}_{\mathbf{V}^{1/2}\mathbf{M}}) \\ &= \mathrm{nzch}(\mathbf{P}_{\mathbf{K}}\mathbf{P}_{\mathbf{L}}),\end{aligned}$$

(c)
$$\begin{aligned}\{\theta_1^2, \dots, \theta_p^2\} &= \mathrm{ch}[\mathbf{X}'\mathbf{X}(\mathbf{X}'\mathbf{VX})^{-1}\mathbf{X}'\mathbf{X} \cdot (\mathbf{X}'\mathbf{V}^{+}\mathbf{X})^{-1}] \\ &= \mathrm{ch}[(\mathrm{cov}(\hat{\boldsymbol{\beta}}))^{-1} \cdot \mathrm{cov}(\tilde{\boldsymbol{\beta}})] \\ &= \mathrm{ch}[(\mathrm{cov}(\hat{\boldsymbol{\beta}}))^{-1}(\mathrm{cov}(\hat{\boldsymbol{\beta}}) - \mathbf{D})] \\ &= \mathrm{ch}[\mathbf{I}_p - \mathbf{X}'\mathbf{X}(\mathbf{K}'\mathbf{K})^{-1}\mathbf{K}'\mathbf{P}_{\mathbf{L}}\mathbf{K}(\mathbf{X}'\mathbf{X})^{-1}] \\ &= \mathrm{ch}[\mathbf{I}_p - (\mathbf{K}'\mathbf{K})^{-1}\mathbf{K}'\mathbf{P}_{\mathbf{L}}\mathbf{K}] = \{1 - \mathrm{ch}[(\mathbf{K}'\mathbf{K})^{-1}\mathbf{K}'\mathbf{P}_{\mathbf{L}}\mathbf{K}]\} \\ &= \{1 - \mathrm{ch}[(\mathbf{X}'\mathbf{VX})^{-1}\mathbf{X}'\mathbf{VM}(\mathbf{MVM})^{-}\mathbf{MVX}]\},\end{aligned}$$

(d) $\theta_i^2 = 1 - \kappa_{p-i+1}^2$, i.e.,

$$\mathrm{cc}_i^2(\hat{\boldsymbol{\beta}}, \tilde{\boldsymbol{\beta}}) = 1 - \mathrm{cc}_{p-i+1}^2(\mathbf{X}'\mathbf{y}, \mathbf{My}), \quad i = 1, \dots, p.$$

11.5 Under a weakly singular model the Watson efficiency can be written as

$$\begin{aligned}\phi &= \frac{|\mathbf{X}'\mathbf{X}|^2}{|\mathbf{X}'\mathbf{VX}| \cdot |\mathbf{X}'\mathbf{V}^{+}\mathbf{X}|} = \frac{|\mathbf{X}'\mathbf{VX} - \mathbf{X}'\mathbf{VM}(\mathbf{MVM})^{-}\mathbf{MVX}|}{|\mathbf{X}'\mathbf{VX}|} \\ &= |\mathbf{I}_p - \mathbf{X}'\mathbf{VM}(\mathbf{MVM})^{-}\mathbf{MVX}(\mathbf{X}'\mathbf{VX})^{-1}| \\ &= |\mathbf{I}_n - \mathbf{P}_{\mathbf{V}^{1/2}\mathbf{X}}\mathbf{P}_{\mathbf{V}^{1/2}\mathbf{M}}| = |\mathbf{I}_n - \mathbf{P}_{\mathbf{V}^{1/2}\mathbf{H}}\mathbf{P}_{\mathbf{V}^{1/2}\mathbf{M}}| \\ &= \prod_{i=1}^{p} \theta_i^2 = \prod_{i=1}^{p} (1 - \kappa_i^2).\end{aligned}$$

11.6 The θ_i^2's are the roots of the equation $\det[\mathrm{cov}(\tilde{\boldsymbol{\beta}}) - \theta^2 \,\mathrm{cov}(\hat{\boldsymbol{\beta}})] = 0$, and thereby they are solutions to $\mathrm{cov}(\tilde{\boldsymbol{\beta}})\mathbf{w} = \theta^2 \,\mathrm{cov}(\hat{\boldsymbol{\beta}})\mathbf{w}$, $\mathbf{w} \neq \mathbf{0}$.

11.7 Let $\mathbf{Gy}$ be the $\mathrm{BLUE}(\mathbf{X}\boldsymbol{\beta})$ and denote $\mathbf{K}_{*} = \mathbf{V}^{1/2}\mathbf{H}$. Then

$$\operatorname{cov}\begin{pmatrix}\mathbf{Hy}\\ \mathbf{My}\end{pmatrix} = \begin{pmatrix}\mathbf{HVH} & \mathbf{HVM}\\ \mathbf{MVH} & \mathbf{MVM}\end{pmatrix} = \begin{pmatrix}\mathbf{K}_*'\mathbf{K}_* & \mathbf{K}_*'\mathbf{L}\\ \mathbf{L}'\mathbf{K}_* & \mathbf{L}'\mathbf{L}\end{pmatrix},$$

$$\operatorname{cov}\begin{pmatrix}\mathbf{Hy}\\ \mathbf{Gy}\end{pmatrix} = \begin{pmatrix}\mathbf{HVH} & \mathbf{GVG}'\\ \mathbf{GVG}' & \mathbf{GVG}'\end{pmatrix} = \begin{pmatrix}\mathbf{K}_*'\mathbf{K}_* & \mathbf{K}_*'(\mathbf{I}-\mathbf{P_L})\mathbf{K}_*\\ \mathbf{K}_*'(\mathbf{I}-\mathbf{P_L})\mathbf{K}_* & \mathbf{K}_*'(\mathbf{I}-\mathbf{P_L})\mathbf{K}_*\end{pmatrix}.$$

Denote $\mathbf{T}_1 = (\mathbf{HVH})^-\mathbf{HVM}(\mathbf{MVM})^-\mathbf{MVH}$, $\mathbf{T}_2 = (\mathbf{HVH})^-\mathbf{GVG}'$. Then

(a) $\operatorname{cc}_+^2(\mathbf{Hy},\mathbf{My}) = \{\kappa_1^2,\dots,\kappa_m^2\}$ $\qquad m = \operatorname{r}(\mathbf{HVM})$

$$= \operatorname{nzch}(\mathbf{T}_1) = \operatorname{nzch}(\mathbf{P_L}\mathbf{P}_{\mathbf{K}_*}) = \operatorname{nzch}(\mathbf{P_L}\mathbf{P_K}) = \operatorname{cc}_+^2(\mathbf{X}'\mathbf{y},\mathbf{My}),$$

(b) $\operatorname{cc}_+^2(\mathbf{Hy},\mathbf{Gy}) = \{\theta_1^2,\dots,\theta_g^2\}$ $\qquad g = \dim\mathscr{C}(\mathbf{X})\cap\mathscr{C}(\mathbf{V})$

$$= \operatorname{nzch}(\mathbf{T}_2) = \operatorname{nzch}[(\mathbf{HVH})^-\mathbf{GVG}'] = \operatorname{nzch}[(\mathbf{K}_*'\mathbf{K}_*)^-\mathbf{K}_*'(\mathbf{I}_n-\mathbf{P_L})\mathbf{K}_*] = \operatorname{nzch}[\mathbf{P}_{\mathbf{K}_*}(\mathbf{I}_n-\mathbf{P_L})] = \operatorname{nzch}[\mathbf{P_K}(\mathbf{I}_n-\mathbf{P_L})],$$

(c) $\operatorname{cc}_1^2(\mathbf{Hy},\mathbf{Gy}) = \max_{\mathbf{a},\mathbf{b}}\operatorname{cor}^2(\mathbf{a}'\mathbf{Hy},\mathbf{b}'\mathbf{Gy})$

$$= \max_{\mathbf{a}}\frac{\mathbf{a}'\mathbf{GVG}'\mathbf{a}}{\mathbf{a}'\mathbf{HVHa}} \qquad \text{max taken subject to } \mathbf{VHa}\neq\mathbf{0}$$

$$= \operatorname{ch}_1[(\mathbf{HVH})^-\mathbf{GVG}'] = \theta_1^2,$$

(d) $\operatorname{cc}_i^2(\mathbf{Hy},\mathbf{My}) = 1-\operatorname{cc}_{h-i+1}^2(\mathbf{Hy},\mathbf{Gy}),\quad i=1,\dots,h,\quad h=\operatorname{r}(\mathbf{VH})$.

11.8 u = # of unit canonical correlations (κ_i's) between $\mathbf{X}'\mathbf{y}$ and $\mathbf{My}$
= # of unit canonical correlations between $\mathbf{Hy}$ and $\mathbf{My}$
$= \dim\mathscr{C}(\mathbf{V}^{1/2}\mathbf{X})\cap\mathscr{C}(\mathbf{V}^{1/2}\mathbf{M}) = \dim\mathscr{C}(\mathbf{VX})\cap\mathscr{C}(\mathbf{VM})$
$= \operatorname{r}(\mathbf{V}) - \dim\mathscr{C}(\mathbf{X})\cap\mathscr{C}(\mathbf{V}) - \dim\mathscr{C}(\mathbf{M})\cap\mathscr{C}(\mathbf{V})$
$= \operatorname{r}(\mathbf{HP_VM})$

11.9 Under $\mathscr{M} = \{\mathbf{y},\mathbf{X}\boldsymbol{\beta},\mathbf{V}\}$, the following statements are equivalent:

(a) $\mathbf{HP_VM} = \mathbf{0}$, (b) $\mathbf{P_VM} = \mathbf{MP_V}$, (c) $\mathscr{C}(\mathbf{P_VH}) \subset \mathscr{C}(\mathbf{H})$,

(d) $\mathscr{C}(\mathbf{VH})\cap\mathscr{C}(\mathbf{VM}) = \{\mathbf{0}\}$, (e) $\mathscr{C}(\mathbf{V}^{1/2}\mathbf{H})\cap\mathscr{C}(\mathbf{V}^{1/2}\mathbf{M}) = \{\mathbf{0}\}$,

(f) $\mathscr{C}(\mathbf{X}) = \mathscr{C}(\mathbf{X})\cap\mathscr{C}(\mathbf{V}) \boxplus \mathscr{C}(\mathbf{X})\cap\mathscr{C}(\mathbf{V})^\perp$,

(g) $u = \dim\mathscr{C}(\mathbf{VH})\cap\mathscr{C}(\mathbf{VM}) = \operatorname{r}(\mathbf{HP_VM}) = 0$, where u is the number of unit canonical correlations between $\mathbf{Hy}$ and $\mathbf{My}$,

(h) $\operatorname{cov}(\mathbf{X}\tilde{\boldsymbol{\beta}}) = \mathbf{P_VX}(\mathbf{X}'\mathbf{V}^+\mathbf{X})^-\mathbf{X}'\mathbf{P_V}$.

11.10 The squared canonical correlations κ_i^2's are the proper eigenvalues of $\mathbf{K}'\mathbf{P_L}\mathbf{K}$ with respect to $\mathbf{K}'\mathbf{K}$, i.e.,

$$\mathbf{HVM}(\mathbf{MVM})^-\mathbf{MVHw} = \kappa^2\mathbf{HVHw}, \quad \mathbf{VHw}\neq\mathbf{0}.$$

The squared canonical correlations θ_i^2's are the proper eigenvalues of $\mathbf{L}'\mathbf{L} = \mathbf{GVG}'$ with respect to $\mathbf{K}'\mathbf{K} = \mathbf{HVH}$:

$$\mathbf{GVG}'\mathbf{w} = \theta^2\mathbf{HVHw} = (1-\kappa^2)\mathbf{HVHw}, \quad \mathbf{VHw} \neq \mathbf{0}.$$

11.11 We can arrange the κ_i^2's as follows:

$$\begin{aligned} &\kappa_1^2 = \cdots = \kappa_u^2 = 1, && u = \mathrm{r}(\mathbf{HP_VM}) \\ &1 > \kappa_{u+1}^2 \geq \cdots \geq \kappa_{u+t}^2 = \kappa_m^2 > 0, && m = \mathrm{r}(\mathbf{HVM}) \\ &\kappa_{u+t+1}^2 = \kappa_{m+1}^2 = \cdots = \kappa_{m+s}^2 = \varrho_h^2 = 0, && \end{aligned}$$

where $\kappa_i^2 = 1 - \theta_{h-i+1}^2$, $i = 1,\dots,h = \mathrm{r}(\mathbf{VH})$, $s = \dim \mathscr{C}(\mathbf{VX}) \cap \mathscr{C}(\mathbf{X})$ and $t = \dim \mathscr{C}(\mathbf{V}) \cap \mathscr{C}(\mathbf{X}) - s$.

11.12 Antieigenvalues. Denoting $\mathbf{x} = \mathbf{V}^{1/2}\mathbf{z}$, the Watson efficiency ϕ can be interpreted as a specific squared cosine:

$$\begin{aligned} \phi &= \frac{(\mathbf{x}'\mathbf{x})^2}{\mathbf{x}'\mathbf{Vx}\cdot\mathbf{x}'\mathbf{V}^{-1}\mathbf{x}} = \frac{(\mathbf{x}'\mathbf{V}^{1/2}\cdot\mathbf{V}^{-1/2}\mathbf{x})^2}{\mathbf{x}'\mathbf{Vx}\cdot\mathbf{x}'\mathbf{V}^{-1}\mathbf{x}} = \cos^2(\mathbf{V}^{1/2}\mathbf{x}, \mathbf{V}^{-1/2}\mathbf{x}) \\ &= \frac{(\mathbf{z}'\mathbf{Vz})^2}{\mathbf{z}'\mathbf{V}^2\mathbf{z}\cdot\mathbf{z}'\mathbf{z}} = \cos^2(\mathbf{Vz},\mathbf{z}). \end{aligned}$$

The vector $\mathbf{z}$ minimizing ϕ [maximizing angle $\angle(\mathbf{Vz},\mathbf{z})$] is called an antieigenvector and the corresponding minimum (τ_1^2) the first antieigenvalue (squared):

$$\tau_1 = \min_{\mathbf{z}\neq\mathbf{0}} \cos(\mathbf{Vz},\mathbf{z}) := \cos\varphi(\mathbf{V}) = \frac{2\sqrt{\lambda_1\lambda_n}}{\lambda_1+\lambda_n} = \frac{\sqrt{\lambda_1\lambda_n}}{(\lambda_1+\lambda_n)/2},$$

where $\varphi(\mathbf{V})$ is the matrix angle of $\mathbf{V}$ and $(\lambda_i, \mathbf{t}_i)$ is the ith eigenpair of $\mathbf{V}$. The first antieigenvector has forms proportional to $\mathbf{z}_1 = \sqrt{\lambda_n}\mathbf{t}_1 \pm \sqrt{\lambda_1}\mathbf{t}_n$. The second antieigenvalue of $\mathbf{V}$ is defined as

$$\tau_2 = \min_{\mathbf{z}\neq\mathbf{0},\,\mathbf{z}'\mathbf{z}_1=0} \cos(\mathbf{Vz},\mathbf{z}) = \frac{2\sqrt{\lambda_2\lambda_{n-1}}}{\lambda_2+\lambda_{n-1}}.$$

11.13 Consider the models $\mathscr{M}_{12} = \{\mathbf{y}, \mathbf{X}_1\boldsymbol{\beta}_1 + \mathbf{X}_2\boldsymbol{\beta}_2, \mathbf{V}\}$, $\mathscr{M}_1 = \{\mathbf{y}, \mathbf{X}_1\boldsymbol{\beta}, \mathbf{V}\}$, and $\mathscr{M}_{1\mathrm{H}} = \{\mathbf{Hy}, \mathbf{X}_1\boldsymbol{\beta}_1, \mathbf{HVH}\}$, where $\mathbf{X}$ is of full column rank and $\mathscr{C}(\mathbf{X}) \subset \mathscr{C}(\mathbf{V})$. Then the corresponding Watson efficiencies are

(a) $$\mathrm{eff}(\hat{\boldsymbol{\beta}} \mid \mathscr{M}_{12}) = \frac{|\mathrm{cov}(\tilde{\boldsymbol{\beta}} \mid \mathscr{M}_{12})|}{|\mathrm{cov}(\hat{\boldsymbol{\beta}} \mid \mathscr{M}_{12})|} = \frac{|\mathbf{X}'\mathbf{X}|^2}{|\mathbf{X}'\mathbf{VX}|\cdot|\mathbf{X}'\mathbf{V}^+\mathbf{X}|},$$

(b) $$\mathrm{eff}(\hat{\boldsymbol{\beta}}_2 \mid \mathscr{M}_{12}) = \frac{|\mathrm{cov}(\tilde{\boldsymbol{\beta}}_2 \mid \mathscr{M}_{12})|}{|\mathrm{cov}(\hat{\boldsymbol{\beta}}_2 \mid \mathscr{M}_{12})|} = \frac{|\mathbf{X}_2'\mathbf{M}_1\mathbf{X}_2|^2}{|\mathbf{X}_2'\mathbf{M}_1\mathbf{VM}_1\mathbf{X}_2|\cdot|\mathbf{X}_2'\dot{\mathbf{M}}_1\mathbf{X}_2|},$$

(c) $$\mathrm{eff}(\hat{\boldsymbol{\beta}}_1 \mid \mathscr{M}_1) = \frac{|\mathbf{X}_1'\mathbf{X}_1|^2}{|\mathbf{X}_1'\mathbf{VX}_1|\cdot|\mathbf{X}_1'\mathbf{V}^+\mathbf{X}_1|},$$

(d) $\mathrm{eff}(\hat{\boldsymbol\beta}_1 \mid \mathscr{M}_{1\mathrm{H}}) = \dfrac{|\mathbf{X}_1'\mathbf{X}_1|^2}{|\mathbf{X}_1'\mathbf{V}\mathbf{X}_1| \cdot |\mathbf{X}_1'(\mathbf{HVH})^-\mathbf{X}_1|}.$

11.14 (Continued ...) The Watson efficiency $\mathrm{eff}(\hat{\boldsymbol\beta} \mid \mathscr{M}_{12})$ can be expressed as

$$\mathrm{eff}(\hat{\boldsymbol\beta} \mid \mathscr{M}_{12}) = \mathrm{eff}(\hat{\boldsymbol\beta}_1 \mid \mathscr{M}_1) \cdot \mathrm{eff}(\hat{\boldsymbol\beta}_2 \mid \mathscr{M}_{12}) \cdot \frac{1}{\mathrm{eff}(\hat{\boldsymbol\beta}_1 \mid \mathscr{M}_{1\mathrm{H}})},$$

where

$$\begin{aligned}\mathrm{eff}(\hat{\boldsymbol\beta}_1 \mid \mathscr{M}_{1\mathrm{H}}) &= \frac{|\mathbf{X}_1'\mathbf{X}_1|^2}{|\mathbf{X}_1'\mathbf{V}\mathbf{X}_1| \cdot |\mathbf{X}_1'(\mathbf{HVH})^-\mathbf{X}_1|} \\ &= |\mathbf{I}_{p_1} - \mathbf{X}_1'\mathbf{V}\mathbf{M}_1\mathbf{X}_2(\mathbf{X}_2'\mathbf{M}_1\mathbf{V}\mathbf{M}_1\mathbf{X}_2)^{-1}\mathbf{X}_2'\mathbf{M}_1\mathbf{V}\mathbf{X}_1'(\mathbf{X}_1'\mathbf{V}\mathbf{X}_1)^{-1}|.\end{aligned}$$

11.15 (Continued ...) The following statements are equivalent:

(a) $\mathrm{eff}(\hat{\boldsymbol\beta} \mid \mathscr{M}_{12}) = \mathrm{eff}(\hat{\boldsymbol\beta}_2 \mid \mathscr{M}_{12})$,

(b) $\mathscr{C}(\mathbf{X}_1) \subset \mathscr{C}(\mathbf{VX})$,

(c) $\mathbf{Hy}$ is linearly sufficient for $\mathbf{X}_1\boldsymbol\beta_1$ under the model $\mathscr{M}_1$.

11.16 (Continued ...) The efficiency factorization multiplier γ is defined as

$$\gamma = \frac{\mathrm{eff}(\hat{\boldsymbol\beta} \mid \mathscr{M}_{12})}{\mathrm{eff}(\hat{\boldsymbol\beta}_1 \mid \mathscr{M}_{12}) \cdot \mathrm{eff}(\hat{\boldsymbol\beta}_2 \mid \mathscr{M}_{12})}.$$

We say that the Watson efficiency factorizes if $\gamma = 1$, i.e., $\mathrm{eff}(\hat{\boldsymbol\beta} \mid \mathscr{M}_{12}) = \mathrm{eff}(\hat{\boldsymbol\beta}_1 \mid \mathscr{M}_{12}) \cdot \mathrm{eff}(\hat{\boldsymbol\beta}_2 \mid \mathscr{M}_{12})$, which happens iff (supposing $\mathbf{X}'\mathbf{X} = \mathbf{I}_p$)

$$|\mathbf{I}_n - \mathbf{P}_{\mathbf{V}^{+1/2}\mathbf{X}_1}\mathbf{P}_{\mathbf{V}^{+1/2}\mathbf{X}_2}| = |\mathbf{I}_n - \mathbf{P}_{\mathbf{V}^{1/2}\mathbf{X}_1}\mathbf{P}_{\mathbf{V}^{1/2}\mathbf{X}_2}|.$$

11.17 Consider a weakly singular linear model $\mathscr{M}_Z = \{\mathbf{y}, \mathbf{Z}\boldsymbol\gamma, \mathbf{V}\} = \{\mathbf{y}, \mathbf{Z}\binom{\boldsymbol\beta}{\delta}, \mathbf{V}\}$, where $\mathbf{Z} = (\mathbf{X} : \mathbf{i}_i)$ has full column rank, and denote $\mathscr{M} = \{\mathbf{y}, \mathbf{X}\boldsymbol\beta, \mathbf{V}\}$, and let $\mathscr{M}_{(i)} = \{\mathbf{y}_{(i)}, \mathbf{X}_{(i)}\boldsymbol\beta, \mathbf{V}_{(i)}\}$ be such a version of $\mathscr{M}$ in which the ith observation is deleted. Assume that $\mathbf{X}'\mathbf{i}_i \neq \mathbf{0}$ $(i = 1, \ldots, n)$, and that OLSE($\boldsymbol\beta$) equals BLUE($\boldsymbol\beta$) under $\mathscr{M}$. Then

$$\hat{\boldsymbol\beta}(\mathscr{M}_{(i)}) = \tilde{\boldsymbol\beta}(\mathscr{M}_{(i)}) \quad \text{for all } i = 1, \ldots, n,$$

iff $\mathbf{V}$ satisfies $\mathbf{MVM} = \lambda^2\mathbf{M}$, for some nonzero scalar λ.

11.18 The Bloomfield–Watson efficiency ψ:

$$\begin{aligned}\psi &= \tfrac{1}{2}\|\mathbf{HV} - \mathbf{VH}\|_F^2 = \|\mathbf{HVM}\|_F^2 = \tfrac{1}{2}\,\mathrm{tr}(\mathbf{HV} - \mathbf{VH})(\mathbf{HV} - \mathbf{VH})' \\ &= \mathrm{tr}(\mathbf{HVMV}) = \mathrm{tr}(\mathbf{HV}^2) - \mathrm{tr}(\mathbf{HV})^2 \leq \tfrac{1}{4}\sum_{i=1}^{p}(\lambda_i - \lambda_{n-i-1})^2,\end{aligned}$$

where the equality is attained in the same situation as the minimum of ϕ_{12}.

11.19 C. R. Rao's criterion for the goodness of OLSE:

$$\eta = \mathrm{tr}[\mathrm{cov}(\mathbf{X}\hat{\boldsymbol{\beta}}) - \mathrm{cov}(\mathbf{X}\tilde{\boldsymbol{\beta}})] = \mathrm{tr}[\mathbf{HVH} - \mathbf{X}(\mathbf{X}'\mathbf{V}^{-1}\mathbf{X})^{-1}\mathbf{X}']$$
$$\leq \sum_{i=1}^{p}(\lambda_i^{1/2} - \lambda_{n-i+1}^{1/2})^2.$$

11.20 Equality of the BLUEs of $\mathbf{X}_2\boldsymbol{\beta}_2$ under two models. Consider the models $\mathscr{M}_{12} = \{\mathbf{y}, \mathbf{X}_1\boldsymbol{\beta}_1 + \mathbf{X}_2\boldsymbol{\beta}_2, \mathbf{V}\}$ and $\underline{\mathscr{M}}_{12} = \{\mathbf{y}, \mathbf{X}_1\boldsymbol{\beta}_1 + \mathbf{X}_2\boldsymbol{\beta}_2, \underline{\mathbf{V}}\}$, where $\mathscr{C}(\mathbf{X}_1) \cap \mathscr{C}(\mathbf{X}_2) = \{\mathbf{0}\}$. Then the following statements are equivalent:

(a) $\{\mathrm{BLUE}(\mathbf{X}_2\boldsymbol{\beta}_2 \mid \mathscr{M}_{12}\} \subset \{\mathrm{BLUE}(\mathbf{X}_2\boldsymbol{\beta}_2 \mid \underline{\mathscr{M}}_{12}\}$,

(b) $\{\mathrm{BLUE}(\mathbf{M}_1\mathbf{X}_2\boldsymbol{\beta}_2 \mid \mathscr{M}_{12}\} \subset \{\mathrm{BLUE}(\mathbf{M}_1\mathbf{X}_2\boldsymbol{\beta}_2 \mid \underline{\mathscr{M}}_{12}\}$,

(c) $\mathscr{C}(\underline{\mathbf{V}}\mathbf{M}) \subset \mathscr{C}(\mathbf{X}_1 : \mathbf{VM})$,

(d) $\mathscr{C}(\mathbf{M}_1\underline{\mathbf{V}}\mathbf{M}) \subset \mathscr{C}(\mathbf{M}_1\mathbf{VM})$,

(e) $\mathscr{C}[\mathbf{M}_1\underline{\mathbf{V}}\mathbf{M}_1(\mathbf{M}_1\mathbf{X}_1)^{\perp}] \subset \mathscr{C}[\mathbf{M}_1\mathbf{VM}_1(\mathbf{M}_1\mathbf{X}_2)^{\perp}]$,

(f) for some $\mathbf{L}_1$ and $\mathbf{L}_2$, the matrix $\mathbf{M}_1\underline{\mathbf{V}}\mathbf{M}_1$ can be expressed as

$$\begin{aligned}\mathbf{M}_1\underline{\mathbf{V}}\mathbf{M}_1 &= \mathbf{M}_1\mathbf{X}_2\mathbf{L}_1\mathbf{X}_2'\mathbf{M}_1 + \mathbf{M}_1\mathbf{VM}_1\mathbf{Q}_{\mathbf{M}_1\mathbf{X}_2}\mathbf{L}_2\mathbf{Q}_{\mathbf{M}_1\mathbf{X}_2}\mathbf{M}_1\mathbf{VM}_1\\ &= \mathbf{M}_1\mathbf{X}_2\mathbf{L}_1\mathbf{X}_2'\mathbf{M}_1 + \mathbf{M}_1\mathbf{VML}_2\mathbf{MVM}_1\end{aligned}$$

12 Linear sufficiency and admissibility

12.1 Under the model $\{\mathbf{y}, \mathbf{X}\boldsymbol{\beta}, \sigma^2\mathbf{V}\}$, a linear statistic $\mathbf{Fy}$ is called linearly sufficient for $\mathbf{X}\boldsymbol{\beta}$ if there exists a matrix $\mathbf{A}$ such that $\mathbf{AFy}$ is the BLUE of $\mathbf{X}\boldsymbol{\beta}$.

12.2 Let $\mathbf{W} = \mathbf{V} + \mathbf{XUU}'\mathbf{X}'$ be an arbitrary nnd matrix satisfying $\mathscr{C}(\mathbf{W}) = \mathscr{C}(\mathbf{X} : \mathbf{V})$; this notation holds throughout this section. Then a statistic $\mathbf{Fy}$ is linearly sufficient for $\mathbf{X}\boldsymbol{\beta}$ iff any of the following equivalent statements holds:

(a) $\mathscr{C}(\mathbf{X}) \subset \mathscr{C}(\mathbf{WF}')$,

(b) $\mathscr{N}(\mathbf{F}) \cap \mathscr{C}(\mathbf{X} : \mathbf{V}) \subset \mathscr{C}(\mathbf{VX}^{\perp})$,

(c) $\mathrm{r}(\mathbf{X} : \mathbf{VF}') = \mathrm{r}(\mathbf{WF}')$,

(d) $\mathscr{C}(\mathbf{X}'\mathbf{F}') = \mathscr{C}(\mathbf{X}')$ and $\mathscr{C}(\mathbf{FX}) \cap \mathscr{C}(\mathbf{FVX}^{\perp}) = \{\mathbf{0}\}$.

(e) The best linear predictor of $\mathbf{y}$ based on $\mathbf{Fy}$, $\mathrm{BLP}(\mathbf{y}; \mathbf{Fy})$, is almost surely equal to a linear function of $\mathbf{Fy}$ which does not depend on $\boldsymbol{\beta}$.

12.3 Let $\mathbf{Fy}$ be linearly sufficient for $\mathbf{X}\boldsymbol{\beta}$ under $\mathscr{M} = \{\mathbf{y}, \mathbf{X}\boldsymbol{\beta}, \sigma^2\mathbf{V}\}$. Then each BLUE of $\mathbf{X}\boldsymbol{\beta}$ under the transformed model $\{\mathbf{Fy}, \mathbf{FX}\boldsymbol{\beta}, \sigma^2\mathbf{FVF}'\}$ is the BLUE of $\mathbf{X}\boldsymbol{\beta}$ under the original model $\mathscr{M}$ and vice versa.

12.4 Let $\mathbf{K}'\boldsymbol{\beta}$ be an estimable parametric function under the model $\{\mathbf{y}, \mathbf{X}\boldsymbol{\beta}, \sigma^2\mathbf{V}\}$. Then the following statements are equivalent:

(a) $\mathbf{Fy}$ is linearly sufficient for $\mathbf{K}'\boldsymbol{\beta}$, (b) $\mathscr{N}(\mathbf{FX} : \mathbf{FVX}^{\perp}) \subset \mathscr{N}(\mathbf{K}' : \mathbf{0})$,

(c) $\mathscr{C}[\mathbf{X}(\mathbf{X}'\mathbf{W}^{-}\mathbf{X})^{-}\mathbf{K}] \subset \mathscr{C}(\mathbf{WF}')$.

12.5 Under the model $\{\mathbf{y}, \mathbf{X}\boldsymbol{\beta}, \sigma^2\mathbf{V}\}$, a linear statistic $\mathbf{Fy}$ is called linearly minimal sufficient for $\mathbf{X}\boldsymbol{\beta}$, if for any other linearly sufficient statistic $\mathbf{Ly}$, there exists a matrix $\mathbf{A}$ such that $\mathbf{Fy} = \mathbf{ALy}$ almost surely.

12.6 The statistic $\mathbf{Fy}$ is linearly minimal sufficient for $\mathbf{X}\boldsymbol{\beta}$ iff $\mathscr{C}(\mathbf{X}) = \mathscr{C}(\mathbf{WF}')$.

12.7 Let $\mathbf{K}'\boldsymbol{\beta}$ be an estimable parametric function under the model $\{\mathbf{y}, \mathbf{X}\boldsymbol{\beta}, \sigma^2\mathbf{V}\}$. Then the following statements are equivalent:

(a) $\mathbf{Fy}$ is linearly minimal sufficient for $\mathbf{K}'\boldsymbol{\beta}$,

(b) $\mathscr{N}(\mathbf{FX} : \mathbf{FVX}^{\perp}) = \mathscr{N}(\mathbf{K}' : \mathbf{0})$,

(c) $\mathscr{C}(\mathbf{K}) = \mathscr{C}(\mathbf{X}'\mathbf{F}')$ and $\mathbf{FVX}^{\perp} = \mathbf{0}$.

12.8 Let $\mathbf{X}_1\boldsymbol{\beta}_1$ be estimable under $\{\mathbf{y}, \mathbf{X}_1\boldsymbol{\beta}_1 + \mathbf{X}_2\boldsymbol{\beta}_2, \mathbf{V}\}$ and denote $\mathbf{W}_1 = \mathbf{V} + \mathbf{X}_1\mathbf{X}_1'$ and $\dot{\mathbf{M}}_2 = \mathbf{M}_2(\mathbf{M}_2\mathbf{W}_1\mathbf{M}_2)^{-}\mathbf{M}_2$. Then $\mathbf{X}_1'\dot{\mathbf{M}}_2\mathbf{y}$ is linearly minimal sufficient for $\mathbf{X}_1\boldsymbol{\beta}_1$.

12.9 Under the model $\{\mathbf{y}, \mathbf{X}\boldsymbol{\beta}, \sigma^2\mathbf{V}\}$, a linear statistic $\mathbf{Fy}$ is called linearly complete for $\mathbf{X}\boldsymbol{\beta}$ if for all matrices $\mathbf{A}$ such that $\mathrm{E}(\mathbf{AFy}) = \mathbf{0}$ it follows that $\mathbf{AFy} = \mathbf{0}$ almost surely.

12.10 A statistic $\mathbf{Fy}$ is linearly complete for $\mathbf{X}\boldsymbol{\beta}$ iff $\mathscr{C}(\mathbf{FV}) \subset \mathscr{C}(\mathbf{FX})$.

12.11 $\mathbf{Fy}$ is linearly complete and linearly sufficient for $\mathbf{X}\boldsymbol{\beta}$ $\iff$ $\mathbf{Fy}$ is minimal linear sufficient $\iff$ $\mathscr{C}(\mathbf{X}) = \mathscr{C}(\mathbf{WF}')$.

12.12 Linear Lehmann–Scheffé theorem. Under the model $\{\mathbf{y}, \mathbf{X}\boldsymbol{\beta}, \sigma^2\mathbf{V}\}$, let $\mathbf{Ly}$ be linear unbiased estimator for $\mathbf{X}\boldsymbol{\beta}$ and let $\mathbf{Fy}$ be linearly complete and linearly sufficient for $\mathbf{X}\boldsymbol{\beta}$. Then the BLUE of $\mathbf{X}\boldsymbol{\beta}$ is almost surely equal to the best linear predictor of $\mathbf{Ly}$ based on $\mathbf{Fy}$, $\mathrm{BLP}(\mathbf{Ly}; \mathbf{Fy})$.

12.13 Admissibility. Consider the linear model $\mathscr{M} = \{\mathbf{y}, \mathbf{X}\boldsymbol{\beta}, \sigma^2\mathbf{V}\}$ and let $\mathbf{K}'\boldsymbol{\beta}$ be an estimable parametric function, $\mathbf{K}' \in \mathbb{R}^{q\times p}$. Denote the set of all linear (homogeneous) estimators of $\mathbf{K}'\boldsymbol{\beta}$ as $\mathrm{LE}_q(\mathbf{y}) = \{\, \mathbf{Fy} : \mathbf{F} \in \mathbb{R}^{q\times n} \,\}$. The mean squared error matrix of $\mathbf{Fy}$ with respect to $\mathbf{K}'\boldsymbol{\beta}$ is defined as

$$\mathrm{MSEM}(\mathbf{Fy}; \mathbf{K}'\boldsymbol{\beta}) = \mathrm{E}(\mathbf{Fy} - \mathbf{K}'\boldsymbol{\beta})(\mathbf{Fy} - \mathbf{K}'\boldsymbol{\beta})',$$

and the quadratic risk of $\mathbf{Fy}$ under $\mathscr{M}$ is

$$\begin{aligned}\text{risk}(\mathbf{Fy};\mathbf{K}'\boldsymbol{\beta}) &= \operatorname{trace}\text{MSEM}(\mathbf{Fy};\mathbf{K}'\boldsymbol{\beta})\\ &= \text{E}(\mathbf{Fy}-\mathbf{K}'\boldsymbol{\beta})'(\mathbf{Fy}-\mathbf{K}'\boldsymbol{\beta})\\ &= \sigma^2\operatorname{trace}(\mathbf{FVF}') + \|(\mathbf{FX}-\mathbf{K}')\boldsymbol{\beta}\|^2\\ &= \operatorname{trace}\operatorname{cov}(\mathbf{Fy}) + \text{bias}^2\,.\end{aligned}$$

A linear estimator $\mathbf{Ay}$ is said to be admissible for $\mathbf{K}'\boldsymbol{\beta}$ among $\text{LE}_q(\mathbf{y})$ under $\mathscr{M}$ if there does not exist $\mathbf{Fy} \in \text{LE}_q(\mathbf{y})$ such that the inequality

$$\text{risk}(\mathbf{Fy};\mathbf{K}'\boldsymbol{\beta}) \leq \text{risk}(\mathbf{Ay};\mathbf{K}'\boldsymbol{\beta})$$

holds for every $(\boldsymbol{\beta},\sigma^2) \in \mathbb{R}^p \times (0,\infty)$ and is strict for at least one point $(\boldsymbol{\beta},\sigma^2)$. The set of admissible estimators of $\mathbf{K}'\boldsymbol{\beta}$ is denoted as $\text{AD}(\mathbf{K}'\boldsymbol{\beta})$.

12.14 Consider the model $\{\mathbf{y},\mathbf{X}\boldsymbol{\beta},\sigma^2\mathbf{V}\}$ and let $\mathbf{K}'\boldsymbol{\beta}$ ($\mathbf{K}' \in \mathbb{R}^{q\times p}$) be estimable, i.e., $\mathbf{K}' = \mathbf{LX}$ for some $\mathbf{L} \in \mathbb{R}^{q\times n}$. Then $\mathbf{Fy} \in \text{AD}(\mathbf{K}'\boldsymbol{\beta})$ iff

$$\mathscr{C}(\mathbf{VF}') \subset \mathscr{C}(\mathbf{X}), \quad \mathbf{FVL}' - \mathbf{FVF}' \geq_{\mathsf{L}} \mathbf{0} \quad \text{and}$$
$$\mathscr{C}[(\mathbf{F}-\mathbf{L})\mathbf{X}] = \mathscr{C}[(\mathbf{F}-\mathbf{L})\mathbf{N}],$$

where $\mathbf{N}$ is a matrix satisfying $\mathscr{C}(\mathbf{N}) = \mathscr{C}(\mathbf{X}) \cap \mathscr{C}(\mathbf{V})$.

12.15 Let $\mathbf{K}'\boldsymbol{\beta}$ be any arbitrary parametric function. Then a necessary condition for $\mathbf{Fy} \in \text{AD}(\mathbf{K}'\boldsymbol{\beta})$ is that $\mathscr{C}(\mathbf{FX} : \mathbf{FV}) \subset \mathscr{C}(\mathbf{K}')$.

12.16 If $\mathbf{Fy}$ is linearly sufficient and admissible for an estimable $\mathbf{K}'\boldsymbol{\beta}$, then $\mathbf{Fy}$ is also linearly minimal sufficient for $\mathbf{K}'\boldsymbol{\beta}$.

13 Best linear unbiased predictor

13.1 BLUP: Best linear unbiased predictor. Consider the linear model with *new* observations:

$$\mathscr{M}_f = \left\{\begin{pmatrix}\mathbf{y}\\ \mathbf{y}_f\end{pmatrix}, \begin{pmatrix}\mathbf{X}\boldsymbol{\beta}\\ \mathbf{X}_f\boldsymbol{\beta}\end{pmatrix}, \sigma^2\begin{pmatrix}\mathbf{V} & \mathbf{V}_{12}\\ \mathbf{V}_{21} & \mathbf{V}_{22}\end{pmatrix}\right\},$$

where $\mathbf{y}_f$ is an unobservable random vector containing new observations (observable in future). Above we have $\text{E}(\mathbf{y}) = \mathbf{X}\boldsymbol{\beta}$, $\text{E}(\mathbf{y}_f) = \mathbf{X}_f\boldsymbol{\beta}$, and

$$\operatorname{cov}(\mathbf{y}) = \sigma^2\mathbf{V}, \quad \operatorname{cov}(\mathbf{y}_f) = \sigma^2\mathbf{V}_{22}, \quad \operatorname{cov}(\mathbf{y},\mathbf{y}_f) = \sigma^2\mathbf{V}_{12}.$$

A linear predictor $\mathbf{Gy}$ is said to be linear unbiased predictor, LUP, for $\mathbf{y}_f$ if $\text{E}(\mathbf{Gy}) = \text{E}(\mathbf{y}_f) = \mathbf{X}_f\boldsymbol{\beta}$ for all $\boldsymbol{\beta} \in \mathbb{R}^p$, i.e., $\text{E}(\mathbf{Gy}-\mathbf{y}_f) = \mathbf{0}$, i.e., $\mathscr{C}(\mathbf{X}_f') \subset \mathscr{C}(\mathbf{X}')$. Then $\mathbf{y}_f$ is said unbiasedly predictable; the difference $\mathbf{Gy}-\mathbf{y}_f$ is the prediction error. A linear unbiased predictor $\mathbf{Gy}$ is the BLUP for $\mathbf{y}_f$ if

$$\operatorname{cov}(\mathbf{Gy}-\mathbf{y}_f) \leq_{\mathsf{L}} \operatorname{cov}(\mathbf{Fy}-\mathbf{y}_f) \quad \text{for all } \mathbf{Fy} \in \{\text{LUP}(\mathbf{y}_f)\}.$$

13.2 A linear predictor $\mathbf{Ay}$ is the BLUP for $\mathbf{y}_f$ iff the equation

$$\mathbf{A}(\mathbf{X} : \mathbf{VX}^{\perp}) = (\mathbf{X}_f : \mathbf{V}_{21}\mathbf{X}^{\perp})$$

is satisfied. This is equivalent to the existence of a matrix $\mathbf{L}$ such that $\mathbf{A}$ satisfies the equation (Pandora's Box for the BLUP)

$$\begin{pmatrix} \mathbf{V} & \mathbf{X} \\ \mathbf{X}' & \mathbf{0} \end{pmatrix} \begin{pmatrix} \mathbf{A}' \\ \mathbf{L} \end{pmatrix} = \begin{pmatrix} \mathbf{V}_{12} \\ \mathbf{X}'_f \end{pmatrix}.$$

13.3 A linear estimator $\boldsymbol{\ell}'\mathbf{y}$ which is unbiased for zero, is called a linear zero function. Every linear zero function can be written as $\mathbf{b}'\mathbf{My}$ for some $\mathbf{b}$. Hence a linear unbiased predictor $\mathbf{Ay}$ is the BLUP for $\mathbf{y}_f$ under $\mathscr{M}_f$ iff

$$\operatorname{cov}(\mathbf{Ay} - \mathbf{y}_f, \boldsymbol{\ell}'\mathbf{y}) = \mathbf{0} \quad \text{for every linear zero function } \boldsymbol{\ell}'\mathbf{y}.$$

13.4 Let $\mathscr{C}(\mathbf{X}'_f) \subset \mathscr{C}(\mathbf{X}')$, i.e., $\mathbf{X}_f\boldsymbol{\beta}$ is estimable (assumed below in all statements). The general solution to 13.2 can be written, for example, as

$$\mathbf{A}_0 = (\mathbf{X}_f : \mathbf{V}_{21}\mathbf{M})(\mathbf{X} : \mathbf{VM})^+ + \mathbf{F}(\mathbf{I}_n - \mathbf{P}_{(\mathbf{X}:\mathbf{VM})}),$$

where $\mathbf{F}$ is free to vary. Even though the multiplier $\mathbf{A}$ may not be unique, the observed value $\mathbf{Ay}$ of the BLUP is unique with probability 1. We can get, for example, the following matrices $\mathbf{A}_i$ such that $\mathbf{A}_i\mathbf{y}$ equals the $\text{BLUP}(\mathbf{y}_f)$:

$$\begin{aligned}
\mathbf{A}_1 &= \mathbf{X}_f\mathbf{B} + \mathbf{V}_{21}\mathbf{W}^-[\mathbf{I}_n - \mathbf{X}(\mathbf{X}'\mathbf{W}^-\mathbf{X})^-\mathbf{X}'\mathbf{W}^-], \\
\mathbf{A}_2 &= \mathbf{X}_f\mathbf{B} + \mathbf{V}_{21}\mathbf{V}^-[\mathbf{I}_n - \mathbf{X}(\mathbf{X}'\mathbf{W}^-\mathbf{X})^-\mathbf{X}'\mathbf{W}^-], \\
\mathbf{A}_3 &= \mathbf{X}_f\mathbf{B} + \mathbf{V}_{21}\mathbf{M}(\mathbf{MVM})^-\mathbf{M}, \\
\mathbf{A}_4 &= \mathbf{X}_f(\mathbf{X}'\mathbf{X})^-\mathbf{X}' + [\mathbf{V}_{21} - \mathbf{X}_f(\mathbf{X}'\mathbf{X})^-\mathbf{X}'\mathbf{V}]\mathbf{M}(\mathbf{MVM})^-\mathbf{M},
\end{aligned}$$

where $\mathbf{W} = \mathbf{V} + \mathbf{XUX}'$ and $\mathbf{B} = (\mathbf{X}'\mathbf{W}^-\mathbf{X})^-\mathbf{X}'\mathbf{W}^-$ with $\mathbf{U}$ satisfying $\mathscr{C}(\mathbf{W}) = \mathscr{C}(\mathbf{X} : \mathbf{V})$.

13.5 The $\text{BLUP}(\mathbf{y}_f)$ can be written as

$$\begin{aligned}
\text{BLUP}(\mathbf{y}_f) &= \mathbf{X}_f\tilde{\boldsymbol{\beta}} + \mathbf{V}_{21}\mathbf{W}^-(\mathbf{y} - \mathbf{X}\tilde{\boldsymbol{\beta}}) \\
&= \mathbf{X}_f\tilde{\boldsymbol{\beta}} + \mathbf{V}_{21}\mathbf{W}^-\tilde{\boldsymbol{\varepsilon}} = \mathbf{X}_f\tilde{\boldsymbol{\beta}} + \mathbf{V}_{21}\mathbf{V}^-\tilde{\boldsymbol{\varepsilon}} \\
&= \mathbf{X}_f\tilde{\boldsymbol{\beta}} + \mathbf{V}_{21}\mathbf{M}(\mathbf{MVM})^-\mathbf{My} = \text{BLUE}(\mathbf{X}_f\boldsymbol{\beta}) + \mathbf{V}_{21}\dot{\mathbf{M}}\mathbf{y} \\
&= \mathbf{X}_f\hat{\boldsymbol{\beta}} + (\mathbf{V}_{21} - \mathbf{X}_f\mathbf{X}^+\mathbf{V})\dot{\mathbf{M}}\mathbf{y} \\
&= \text{OLSE}(\mathbf{X}_f\boldsymbol{\beta}) + (\mathbf{V}_{21} - \mathbf{X}_f\mathbf{X}^+\mathbf{V})\dot{\mathbf{M}}\mathbf{y},
\end{aligned}$$

where $\tilde{\boldsymbol{\varepsilon}} = \mathbf{y} - \mathbf{X}\tilde{\boldsymbol{\beta}}$ is the vector of the BLUE's residual:

$$\mathbf{X}\tilde{\boldsymbol{\beta}} = \mathbf{y} - \mathbf{VM}(\mathbf{MVM})^-\mathbf{My} = \mathbf{y} - \mathbf{V}\dot{\mathbf{M}}\mathbf{y}, \quad \tilde{\boldsymbol{\varepsilon}} = \mathbf{V}\dot{\mathbf{M}}\mathbf{y} = \mathbf{W}\dot{\mathbf{M}}\mathbf{y}.$$

13.6 $\text{BLUP}(\mathbf{y}_f) = \text{BLUE}(\mathbf{X}_f\boldsymbol{\beta}) \iff \mathscr{C}(\mathbf{V}_{12}) \subset \mathscr{C}(\mathbf{X})$

13.7 The following statements are equivalent:

(a) $\mathrm{BLUP}(\mathbf{y}_f) = \mathrm{OLSE}(\mathbf{X}_f\boldsymbol{\beta}) = \mathbf{X}_f\hat{\boldsymbol{\beta}}$ for a fixed $\mathbf{X}_f = \mathbf{LX}$,

(b) $\mathscr{C}[\mathbf{V}_{21} - \mathbf{V}(\mathbf{X}')^{+}\mathbf{X}'_f] \subset \mathscr{C}(\mathbf{X})$,

(c) $\mathscr{C}(\mathbf{V}_{21} - \mathbf{VHL}') \subset \mathscr{C}(\mathbf{X})$.

13.8 The following statements are equivalent:

(a) $\mathrm{BLUP}(\mathbf{y}_f) = \mathrm{OLSE}(\mathbf{X}_f\boldsymbol{\beta}) = \mathbf{X}_f\hat{\boldsymbol{\beta}}$ for all $\mathbf{X}_f$ of the form $\mathbf{X}_f = \mathbf{LX}$,

(b) $\mathscr{C}(\mathbf{VX}) \subset \mathscr{C}(\mathbf{X})$ and $\mathscr{C}(\mathbf{V}_{12}) \subset \mathscr{C}(\mathbf{X})$,

(c) $\mathrm{OLSE}(\mathbf{X}\boldsymbol{\beta}) = \mathrm{BLUE}(\mathbf{X}\boldsymbol{\beta})$ and $\mathrm{BLUP}(\mathbf{y}_f) = \mathrm{BLUE}(\mathbf{X}_f\boldsymbol{\beta})$.

13.9 Let $\mathbf{X}_f\boldsymbol{\beta}$ be estimable so that $\mathbf{X}_f = \mathbf{LX}$ for some $\mathbf{L}$, and denote $\mathbf{Gy} = \mathrm{BLUE}(\mathbf{X}\boldsymbol{\beta})$. Then

$$\begin{aligned}
\operatorname{cov}[\mathrm{BLUP}(\mathbf{y}_f)] &= \operatorname{cov}(\mathbf{X}_f\tilde{\boldsymbol{\beta}}) + \operatorname{cov}[\mathbf{V}_{21}\mathbf{V}^{-}(\mathbf{y} - \mathbf{X}\tilde{\boldsymbol{\beta}})] \\
&= \operatorname{cov}(\mathbf{LX}\tilde{\boldsymbol{\beta}}) + \operatorname{cov}[\mathbf{V}_{21}\mathbf{V}^{-}(\mathbf{y} - \mathbf{X}\tilde{\boldsymbol{\beta}})] \\
&= \operatorname{cov}(\mathbf{LGy}) + \operatorname{cov}[\mathbf{V}_{21}\mathbf{V}^{-}(\mathbf{y} - \mathbf{Gy})] \\
&= \operatorname{cov}(\mathbf{LGy}) + \operatorname{cov}(\mathbf{V}_{21}\dot{\mathbf{M}}\mathbf{y}) \\
&= \mathbf{L}\operatorname{cov}(\mathbf{Gy})\mathbf{L}' + \mathbf{V}_{21}\dot{\mathbf{M}}\mathbf{V}_{12} \\
&= \mathbf{L}\operatorname{cov}(\mathbf{Gy})\mathbf{L}' + \mathbf{V}_{21}\mathbf{V}^{-}[\mathbf{V} - \operatorname{cov}(\mathbf{Gy})]\mathbf{V}^{-}\mathbf{V}_{12}.
\end{aligned}$$

13.10 If the covariance matrix of $(\mathbf{y}', \mathbf{y}'_f)'$ has an intraclass correlation structure and $\mathbf{1}_n \in \mathscr{C}(\mathbf{X})$, $\mathbf{1}_m \in \mathscr{C}(\mathbf{X}_f)$, then

$$\mathrm{BLUP}(\mathbf{y}_f) = \mathrm{OLSE}(\mathbf{X}_f\boldsymbol{\beta}) = \mathbf{X}_f\hat{\boldsymbol{\beta}} = \mathbf{X}_f(\mathbf{X}'\mathbf{X})^{-}\mathbf{X}'\mathbf{y} = \mathbf{X}_f\mathbf{X}^{+}\mathbf{y}.$$

13.11 If the covariance matrix of $(\mathbf{y}', y_f)'$ has an AR(1)-structure $\{\varrho^{|i-j|}\}$, then

$$\mathrm{BLUP}(y_f) = \mathbf{x}'_f\tilde{\boldsymbol{\beta}} + \varrho\mathbf{i}'_n\tilde{\boldsymbol{\varepsilon}} = \mathbf{x}'_f\tilde{\boldsymbol{\beta}} + \varrho\tilde{e}_n,$$

where $\mathbf{x}'_f$ corresponds to $\mathbf{X}_f$ and $\tilde{\boldsymbol{\varepsilon}}$ is the vector of the BLUE's residual.

13.12 Consider the models $\mathscr{M}_f$ and $\underline{\mathscr{M}}_f$, where $\mathbf{X}_f\boldsymbol{\beta}$ is a given estimable parametric function such that $\mathscr{C}(\mathbf{X}'_f) \subset \mathscr{C}(\mathbf{X}')$ and $\mathscr{C}(\underline{\mathbf{X}}'_f) \subset \mathscr{C}(\underline{\mathbf{X}}')$:

$$\mathscr{M}_f = \left\{\begin{pmatrix}\mathbf{y}\\ \mathbf{y}_f\end{pmatrix}, \begin{pmatrix}\mathbf{X}\\ \mathbf{X}_f\end{pmatrix}\boldsymbol{\beta}, \begin{pmatrix}\mathbf{V}_{11} & \mathbf{V}_{12}\\ \mathbf{V}_{21} & \mathbf{V}_{22}\end{pmatrix}\right\},$$

$$\underline{\mathscr{M}}_f = \left\{\begin{pmatrix}\mathbf{y}\\ \mathbf{y}_f\end{pmatrix}, \begin{pmatrix}\underline{\mathbf{X}}\\ \underline{\mathbf{X}}_f\end{pmatrix}\boldsymbol{\beta}, \begin{pmatrix}\underline{\mathbf{V}}_{11} & \underline{\mathbf{V}}_{12}\\ \underline{\mathbf{V}}_{21} & \underline{\mathbf{V}}_{22}\end{pmatrix}\right\}.$$

Then every representation of the BLUP for $\mathbf{y}_f$ under the model $\mathscr{M}_f$ is also a BLUP for $\mathbf{y}_f$ under the model $\underline{\mathscr{M}}_f$ if and only if

$$\mathscr{C}\begin{pmatrix}\underline{\mathbf{X}} & \underline{\mathbf{V}}_{11}\underline{\mathbf{M}}\\ \underline{\mathbf{X}}_f & \underline{\mathbf{V}}_{21}\underline{\mathbf{M}}\end{pmatrix} \subset \mathscr{C}\begin{pmatrix}\mathbf{X} & \mathbf{V}_{11}\mathbf{M}\\ \mathbf{X}_f & \mathbf{V}_{21}\mathbf{M}\end{pmatrix}.$$

13.13 Under the model $\mathscr{M}_f$, a linear statistic $\mathbf{Fy}$ is called linearly prediction sufficient for $\mathbf{y}_f$ if there exists a matrix $\mathbf{A}$ such that $\mathbf{AFy}$ is the BLUP for $\mathbf{y}_f$. Moreover, $\mathbf{Fy}$ is called linearly minimal prediction sufficient for $\mathbf{y}_f$ if for any other linearly prediction sufficient statistics $\mathbf{Sy}$, there exists a matrix $\mathbf{A}$ such that $\mathbf{Fy} = \mathbf{ASy}$ almost surely. Under the model $\mathscr{M}_f$, $\mathbf{Fy}$ is linearly prediction sufficient for $\mathbf{y}_f$ iff

$$\mathscr{N}(\mathbf{FX} : \mathbf{FVX}^{\perp}) \subset \mathscr{N}(\mathbf{X}_f : \mathbf{V}_{21}\mathbf{X}^{\perp}),$$

and $\mathbf{Fy}$ is linearly minimal prediction sufficient iff the equality holds above.

13.14 Let $\mathbf{Fy}$ be linearly prediction sufficient for $\mathbf{y}_f$ under $\mathscr{M}_f$. Then every representation of the BLUP for $\mathbf{y}_f$ under the transformed model

$$\mathscr{M}_f^* = \left\{ \begin{pmatrix} \mathbf{Fy} \\ \mathbf{y}_f \end{pmatrix}, \begin{pmatrix} \mathbf{FX}\boldsymbol{\beta} \\ \mathbf{X}_f\boldsymbol{\beta} \end{pmatrix}, \sigma^2 \begin{pmatrix} \mathbf{FVF}' & \mathbf{FV}_{12} \\ \mathbf{V}_{21}\mathbf{F}' & \mathbf{V}_{22} \end{pmatrix} \right\}$$

is also the BLUP under $\mathscr{M}_f$ and vice versa.

14 Mixed model

14.1 The mixed model: $\mathbf{y} = \mathbf{X}\boldsymbol{\beta} + \mathbf{Z}\boldsymbol{\gamma} + \boldsymbol{\varepsilon}$, where $\mathbf{y}$ is an observable and $\boldsymbol{\varepsilon}$ an unobservable random vector, $\mathbf{X}_{n\times p}$ and $\mathbf{Z}_{n\times q}$ are known matrices, $\boldsymbol{\beta}$ is a vector of unknown parameters having fixed values (fixed effects), $\boldsymbol{\gamma}$ is an unobservable random vector (random effects) such that $\mathrm{E}(\boldsymbol{\gamma}) = \mathbf{0}$, $\mathrm{E}(\boldsymbol{\varepsilon}) = \mathbf{0}$, and $\mathrm{cov}(\boldsymbol{\gamma}) = \sigma^2\mathbf{D}$, $\mathrm{cov}(\boldsymbol{\varepsilon}) = \sigma^2\mathbf{R}$, $\mathrm{cov}(\boldsymbol{\gamma}, \boldsymbol{\varepsilon}) = \mathbf{0}$. We may denote briefly

$$\mathscr{M}_{\text{mix}} = \{\mathbf{y},\, \mathbf{X}\boldsymbol{\beta} + \mathbf{Z}\boldsymbol{\gamma},\, \sigma^2\mathbf{D},\, \sigma^2\mathbf{R}\}.$$

Then $\mathrm{E}(\mathbf{y}) = \mathbf{X}\boldsymbol{\beta}$, $\mathrm{cov}(\mathbf{y}) = \mathrm{cov}(\mathbf{Z}\boldsymbol{\gamma} + \boldsymbol{\varepsilon}) = \sigma^2(\mathbf{ZDZ}' + \mathbf{R}) := \sigma^2\boldsymbol{\Sigma}$, and

$$\mathrm{cov}\begin{pmatrix} \mathbf{y} \\ \boldsymbol{\gamma} \end{pmatrix} = \sigma^2 \begin{pmatrix} \mathbf{ZDZ}' + \mathbf{R} & \mathbf{ZD} \\ \mathbf{DZ}' & \mathbf{D} \end{pmatrix} = \sigma^2 \begin{pmatrix} \boldsymbol{\Sigma} & \mathbf{ZD} \\ \mathbf{DZ}' & \mathbf{D} \end{pmatrix}.$$

14.2 The mixed model can be presented as a version of the model with "new observations"; the new observations being now in $\boldsymbol{\gamma}$: $\mathbf{y} = \mathbf{X}\boldsymbol{\beta} + \mathbf{Z}\boldsymbol{\gamma} + \boldsymbol{\varepsilon}$, $\boldsymbol{\gamma} = \mathbf{0}\cdot\boldsymbol{\beta} + \boldsymbol{\varepsilon}_f$, where $\mathrm{cov}(\boldsymbol{\varepsilon}_f) = \mathrm{cov}(\boldsymbol{\gamma}) = \sigma^2\mathbf{D}$, $\mathrm{cov}(\boldsymbol{\gamma}, \boldsymbol{\varepsilon}) = \mathrm{cov}(\boldsymbol{\varepsilon}_f, \boldsymbol{\varepsilon}) = \mathbf{0}$:

$$\mathscr{M}_{\text{mix-new}} = \left\{ \begin{pmatrix} \mathbf{y} \\ \boldsymbol{\gamma} \end{pmatrix}, \begin{pmatrix} \mathbf{X} \\ \mathbf{0} \end{pmatrix}\boldsymbol{\beta}, \sigma^2 \begin{pmatrix} \mathbf{ZDZ}' + \mathbf{R} & \mathbf{ZD} \\ \mathbf{DZ}' & \mathbf{D} \end{pmatrix} \right\}.$$

14.3 The linear predictor $\mathbf{Ay}$ is the BLUP for $\boldsymbol{\gamma}$ under the mixed model $\mathscr{M}_{\text{mix}}$ iff

$$\mathbf{A}(\mathbf{X} : \boldsymbol{\Sigma}\mathbf{M}) = (\mathbf{0} : \mathbf{DZ}'\mathbf{M}),$$

where $\boldsymbol{\Sigma} = \mathbf{ZDZ}' + \mathbf{R}$. In terms of Pandora's Box, $\mathbf{Ay} = \mathrm{BLUP}(\boldsymbol{\gamma})$ iff there exists a matrix $\mathbf{L}$ such that $\mathbf{B}$ satisfies the equation

$$\begin{pmatrix} \boldsymbol{\Sigma} & \mathbf{X} \\ \mathbf{X}' & \mathbf{0} \end{pmatrix} \begin{pmatrix} \mathbf{B}' \\ \mathbf{L} \end{pmatrix} = \begin{pmatrix} \mathbf{ZD} \\ \mathbf{0} \end{pmatrix}.$$

The linear estimator $\mathbf{By}$ is the BLUE for $\mathbf{X}\boldsymbol{\beta}$ under the model $\mathscr{M}_{\text{mix}}$ iff

$$\mathbf{B}(\mathbf{X} : \boldsymbol{\Sigma}\mathbf{M}) = (\mathbf{X} : \mathbf{0}).$$

14.4 Under the mixed model $\mathscr{M}_{\text{mix}}$ we have

(a) $\text{BLUE}(\mathbf{X}\boldsymbol{\beta}) = \mathbf{X}\tilde{\boldsymbol{\beta}} = \mathbf{X}(\mathbf{X}'\mathbf{W}^{-}\mathbf{X})^{-}\mathbf{X}'\mathbf{W}^{-}\mathbf{y}$,

(b) $\text{BLUP}(\boldsymbol{\gamma}) = \tilde{\boldsymbol{\gamma}} = \mathbf{DZ}'\mathbf{W}^{-}(\mathbf{y} - \mathbf{X}\tilde{\boldsymbol{\beta}}) = \mathbf{DZ}'\mathbf{W}^{-}\tilde{\boldsymbol{\varepsilon}} = \mathbf{DZ}'\boldsymbol{\Sigma}^{-}\tilde{\boldsymbol{\varepsilon}}$
$= \mathbf{DZ}'\mathbf{M}(\mathbf{M}\boldsymbol{\Sigma}\mathbf{M})^{-}\mathbf{My} = \mathbf{DZ}'\dot{\mathbf{M}}_{\Sigma}\mathbf{y}$,

where $\mathbf{W} = \boldsymbol{\Sigma} + \mathbf{XUX}'$, $\mathscr{C}(\mathbf{W}) = \mathscr{C}(\mathbf{X} : \boldsymbol{\Sigma})$ and $\dot{\mathbf{M}}_{\Sigma} = \mathbf{Z}'\mathbf{M}(\mathbf{M}\boldsymbol{\Sigma}\mathbf{M})^{-}\mathbf{M}$.

14.5 Henderson's mixed model equations are defined as

$$\begin{pmatrix} \mathbf{X}'\mathbf{R}^{-1}\mathbf{X} & \mathbf{X}'\mathbf{R}^{-1}\mathbf{Z} \\ \mathbf{Z}'\mathbf{R}^{-1}\mathbf{X} & \mathbf{Z}'\mathbf{R}^{-1}\mathbf{Z} + \mathbf{D}^{-1} \end{pmatrix} \begin{pmatrix} \boldsymbol{\beta} \\ \boldsymbol{\gamma} \end{pmatrix} = \begin{pmatrix} \mathbf{X}'\mathbf{R}^{-1}\mathbf{y} \\ \mathbf{Z}'\mathbf{R}^{-1}\mathbf{y} \end{pmatrix}.$$

They are obtained by minimizing the following quadratic form $f(\boldsymbol{\beta}, \boldsymbol{\gamma})$ with respect to $\boldsymbol{\beta}$ and $\boldsymbol{\gamma}$ (keeping also $\boldsymbol{\gamma}$ as a non-random vector):

$$f(\boldsymbol{\beta}, \boldsymbol{\gamma}) = \begin{pmatrix} \mathbf{y} - \mathbf{X}\boldsymbol{\beta} - \mathbf{Z}\boldsymbol{\gamma} \\ \boldsymbol{\gamma} \end{pmatrix}' \begin{pmatrix} \mathbf{R} & \mathbf{0} \\ \mathbf{0} & \mathbf{D} \end{pmatrix}^{-1} \begin{pmatrix} \mathbf{y} - \mathbf{X}\boldsymbol{\beta} - \mathbf{Z}\boldsymbol{\gamma} \\ \boldsymbol{\gamma} \end{pmatrix}.$$

If $\boldsymbol{\beta}_*$ and $\boldsymbol{\gamma}_*$ are solutions to the mixed model equations, then $\mathbf{X}\boldsymbol{\beta}_* = \text{BLUE}(\mathbf{X}\boldsymbol{\beta}) = \mathbf{X}\tilde{\boldsymbol{\beta}}$ and $\boldsymbol{\gamma}_* = \text{BLUP}(\boldsymbol{\gamma})$.

14.6 Let us denote

$$\mathbf{y}_{\#} = \begin{pmatrix} \mathbf{y} \\ \mathbf{0} \end{pmatrix}, \quad \mathbf{X}_* = \begin{pmatrix} \mathbf{X} & \mathbf{Z} \\ \mathbf{0} & \mathbf{I}_q \end{pmatrix}, \quad \mathbf{V}_* = \begin{pmatrix} \mathbf{R} & \mathbf{0} \\ \mathbf{0} & \mathbf{D} \end{pmatrix}, \quad \boldsymbol{\alpha} = \begin{pmatrix} \boldsymbol{\beta} \\ \boldsymbol{\gamma} \end{pmatrix}.$$

Then $f(\boldsymbol{\beta}, \boldsymbol{\gamma})$ in 14.5 can be expressed as

$$f(\boldsymbol{\beta}, \boldsymbol{\gamma}) = (\mathbf{y}_{\#} - \mathbf{X}_*\boldsymbol{\alpha})'\mathbf{V}_*^{-1}(\mathbf{y}_{\#} - \mathbf{X}_*\boldsymbol{\alpha}),$$

and the minimum of $f(\boldsymbol{\beta}, \boldsymbol{\gamma})$ is attained when $\boldsymbol{\alpha}$ has value

$$\tilde{\boldsymbol{\alpha}} = \begin{pmatrix} \tilde{\boldsymbol{\beta}} \\ \tilde{\boldsymbol{\gamma}} \end{pmatrix} = (\mathbf{X}_*'\mathbf{V}_*^{-1}\mathbf{X}_*)^{-1}\mathbf{X}_*'\mathbf{V}_*^{-1}\mathbf{y}_*; \quad \mathbf{X}_*\tilde{\boldsymbol{\alpha}} = \begin{pmatrix} \mathbf{X}\tilde{\boldsymbol{\beta}} + \mathbf{Z}\tilde{\boldsymbol{\gamma}} \\ \tilde{\boldsymbol{\gamma}} \end{pmatrix}.$$

14.7 If $\mathbf{V}_*$ is singular, then minimizing the quadratic form

$$f(\boldsymbol{\beta}, \boldsymbol{\gamma}) = (\mathbf{y}_{\#} - \mathbf{X}_*\boldsymbol{\alpha})'\mathbf{W}_*^{-}(\mathbf{y}_{\#} - \mathbf{X}_*\boldsymbol{\alpha}),$$

where $\mathbf{W}_* = \mathbf{V}_* + \mathbf{X}_*\mathbf{X}_*'$, yields $\tilde{\boldsymbol{\beta}}$ and $\tilde{\boldsymbol{\gamma}}$.

14.8 One choice for $\mathbf{X}_*^{\perp}$ is $\mathbf{X}_*^{\perp} = \begin{pmatrix} \mathbf{I}_n \\ -\mathbf{Z}' \end{pmatrix}\mathbf{M} = \begin{pmatrix} \mathbf{M} \\ -\mathbf{Z}'\mathbf{M} \end{pmatrix}.$

14.9 Let us consider the fixed effects partitioned model

$$\mathscr{F}: \mathbf{y} = \mathbf{X}\boldsymbol{\beta} + \mathbf{Z}\boldsymbol{\gamma} + \boldsymbol{\varepsilon}, \quad \operatorname{cov}(\mathbf{y}) = \operatorname{cov}(\boldsymbol{\varepsilon}) = \mathbf{R},$$

where both $\boldsymbol{\beta}$ and $\boldsymbol{\gamma}$ are fixed (but of course unknown) coefficients, and supplement $\mathscr{F}$ with the stochastic restrictions $\mathbf{y}_0 = \boldsymbol{\gamma} + \boldsymbol{\varepsilon}_0$, $\operatorname{cov}(\boldsymbol{\varepsilon}_0) = \mathbf{D}$. This supplementation can be expressed as the partitioned model:

$$\mathscr{F}_* = \{\mathbf{y}_*, \mathbf{X}_*\boldsymbol{\alpha}, \mathbf{V}_*\} = \left\{ \begin{pmatrix} \mathbf{y} \\ \mathbf{y}_0 \end{pmatrix}, \begin{pmatrix} \mathbf{X} & \mathbf{Z} \\ \mathbf{0} & \mathbf{I}_q \end{pmatrix} \begin{pmatrix} \boldsymbol{\beta} \\ \boldsymbol{\gamma} \end{pmatrix}, \begin{pmatrix} \mathbf{R} & \mathbf{0} \\ \mathbf{0} & \mathbf{D} \end{pmatrix} \right\}.$$

Model $\mathscr{F}_*$ is a partitioned linear model $\mathscr{F} = \{\mathbf{y}, \mathbf{X}\boldsymbol{\beta} + \mathbf{Z}\boldsymbol{\gamma}, \mathbf{R}\}$ supplemented with stochastic restrictions on $\boldsymbol{\gamma}$.

14.10 The estimator $\mathbf{B}\mathbf{y}_*$ is the BLUE for $\mathbf{X}_*\boldsymbol{\alpha}$ under the model $\mathscr{F}_*$ iff $\mathbf{B}$ satisfies the equation $\mathbf{B}(\mathbf{X}_* : \mathbf{V}_*\mathbf{X}_*^{\perp}) = (\mathbf{X}_* : \mathbf{0})$, i.e.,

(a) $$\begin{pmatrix} \mathbf{B}_{11} & \mathbf{B}_{12} \\ \mathbf{B}_{21} & \mathbf{B}_{22} \end{pmatrix} \begin{pmatrix} \mathbf{X} & \mathbf{Z} & \mathbf{RM} \\ \mathbf{0} & \mathbf{I}_q & -\mathbf{DZ}'\mathbf{M} \end{pmatrix} = \begin{pmatrix} \mathbf{X} & \mathbf{Z} & \mathbf{0} \\ \mathbf{0} & \mathbf{I}_q & \mathbf{0} \end{pmatrix}.$$

14.11 Let $\mathbf{B}\mathbf{y}_*$ be the BLUE of $\mathbf{X}_*\boldsymbol{\alpha}$ under the augmented model $\mathscr{F}_*$, i.e.,

$$\begin{aligned} \mathbf{B}\mathbf{y}_* &= \begin{pmatrix} \mathbf{B}_{11} & \mathbf{B}_{12} \\ \mathbf{B}_{21} & \mathbf{B}_{22} \end{pmatrix} \mathbf{y}_* = \begin{pmatrix} \mathbf{B}_{11} \\ \mathbf{B}_{21} \end{pmatrix} \mathbf{y} + \begin{pmatrix} \mathbf{B}_{12} \\ \mathbf{B}_{22} \end{pmatrix} \mathbf{y}_0 \\ &= \operatorname{BLUE}(\mathbf{X}_*\boldsymbol{\alpha} \mid \mathscr{F}_*). \end{aligned}$$

Then it is necessary and sufficient that $\mathbf{B}_{21}\mathbf{y}$ is the BLUP of $\boldsymbol{\gamma}$ and $(\mathbf{B}_{11} - \mathbf{Z}\mathbf{B}_{21})\mathbf{y}$ is the BLUE of $\mathbf{X}\boldsymbol{\beta}$ under the mixed model $\mathscr{M}_{\text{mix}}$; in other words, (a) in 14.10 holds iff

$$\begin{pmatrix} \mathbf{I}_n & -\mathbf{Z} \\ \mathbf{0} & \mathbf{I}_q \end{pmatrix} \mathbf{B} \begin{pmatrix} \mathbf{y} \\ \mathbf{0} \end{pmatrix} = \begin{pmatrix} \mathbf{B}_{11} - \mathbf{Z}\mathbf{B}_{21} \\ \mathbf{B}_{21} \end{pmatrix} \mathbf{y} = \begin{pmatrix} \operatorname{BLUE}(\mathbf{X}\boldsymbol{\beta} \mid \mathscr{M}) \\ \operatorname{BLUP}(\boldsymbol{\gamma} \mid \mathscr{M}) \end{pmatrix},$$

or equivalently

$$\mathbf{B} \begin{pmatrix} \mathbf{y} \\ \mathbf{0} \end{pmatrix} = \begin{pmatrix} \mathbf{B}_{11} \\ \mathbf{B}_{21} \end{pmatrix} \mathbf{y} = \begin{pmatrix} \operatorname{BLUE}(\mathbf{X}\boldsymbol{\beta} \mid \mathscr{M}) + \operatorname{BLUP}(\mathbf{Z}\boldsymbol{\gamma} \mid \mathscr{M}) \\ \operatorname{BLUP}(\boldsymbol{\gamma} \mid \mathscr{M}) \end{pmatrix}.$$

14.12 Assume that $\mathbf{B}$ satisfies (a) in 14.10. Then all representations of BLUEs of $\mathbf{X}\boldsymbol{\beta}$ and BLUPs of $\boldsymbol{\gamma}$ in the mixed model $\mathscr{M}_{\text{mix}}$ can be generated through

$$\begin{pmatrix} \mathbf{B}_{11} - \mathbf{Z}\mathbf{B}_{21} \\ \mathbf{B}_{21} \end{pmatrix} \mathbf{y} \quad \text{by varying } \mathbf{B}_{11} \text{ and } \mathbf{B}_{21}.$$

14.13 Consider two mixed models:

$$\mathscr{M}_1 = \{\mathbf{y}, \mathbf{X}\boldsymbol{\beta} + \mathbf{Z}\boldsymbol{\gamma}, \mathbf{D}_1, \mathbf{R}_1\}, \quad \mathscr{M}_2 = \{\mathbf{y}, \mathbf{X}\boldsymbol{\beta} + \mathbf{Z}\boldsymbol{\gamma}, \mathbf{D}_2, \mathbf{R}_2\},$$

and denote $\boldsymbol{\Sigma}_i = \mathbf{Z}\mathbf{D}_i\mathbf{Z}' + \mathbf{R}_i$. Then every representation of the $\operatorname{BLUP}(\boldsymbol{\gamma} \mid \mathscr{M}_1)$ continues to be $\operatorname{BLUP}(\boldsymbol{\gamma} \mid \mathscr{M}_2)$ iff any of the following equivalent conditions holds:

(a) $\mathscr{C}\begin{pmatrix}\mathbf{\Sigma}_2\mathbf{M}\\ \mathbf{D}_2\mathbf{Z}'\mathbf{M}\end{pmatrix} \subset \mathscr{C}\begin{pmatrix}\mathbf{X} & \mathbf{\Sigma}_1\mathbf{M}\\ \mathbf{0} & \mathbf{D}_1\mathbf{Z}'\mathbf{M}\end{pmatrix}$,

(b) $\mathscr{C}\begin{pmatrix}\mathbf{R}_2\mathbf{M}\\ \mathbf{D}_2\mathbf{Z}'\mathbf{M}\end{pmatrix} \subset \mathscr{C}\begin{pmatrix}\mathbf{X} & \mathbf{R}_1\mathbf{M}\\ \mathbf{0} & \mathbf{D}_1\mathbf{Z}'\mathbf{M}\end{pmatrix}$,

(c) $\mathscr{C}\begin{pmatrix}\mathbf{M}\mathbf{\Sigma}_2\mathbf{M}\\ \mathbf{D}_2\mathbf{Z}'\mathbf{M}\end{pmatrix} \subset \mathscr{C}\begin{pmatrix}\mathbf{M}\mathbf{\Sigma}_1\mathbf{M}\\ \mathbf{D}_1\mathbf{Z}'\mathbf{M}\end{pmatrix}$,

(d) $\mathscr{C}\begin{pmatrix}\mathbf{M}\mathbf{R}_2\mathbf{M}\\ \mathbf{D}_2\mathbf{Z}'\mathbf{M}\end{pmatrix} \subset \mathscr{C}\begin{pmatrix}\mathbf{M}\mathbf{R}_1\mathbf{M}\\ \mathbf{D}_1\mathbf{Z}'\mathbf{M}\end{pmatrix}$.

14.14 (Continued ...) Both the BLUE($\mathbf{X}\boldsymbol{\beta} \mid \mathscr{M}_1$) continues to be the BLUE($\mathbf{X}\boldsymbol{\beta} \mid \mathscr{M}_2$) and the BLUP($\boldsymbol{\gamma} \mid \mathscr{M}_1$) continues to be the BLUP($\boldsymbol{\gamma} \mid \mathscr{M}_2$) iff any of the following equivalent conditions holds:

(a) $\mathscr{C}\begin{pmatrix}\mathbf{\Sigma}_2\mathbf{M}\\ \mathbf{D}_2\mathbf{Z}'\mathbf{M}\end{pmatrix} \subset \mathscr{C}\begin{pmatrix}\mathbf{\Sigma}_1\mathbf{M}\\ \mathbf{D}_1\mathbf{Z}'\mathbf{M}\end{pmatrix}$,

(b) $\mathscr{C}\begin{pmatrix}\mathbf{R}_2\mathbf{M}\\ \mathbf{D}_2\mathbf{Z}'\mathbf{M}\end{pmatrix} \subset \mathscr{C}\begin{pmatrix}\mathbf{R}_1\mathbf{M}\\ \mathbf{D}_1\mathbf{Z}'\mathbf{M}\end{pmatrix}$,

(c) $\mathscr{C}(\mathbf{V}_{*2}\mathbf{X}_*^{\perp}) \subset \mathscr{C}(\mathbf{V}_{*1}\mathbf{X}_*^{\perp})$, where

$$\mathbf{V}_{*i} = \begin{pmatrix}\mathbf{R}_i & \mathbf{0}\\ \mathbf{0} & \mathbf{D}_i\end{pmatrix} \quad \text{and} \quad \mathbf{X}_*^{\perp} = \begin{pmatrix}\mathbf{I}_n\\ -\mathbf{Z}'\end{pmatrix}\mathbf{M}.$$

(d) The matrix $\mathbf{V}_{*2}$ can be expressed as

$$\mathbf{V}_{*2} = \mathbf{X}_*\mathbf{N}_1\mathbf{X}_* + \mathbf{V}_{*1}\mathbf{X}_*^{\perp}\mathbf{N}_2(\mathbf{X}_*^{\perp})'\mathbf{V}_{*1} \quad \text{for some } \mathbf{N}_1, \mathbf{N}_2.$$

14.15 Consider the partitioned linear fixed effects model

$$\mathscr{F} = \{\mathbf{y},\, \mathbf{X}_1\boldsymbol{\beta}_1 + \mathbf{X}_2\boldsymbol{\beta}_2,\, \mathbf{R}\} = \{\mathbf{y}, \mathbf{X}\boldsymbol{\beta}, \mathbf{R}\}.$$

Let $\mathscr{M}$ be a linear mixed effects model $\mathbf{y} = \mathbf{X}_1\boldsymbol{\beta}_1 + \mathbf{X}_2\boldsymbol{\gamma}_2 + \boldsymbol{\varepsilon}$, where $\operatorname{cov}(\boldsymbol{\gamma}_2) = \mathbf{D}$, $\operatorname{cov}(\boldsymbol{\varepsilon}) = \mathbf{R}$, which we denote as

$$\mathscr{M} = \{\mathbf{y}, \mathbf{X}_1\boldsymbol{\beta}_1, \mathbf{\Sigma}\} = \{\mathbf{y},\, \mathbf{X}_1\boldsymbol{\beta}_1,\, \mathbf{X}_2\mathbf{D}\mathbf{X}_2' + \mathbf{R}\}.$$

Then, denoting $\mathbf{\Sigma} = \mathbf{X}_2\mathbf{D}\mathbf{X}_2' + \mathbf{R}$, the following statements hold:

(a) There exists a matrix $\mathbf{L}$ such that $\mathbf{Ly}$ is the BLUE of $\mathbf{M}_2\mathbf{X}_1\boldsymbol{\beta}_1$ under the models $\mathscr{F}$ and $\mathscr{M}$ iff $\mathscr{N}(\mathbf{X}_1 : \mathbf{X}_2 : \mathbf{R}\mathbf{X}_1^{\perp}) \subset \mathscr{N}(\mathbf{M}_2\mathbf{X}_1 : \mathbf{0} : \mathbf{0})$.

(b) Every representation of the BLUE of $\mathbf{M}_2\mathbf{X}_1\boldsymbol{\beta}_1$ under $\mathscr{F}$ is also the BLUE of $\mathbf{M}_2\mathbf{X}_1\boldsymbol{\beta}_1$ under $\mathscr{M}$ iff $\mathscr{C}(\mathbf{X}_2 : \mathbf{R}\mathbf{X}^{\perp}) \supset \mathscr{C}(\mathbf{\Sigma}\mathbf{X}_1^{\perp})$.

(c) Every representation of the BLUE of $\mathbf{M}_2\mathbf{X}_1\boldsymbol{\beta}_1$ under $\mathscr{M}$ is also the BLUE of $\mathbf{M}_2\mathbf{X}_1\boldsymbol{\beta}_1$ under $\mathscr{F}$ iff $\mathscr{C}(\mathbf{X}_2 : \mathbf{R}\mathbf{X}^{\perp}) \subset \mathscr{C}(\mathbf{\Sigma}\mathbf{X}_1^{\perp})$.

15 Multivariate linear model

15.1 Instead of one response variable y, consider d response variables $y_1, \dots, y_d$. Let the n observed values of these d variables be in the data matrix $\mathbf{Y}_{n\times d}$ while $\mathbf{X}_{n\times p}$ is the usual model matrix:

$$\mathbf{Y} = (\mathbf{y}_1 : \dots : \mathbf{y}_d) = \begin{pmatrix} \mathbf{y}'_{(1)} \\ \vdots \\ \mathbf{y}'_{(n)} \end{pmatrix}, \quad \mathbf{X} = (\mathbf{x}_1 : \dots : \mathbf{x}_p) = \begin{pmatrix} \mathbf{x}'_{(1)} \\ \vdots \\ \mathbf{x}'_{(n)} \end{pmatrix}.$$

Denote $\mathbb{B} = (\boldsymbol{\beta}_1 : \dots : \boldsymbol{\beta}_d) \in \mathbb{R}^{p\times d}$, and assume that

$$\mathbf{y}_j = \mathbf{X}\boldsymbol{\beta}_j + \boldsymbol{\varepsilon}_j, \quad \operatorname{cov}(\boldsymbol{\varepsilon}_j) = \sigma_j^2\mathbf{I}_n, \quad \operatorname{E}(\boldsymbol{\varepsilon}_j) = \mathbf{0}, \quad j = 1, \dots, d,$$

$$(\mathbf{y}_1 : \dots : \mathbf{y}_d) = \mathbf{X}(\boldsymbol{\beta}_1 : \dots : \boldsymbol{\beta}_d) + (\boldsymbol{\varepsilon}_1 : \dots : \boldsymbol{\varepsilon}_d), \quad \mathbf{Y} = \mathbf{X}\mathbb{B} + \boldsymbol{\mathcal{E}}.$$

The columns of $\mathbf{Y}$, $\mathbf{y}_1, \dots, \mathbf{y}_d$, are n-dimensional random vectors such that

$$\mathbf{y}_j \sim (\mathbf{X}\boldsymbol{\beta}_j, \sigma_j^2\mathbf{I}_n), \quad \operatorname{cov}(\mathbf{y}_j, \mathbf{y}_k) = \sigma_{jk}\mathbf{I}_n, \quad j,k = 1, \dots, d.$$

Transposing $\mathbf{Y} = \mathbf{X}\mathbb{B} + \boldsymbol{\mathcal{E}}$ yields

$$(\mathbf{y}_{(1)} : \dots : \mathbf{y}_{(n)}) = \mathbb{B}'(\mathbf{x}_{(1)} : \dots : \mathbf{x}_{(n)}) + (\boldsymbol{\varepsilon}_{(1)} : \dots : \boldsymbol{\varepsilon}_{(n)}),$$

where the (transposed) rows of $\mathbf{Y}$, i.e., $\mathbf{y}_{(1)}, \dots, \mathbf{y}_{(n)}$, are independent d-dimensional random vectors such that

$$\mathbf{y}_{(i)} \sim (\mathbb{B}'\mathbf{x}_{(i)}, \boldsymbol{\Sigma}), \quad \operatorname{cov}(\mathbf{y}_{(i)}, \mathbf{y}_{(j)}) = \mathbf{0}, \quad i \neq j.$$

Notice that in this setup rows (observations) are independent but columns (variables) may correlate. We denote this model shortly as $\mathscr{A} = \{\mathbf{Y}, \mathbf{X}\mathbb{B}, \boldsymbol{\Sigma}\}$.

15.2 Denoting

$$\mathbf{y}_* = \operatorname{vec}(\mathbf{Y}) = \begin{pmatrix} \mathbf{y}_1 \\ \vdots \\ \mathbf{y}_d \end{pmatrix}, \quad \operatorname{E}(\mathbf{y}_*) = \begin{pmatrix} \mathbf{X} & \dots & \mathbf{0} \\ \vdots & \ddots & \vdots \\ \mathbf{0} & \dots & \mathbf{X} \end{pmatrix} \begin{pmatrix} \boldsymbol{\beta}_1 \\ \vdots \\ \boldsymbol{\beta}_d \end{pmatrix}$$
$$= (\mathbf{I}_d \otimes \mathbf{X}) \operatorname{vec}(\mathbb{B}),$$

we get

$$\operatorname{cov}(\mathbf{y}_*) = \begin{pmatrix} \sigma_1^2\mathbf{I}_n & \sigma_{12}\mathbf{I}_n & \dots & \sigma_{1d}\mathbf{I}_n \\ \vdots & \vdots & & \vdots \\ \sigma_{d1}\mathbf{I}_n & \sigma_{d2}\mathbf{I}_n & \dots & \sigma_d^2\mathbf{I}_n \end{pmatrix} = \boldsymbol{\Sigma} \otimes \mathbf{I}_n,$$

and hence the multivariate model can be rewritten as a univariate model

$$\mathscr{B} = \{\operatorname{vec}(\mathbf{Y}), (\mathbf{I}_d \otimes \mathbf{X}) \operatorname{vec}(\mathbb{B}), \boldsymbol{\Sigma} \otimes \mathbf{I}_n\} := \{\mathbf{y}_*, \mathbf{X}_*\boldsymbol{\beta}_*, \mathbf{V}_*\}.$$

15.3 By analogy of the univariate model, we can estimate $\mathbf{X}\mathbb{B}$ by minimizing

$$\operatorname{tr}(\mathbf{Y} - \mathbf{X}\mathbb{B})'(\mathbf{Y} - \mathbf{X}\mathbb{B}) = \|\mathbf{Y} - \mathbf{X}\mathbb{B}\|_F^2.$$

The resulting $\mathbf{X}\hat{\mathbb{B}} = \mathbf{HY}$ is the OLSE of $\mathbf{X}\mathbb{B}$. Moreover, we have

$$(\mathbf{Y}-\mathbf{X}\mathbb{B})'(\mathbf{Y}-\mathbf{X}\mathbb{B}) \geq_{\mathsf{L}} (\mathbf{Y}-\mathbf{HY})'(\mathbf{Y}-\mathbf{HY}) = \mathbf{Y}'\mathbf{MY} := \mathbf{E}_{\text{res}} \quad \forall\, \mathbb{B},$$

where the equality holds iff $\mathbf{X}\mathbb{B} = \mathbf{X}\hat{\mathbb{B}} = \mathbf{P_X Y} = \mathbf{HY}$.

15.4 Under multinormality, $\mathbf{Y}'\mathbf{MY} \sim \mathrm{W}_d[n - \mathrm{r}(\mathbf{X}), \boldsymbol{\Sigma}]$, and if $\mathrm{r}(\mathbf{X}) = p$, we have $\mathrm{MLE}(\mathbb{B}) = \hat{\mathbb{B}} = (\mathbf{X}'\mathbf{X})^{-1}\mathbf{X}'\mathbf{Y}$, $\mathrm{MLE}(\boldsymbol{\Sigma}) = \frac{1}{n}\mathbf{Y}'\mathbf{MY} = \frac{1}{n}\mathbf{E}_{\text{res}}$.

15.5 $\mathscr{C}(\mathbf{V}_*\mathbf{X}_*) \subset \mathscr{C}(\mathbf{X}_*) \implies \mathrm{BLUE}(\mathbf{X}_*\boldsymbol{\beta}_*) = \mathrm{OLSE}(\mathbf{X}_*\boldsymbol{\beta}_*)$

15.6 Consider a linear hypothesis $H\colon \mathbf{K}'\mathbb{B} = \mathbf{D}$, where $\mathbf{K} \in \mathbb{R}^{p\times q}_q$ and $\mathbf{D} \in \mathbb{R}^{q\times d}$ are such that $\mathbf{K}'\mathbb{B}$ is estimable, Then the minimum $\mathbf{E}_H$, say, of

$$(\mathbf{Y}-\mathbf{X}\mathbb{B})'(\mathbf{Y}-\mathbf{X}\mathbb{B}) \quad \text{subject to} \quad \mathbf{K}'\mathbb{B} = \mathbf{D}$$

occurs (in the Löwner sense) when $\mathbf{X}\mathbb{B}$ equals

$$\mathbf{X}\hat{\mathbb{B}}_H = \mathbf{X}\hat{\mathbb{B}} - \mathbf{X}(\mathbf{X}'\mathbf{X})^{-}\mathbf{K}[\mathbf{K}'(\mathbf{X}'\mathbf{X})^{-}\mathbf{K}]^{-1}(\mathbf{K}'\hat{\mathbb{B}} - \mathbf{D}),$$

and so we have, corresponding to 7.17 (p. 35),

$$\mathbf{E}_H - \mathbf{E}_{\text{res}} = (\mathbf{K}'\hat{\mathbb{B}} - \mathbf{D})'[\mathbf{K}'(\mathbf{X}'\mathbf{X})^{-}\mathbf{K}]^{-1}(\mathbf{K}'\hat{\mathbb{B}} - \mathbf{D}).$$

The hypothesis testing in multivariate model can be based on some appropriate function of $(\mathbf{E}_H - \mathbf{E}_{\text{res}})\mathbf{E}_{\text{res}}^{-1}$, or on some closely related matrix.

15.7 Consider two independent samples $\mathbf{Y}_1' = (\mathbf{y}_{(11)} : \ldots : \mathbf{y}_{(1n_1)})$ and $\mathbf{Y}_2' = (\mathbf{y}_{(21)} : \ldots : \mathbf{y}_{(2n_2)})$ so that each $\mathbf{y}_{(1i)} \sim \mathrm{N}_d(\boldsymbol{\mu}_1, \boldsymbol{\Sigma})$ and $\mathbf{y}_{(2i)} \sim \mathrm{N}_d(\boldsymbol{\mu}_2, \boldsymbol{\Sigma})$. Then the test statistics for the hypothesis $\boldsymbol{\mu}_1 = \boldsymbol{\mu}_2$ can be be based, e.g., on the

$$\begin{aligned}\mathrm{ch}_1[(\mathbf{E}_H - \mathbf{E}_{\text{res}})\mathbf{E}_{\text{res}}^{-1}] &= \mathrm{ch}_1\left[\frac{n_1 n_2}{n_1+n_2}(\bar{\mathbf{y}}_1 - \bar{\mathbf{y}}_2)(\bar{\mathbf{y}}_1 - \bar{\mathbf{y}}_2)'\mathbf{E}_{\text{res}}^{-1}\right] \\ &= \alpha(\bar{\mathbf{y}}_1 - \bar{\mathbf{y}}_2)'\mathbf{S}_*^{-1}(\bar{\mathbf{y}}_1 - \bar{\mathbf{y}}_2),\end{aligned}$$

where $\alpha = \frac{(n_1+n_2-2)n_1 n_2}{n_1+n_2}$, $\mathbf{S}_* = \frac{1}{n_1+n_2-2}\mathbf{E}_{\text{res}} = \frac{1}{n_1+n_2-2}\mathbf{T}_*$, and

$$\begin{aligned}\mathbf{E}_{\text{res}} &= \mathbf{Y}'(\mathbf{I}_n - \mathbf{H})\mathbf{Y} = \mathbf{Y}_1'\mathbf{C}_{n_1}\mathbf{Y}_1 + \mathbf{Y}_2'\mathbf{C}_{n_2}\mathbf{Y}_2 = \mathbf{T}_*, \\ \mathbf{E}_H - \mathbf{E}_{\text{res}} &= \mathbf{Y}'(\mathbf{I}_n - \mathbf{J}_n)\mathbf{Y} - \mathbf{Y}'(\mathbf{I}_n - \mathbf{H})\mathbf{Y} = \mathbf{Y}'(\mathbf{H} - \mathbf{J}_n)\mathbf{Y} \\ &= n_1(\bar{\mathbf{y}}_1 - \bar{\mathbf{y}})(\bar{\mathbf{y}}_1 - \bar{\mathbf{y}})' + n_2(\bar{\mathbf{y}}_2 - \bar{\mathbf{y}})(\bar{\mathbf{y}}_2 - \bar{\mathbf{y}})' \\ &= \frac{n_1 n_2}{n_1+n_2}(\bar{\mathbf{y}}_1 - \bar{\mathbf{y}}_2)(\bar{\mathbf{y}}_1 - \bar{\mathbf{y}}_2)' := \mathbf{E}_{\text{Between}}, \\ \mathbf{E}_H &= \mathbf{E}_{\text{Between}} + \mathbf{E}_{\text{res}}.\end{aligned}$$

Hence one appropriate test statistics appears to be a function of the squared Mahalanobis distance

$$\mathrm{MHLN}^2(\bar{\mathbf{y}}_1, \bar{\mathbf{y}}_2, \mathbf{S}_*) = (\bar{\mathbf{y}}_1 - \bar{\mathbf{y}}_2)'\mathbf{S}_*^{-1}(\bar{\mathbf{y}}_1 - \bar{\mathbf{y}}_2).$$

One such statistics is $\frac{n_1 n_2}{n_1+n_2}(\bar{\mathbf{y}}_1 - \bar{\mathbf{y}}_2)'\mathbf{S}_*^{-1}(\bar{\mathbf{y}}_1 - \bar{\mathbf{y}}_2) = T^2 =$ Hotelling's T^2.

16 Principal components, discriminant analysis, factor analysis

16.1 Principal component analysis, PCA. Let a p-dimensional random vector $\mathbf{z}$ have $\mathrm{E}(\mathbf{z}) = \mathbf{0}$ and $\mathrm{cov}(\mathbf{z}) = \mathbf{\Sigma}$. Consider the eigenvalue decomposition of $\mathbf{\Sigma}$: $\mathbf{\Sigma} = \mathbf{T\Lambda T}'$, $\mathbf{\Lambda} = \mathrm{diag}(\lambda_1, \dots, \lambda_n)$, $\mathbf{T}'\mathbf{T} = \mathbf{I}_n$, $\mathbf{T} = (\mathbf{t}_1 : \dots : \mathbf{t}_n)$, where $\lambda_1 \geq \lambda_2 \geq \dots \geq \lambda_n \geq 0$ are the ordered eigenvalues of $\mathbf{\Sigma}$. Then the random variable $\mathbf{t}_i'\mathbf{z}$, which is the ith element of the random vector $\mathbf{T}'\mathbf{z}$, is the ith (population) principal component of $\mathbf{z}$.

16.2 Denote $\mathbf{T}_{(i)} = (\mathbf{t}_1 : \dots : \mathbf{t}_i)$. Then, in the above notation,

- $\max_{\mathbf{b}'\mathbf{b}=1} \mathrm{var}(\mathbf{b}'\mathbf{z}) = \mathrm{var}(\mathbf{t}_1'\mathbf{z}) = \lambda_1,$
- $\max_{\substack{\mathbf{b}'\mathbf{b}=1 \\ \mathbf{T}_{(i-1)}'\mathbf{b}=\mathbf{0}}} \mathrm{var}(\mathbf{b}'\mathbf{z}) = \mathrm{var}(\mathbf{t}_i'\mathbf{z}) = \lambda_i,$

i.e., $\mathbf{t}_i'\mathbf{z}$ has maximum variance of all normalized linear combinations uncorrelated with the elements of $\mathbf{T}_{(i-1)}'\mathbf{z}$.

16.3 Predictive approach to PCA. Let $\mathbf{A}_{p\times k}$ have orthonormal columns, and consider the best linear predictor of $\mathbf{z}$ on the basis of $\mathbf{A}'\mathbf{z}$. Then, the Euclidean norm of the covariance matrix of the prediction error, see 6.14 (p. 29),

$$\|\mathbf{\Sigma} - \mathbf{\Sigma A}(\mathbf{A}'\mathbf{\Sigma A})^{-1}\mathbf{A}'\mathbf{\Sigma}\|_F = \|\mathbf{\Sigma}^{1/2}(\mathbf{I}_d - \mathbf{P}_{\mathbf{\Sigma}^{1/2}\mathbf{A}})\mathbf{\Sigma}^{1/2}\|_F,$$

is a minimum when $\mathbf{A}$ is chosen as $\mathbf{T}_{(k)} = (\mathbf{t}_1 : \dots : \mathbf{t}_k)$. Moreover, minimizing the trace of the covariance matrix of the prediction error, i.e., maximizing $\mathrm{tr}(\mathbf{P}_{\mathbf{\Sigma}^{1/2}\mathbf{A}}\mathbf{\Sigma})$ yields the same result.

16.4 Sample principal components, geometric interpretation. Let $\tilde{\mathbf{X}}$ be an $n \times p$ centered data matrix. How should $\mathbf{G}_{n\times k}$ be chosen if we wish to minimize the sum of orthogonal distances (squared) of the observations $\tilde{\mathbf{x}}_{(i)}$ from $\mathscr{C}(\mathbf{G})$? The function to be minimized is

$$\|\mathbf{E}\|_F^2 = \|\tilde{\mathbf{X}} - \tilde{\mathbf{X}}\mathbf{P}_\mathbf{G}\|_F^2 = \mathrm{tr}(\tilde{\mathbf{X}}'\tilde{\mathbf{X}}) - \mathrm{tr}(\mathbf{P}_\mathbf{G}\tilde{\mathbf{X}}'\tilde{\mathbf{X}}),$$

and the solution is $\dot{\mathbf{G}} = \mathbf{T}_{(k)} = (\mathbf{t}_1 : \dots : \mathbf{t}_k)$, where $\mathbf{t}_k$ are the first k orthonormal eigenvectors of $\tilde{\mathbf{X}}'\tilde{\mathbf{X}}$ (which are the same as those of the corresponding covariance matrix). The new projected observations are the columns of matrix $\mathbf{T}_{(k)}\mathbf{T}_{(k)}'\tilde{\mathbf{X}}' = \mathbf{P}_{\mathbf{T}_{(k)}}\tilde{\mathbf{X}}'$. In particular, if $k = 1$, the new projected observations are the columns of

$$\mathbf{t}_1\mathbf{t}_1'\tilde{\mathbf{X}}' = \mathbf{t}_1(\mathbf{t}_1'\tilde{\mathbf{X}}') = \mathbf{t}_1(\mathbf{t}_1'\tilde{\mathbf{x}}_{(1)}, \dots, \mathbf{t}_1'\tilde{\mathbf{x}}_{(n)}) := \mathbf{t}_1\mathbf{s}_1',$$

where the vector $\mathbf{s}_1 = \tilde{\mathbf{X}}\mathbf{t}_1 = (\mathbf{t}_1'\tilde{\mathbf{x}}_{(1)}, \dots, \mathbf{t}_n'\tilde{\mathbf{x}}_{(n)})' \in \mathbb{R}^n$ comprises the values (scores) of the first (sample) principal component; the ith individual has the score $\mathbf{t}_i'\tilde{\mathbf{x}}_{(i)}$ on the first principal component. The scores are the (signed) lengths of the new projected observations.

16.5 PCA and the matrix approximation. The matrix approximation of the centered data matrix $\tilde{\mathbf{X}}$ by a matrix of a lower rank yields the PCA. The scores can be obtained directly from the SVD of $\tilde{\mathbf{X}}$: $\tilde{\mathbf{X}} = \mathbf{U}\boldsymbol{\Delta}\mathbf{T}'$. With $k = 1$, the scores of the jth principal component are in the jth column of $\mathbf{U}\boldsymbol{\Delta} = \tilde{\mathbf{X}}\mathbf{T}$. The columns of matrix $\mathbf{T}'\tilde{\mathbf{X}}'$ represent the new rotated observations.

16.6 Discriminant analysis. Let $\mathbf{x}$ denote a d-variate random vector with $\mathrm{E}(\mathbf{x}) = \boldsymbol{\mu}_1$ in population 1 and $\mathrm{E}(\mathbf{x}) = \boldsymbol{\mu}_2$ in population 2, and $\mathrm{cov}(\mathbf{x}) = \boldsymbol{\Sigma}$ (pd) in both populations. If

$$\max_{\mathbf{a}\neq\mathbf{0}} \frac{[\mathbf{a}'(\boldsymbol{\mu}_1 - \boldsymbol{\mu}_2)]^2}{\mathrm{var}(\mathbf{a}'\mathbf{x})} = \max_{\mathbf{a}\neq\mathbf{0}} \frac{[\mathbf{a}'(\boldsymbol{\mu}_1 - \boldsymbol{\mu}_2)]^2}{\mathbf{a}'\boldsymbol{\Sigma}\mathbf{a}} = \frac{[\mathbf{a}_*'(\boldsymbol{\mu}_1 - \boldsymbol{\mu}_2)]^2}{\mathbf{a}_*'\boldsymbol{\Sigma}\mathbf{a}_*},$$

then $\mathbf{a}_*'\mathbf{x}$ is called a linear discriminant function for the two populations. Moreover,

$$\max_{\mathbf{a}\neq\mathbf{0}} \frac{[\mathbf{a}'(\boldsymbol{\mu}_1 - \boldsymbol{\mu}_2)]^2}{\mathbf{a}'\boldsymbol{\Sigma}\mathbf{a}} = (\boldsymbol{\mu}_1-\boldsymbol{\mu}_2)'\boldsymbol{\Sigma}^{-1}(\boldsymbol{\mu}_1-\boldsymbol{\mu}_2) = \mathrm{MHLN}^2(\boldsymbol{\mu}_1, \boldsymbol{\mu}_2, \boldsymbol{\Sigma}),$$

and any linear combination $\mathbf{a}_*'\mathbf{x}$ with $\mathbf{a}_* = b \cdot \boldsymbol{\Sigma}^{-1}(\boldsymbol{\mu}_1 - \boldsymbol{\mu}_2)$, where $b \neq 0$ is an arbitrary constant, is a linear discriminant function for the two populations. In other words, finding that linear combination $\mathbf{a}'\mathbf{x}$ for which $\mathbf{a}'\boldsymbol{\mu}_1$ and $\mathbf{a}'\boldsymbol{\mu}_2$ are as distant from each other as possible—distance measure in terms of standard deviation of $\mathbf{a}'\mathbf{x}$, yields a linear discriminant function $\mathbf{a}_*'\mathbf{x}$.

16.7 Let $\mathbf{U}_1' = (\mathbf{u}_{(11)} : \ldots : \mathbf{u}_{(1n_1)})$ and $\mathbf{U}_2' = (\mathbf{u}_{(21)} : \ldots : \mathbf{u}_{(2n_2)})$ be independent random samples from d-dimensional populations $(\boldsymbol{\mu}_1, \boldsymbol{\Sigma})$ and $(\boldsymbol{\mu}_2, \boldsymbol{\Sigma})$, respectively. Denote, cf. 5.46 (p. 26), 15.7 (p. 73),

$$\mathbf{T}_i = \mathbf{U}_i'\mathbf{C}_{n_i}\mathbf{U}_i, \quad \mathbf{T}_* = \mathbf{T}_1 + \mathbf{T}_2,$$

$$\mathbf{S}_* = \frac{1}{n_1 + n_2 - 2}\mathbf{T}_*, \quad n = n_1 + n_2,$$

and $\bar{\mathbf{u}}_i = \frac{1}{n_i}\mathbf{U}_i'\mathbf{1}_{n_i}$, $\bar{\mathbf{u}} = \frac{1}{n}(\mathbf{U}_1' : \mathbf{U}_2')\mathbf{1}_n$. Let $\mathbf{S}_*$ be pd. Then

$$\begin{aligned}
&\max_{\mathbf{a}\neq\mathbf{0}} \frac{[\mathbf{a}'(\bar{\mathbf{u}}_1 - \bar{\mathbf{u}}_2)]^2}{\mathbf{a}'\mathbf{S}_*\mathbf{a}} \\
&= (\bar{\mathbf{u}}_1 - \bar{\mathbf{u}}_2)'\mathbf{S}_*^{-1}(\bar{\mathbf{u}}_1 - \bar{\mathbf{u}}_2) = \mathrm{MHLN}^2(\bar{\mathbf{u}}_1, \bar{\mathbf{u}}_2, \mathbf{S}_*) \\
&= (n-2)\max_{\mathbf{a}\neq\mathbf{0}} \frac{\mathbf{a}'(\bar{\mathbf{u}}_1 - \bar{\mathbf{u}}_2)(\bar{\mathbf{u}}_1 - \bar{\mathbf{u}}_2)'\mathbf{a}}{\mathbf{a}'\mathbf{T}_*\mathbf{a}} = (n-2)\max_{\mathbf{a}\neq\mathbf{0}} \frac{\mathbf{a}'\mathbf{B}\mathbf{a}}{\mathbf{a}'\mathbf{T}_*\mathbf{a}},
\end{aligned}$$

where $\mathbf{B} = (\bar{\mathbf{u}}_1 - \bar{\mathbf{u}}_2)(\bar{\mathbf{u}}_1 - \bar{\mathbf{u}}_2)' = \frac{n}{n_1 n_2}\sum_{i=1}^{2} n_i(\bar{\mathbf{u}}_i - \bar{\mathbf{u}})(\bar{\mathbf{u}}_i - \bar{\mathbf{u}})'$. The vector of coefficients of the discriminant function is given by $\mathbf{a}_* = \mathbf{S}_*^{-1}(\bar{\mathbf{u}}_1 - \bar{\mathbf{u}}_2)$ or any vector proportional to it.

16.8 The model for factor analysis is $\mathbf{x} = \boldsymbol{\mu} + \mathbf{A}\mathbf{f} + \boldsymbol{\varepsilon}$, where $\mathbf{x}$ is an observable random vector of p components; $\mathrm{E}(\mathbf{x}) = \boldsymbol{\mu}$ and $\mathrm{cov}(\mathbf{x}) = \boldsymbol{\Sigma}$. Vector $\mathbf{f}$ is an

m-dimensional random vector, $m \leq p$, whose elements are called (common) factors. The elements of $\boldsymbol{\varepsilon}$ are called specific or unique factors. The matrix $\mathbf{A}_{p \times m}$ is the unknown matrix of factor loadings. Moreover,

$$\mathrm{E}(\boldsymbol{\varepsilon}) = \mathbf{0}, \quad \operatorname{cov}(\boldsymbol{\varepsilon}) = \boldsymbol{\Psi} = \operatorname{diag}(\psi_1^2, \dots, \psi_p^2),$$
$$\mathrm{E}(\mathbf{f}) = \mathbf{0}, \quad \operatorname{cov}(\mathbf{f}) = \boldsymbol{\Phi} = \mathbf{I}_m, \quad \operatorname{cov}(\mathbf{f}, \boldsymbol{\varepsilon}) = \mathbf{0}.$$

The fundamental equation for factor analysis is $\operatorname{cov}(\mathbf{x}) = \operatorname{cov}(\mathbf{Af}) + \operatorname{cov}(\boldsymbol{\varepsilon})$, i.e., $\boldsymbol{\Sigma} = \mathbf{AA}' + \boldsymbol{\Psi}$. Then

$$\operatorname{cov}\begin{pmatrix}\mathbf{x}\\ \mathbf{f}\end{pmatrix} = \begin{pmatrix}\boldsymbol{\Sigma} & \mathbf{A}\\ \mathbf{A}' & \mathbf{I}_m\end{pmatrix} \quad \text{and} \quad \operatorname{cov}(\mathbf{f} - \mathbf{Lx}) \geq_{\mathsf{L}} \operatorname{cov}(\mathbf{f} - \mathbf{A}'\boldsymbol{\Sigma}^{-1}\mathbf{x}) \quad \text{for all } \mathbf{L}.$$

The matrix $\mathbf{L}_* = \mathbf{A}'\boldsymbol{\Sigma}^{-1}$ can be represented as

$$\mathbf{L}_* = \mathbf{A}'\boldsymbol{\Sigma}^{-1} = (\mathbf{A}'\boldsymbol{\Psi}^{-1}\mathbf{A} + \mathbf{I}_m)^{-1}\mathbf{A}'\boldsymbol{\Psi}^{-1},$$

and the individual factor scores can be obtained by $\mathbf{L}_*(\mathbf{x} - \boldsymbol{\mu})$.

17 Canonical correlations

17.1 Let $\mathbf{z} = \binom{\mathbf{x}}{\mathbf{y}}$ denote a d-dimensional random vector with (possibly singular) covariance matrix $\boldsymbol{\Sigma}$. Let $\mathbf{x}$ and $\mathbf{y}$ have dimensions d_1 and $d_2 = d - d_1$, respectively. Denote

$$\operatorname{cov}\begin{pmatrix}\mathbf{x}\\ \mathbf{y}\end{pmatrix} = \boldsymbol{\Sigma} = \begin{pmatrix}\boldsymbol{\Sigma}_{\mathbf{xx}} & \boldsymbol{\Sigma}_{\mathbf{xy}}\\ \boldsymbol{\Sigma}_{\mathbf{yx}} & \boldsymbol{\Sigma}_{\mathbf{yy}}\end{pmatrix}, \quad \mathrm{r}(\boldsymbol{\Sigma}_{\mathbf{xy}}) = m, \quad \mathrm{r}(\boldsymbol{\Sigma}_{\mathbf{xx}}) = h \leq \mathrm{r}(\boldsymbol{\Sigma}_{\mathbf{yy}}).$$

Let ϱ_1^2 be the maximum value of $\operatorname{cor}^2(\mathbf{a}'\mathbf{x}, \mathbf{b}'\mathbf{y})$, and let $\mathbf{a} = \mathbf{a}_1$ and $\mathbf{b} = \mathbf{b}_1$ be the corresponding maximizing values of $\mathbf{a}$ and $\mathbf{b}$. Then the positive square root $\sqrt{\varrho_1^2}$ is called the first (population) canonical correlation between $\mathbf{x}$ and $\mathbf{y}$, denoted as $\operatorname{cc}_1(\mathbf{x}, \mathbf{y}) = \varrho_1$, and $u_1 = \mathbf{a}_1'\mathbf{x}$ and $v_1 = \mathbf{b}_1'\mathbf{y}$ are called the first (population) canonical variables.

17.2 Let ϱ_2^2 be the maximum value of $\operatorname{cor}^2(\mathbf{a}'\mathbf{x}, \mathbf{b}'\mathbf{y})$, where $\mathbf{a}'\mathbf{x}$ is uncorrelated with $\mathbf{a}_1'\mathbf{x}$ and $\mathbf{b}'\mathbf{y}$ is uncorrelated with $\mathbf{b}_1'\mathbf{y}$, and let $u_2 = \mathbf{a}_2'\mathbf{x}$ and $v_2 = \mathbf{b}_2'\mathbf{y}$ be the maximizing values. The positive square root $\sqrt{\varrho_2^2}$ is called the second canonical correlation, denoted as $\operatorname{cc}_2(\mathbf{x}, \mathbf{y}) = \varrho_2$, and u_2 and v_2 are called the second canonical variables. Continuing in this manner, we obtain h pairs of canonical variables $\mathbf{u} = (u_1, u_2, \dots, u_h)'$ and $\mathbf{v} = (v_1, v_2, \dots, v_h)'$.

17.3 We have

$$\operatorname{cor}^2(\mathbf{a}'\mathbf{x}, \mathbf{b}'\mathbf{y}) = \frac{(\mathbf{a}'\boldsymbol{\Sigma}_{\mathbf{xy}}\mathbf{b})^2}{\mathbf{a}'\boldsymbol{\Sigma}_{\mathbf{xx}}\mathbf{a} \cdot \mathbf{b}'\boldsymbol{\Sigma}_{\mathbf{yy}}\mathbf{b}} = \frac{(\mathbf{a}_*'\boldsymbol{\Sigma}_{\mathbf{xx}}^{+1/2}\boldsymbol{\Sigma}_{\mathbf{xy}}\boldsymbol{\Sigma}_{\mathbf{yy}}^{+1/2}\mathbf{b}_*)^2}{\mathbf{a}_*'\mathbf{a}_* \cdot \mathbf{b}_*'\mathbf{b}_*},$$

where $\mathbf{a}_* = \boldsymbol{\Sigma}_{\mathbf{xx}}^{1/2}\mathbf{a}$, $\mathbf{b}_* = \boldsymbol{\Sigma}_{\mathbf{yy}}^{1/2}\mathbf{b}$. In view of 23.3 (p. 106), we get

$$\max_{\mathbf{a},\mathbf{b}} \operatorname{cor}^2(\mathbf{a}'\mathbf{x}, \mathbf{b}'\mathbf{y}) = \operatorname{sg}_1^2(\boldsymbol{\Sigma}_{\mathbf{xx}}^{+1/2}\boldsymbol{\Sigma}_{\mathbf{xy}}\boldsymbol{\Sigma}_{\mathbf{yy}}^{+1/2})$$
$$= \operatorname{ch}_1(\boldsymbol{\Sigma}_{\mathbf{xx}}^{+}\boldsymbol{\Sigma}_{\mathbf{xy}}\boldsymbol{\Sigma}_{\mathbf{yy}}^{+}\boldsymbol{\Sigma}_{\mathbf{yx}}) = \operatorname{cc}_1^2(\mathbf{x},\mathbf{y}) = \varrho_1^2.$$

17.4 The minimal angle between the subspaces $\mathcal{A} = \mathscr{C}(\mathbf{A})$ and $\mathcal{B} = \mathscr{C}(\mathbf{B})$ is defined to be the number $0 \le \theta_{\min} \le \pi/2$ for which

$$\cos^2\theta_{\min} = \max_{\substack{\mathbf{u}\in\mathcal{A},\, \mathbf{v}\in\mathcal{B}\\ \mathbf{u}\neq\mathbf{0},\, \mathbf{v}\neq\mathbf{0}}} \cos^2(\mathbf{u},\mathbf{v}) = \max_{\substack{\mathbf{A}\boldsymbol{\alpha}\neq\mathbf{0}\\ \mathbf{B}\boldsymbol{\beta}\neq\mathbf{0}}} \frac{(\boldsymbol{\alpha}'\mathbf{A}'\mathbf{B}\boldsymbol{\beta})^2}{\boldsymbol{\alpha}'\mathbf{A}'\mathbf{A}\boldsymbol{\alpha}\cdot\boldsymbol{\beta}'\mathbf{B}'\mathbf{B}\boldsymbol{\beta}}.$$

17.5 Let $\mathbf{A}_{n\times a}$ and $\mathbf{B}_{n\times b}$ be given matrices. Then

$$\max_{\substack{\mathbf{A}\boldsymbol{\alpha}\neq\mathbf{0}\\ \mathbf{B}\boldsymbol{\beta}\neq\mathbf{0}}} \frac{(\boldsymbol{\alpha}'\mathbf{A}'\mathbf{B}\boldsymbol{\beta})^2}{\boldsymbol{\alpha}'\mathbf{A}'\mathbf{A}\boldsymbol{\alpha}\cdot\boldsymbol{\beta}'\mathbf{B}'\mathbf{B}\boldsymbol{\beta}} = \frac{(\boldsymbol{\alpha}_1'\mathbf{A}'\mathbf{B}\boldsymbol{\beta}_1)^2}{\boldsymbol{\alpha}_1'\mathbf{A}'\mathbf{A}\boldsymbol{\alpha}_1\cdot\boldsymbol{\beta}_1'\mathbf{B}'\mathbf{B}\boldsymbol{\beta}_2}$$
$$= \frac{\boldsymbol{\alpha}_1'\mathbf{A}'\mathbf{P}_\mathbf{B}\mathbf{A}\boldsymbol{\alpha}_1}{\boldsymbol{\alpha}_1'\mathbf{A}'\mathbf{A}\boldsymbol{\alpha}_1} = \operatorname{ch}_1(\mathbf{P}_\mathbf{A}\mathbf{P}_\mathbf{B}) = \lambda_1^2,$$

where $(\lambda_1^2, \boldsymbol{\alpha}_1)$ is the first proper eigenpair for $(\mathbf{A}'\mathbf{P}_\mathbf{B}\mathbf{A}, \mathbf{A}'\mathbf{A})$ satisfying

$$\mathbf{A}'\mathbf{P}_\mathbf{B}\mathbf{A}\boldsymbol{\alpha}_1 = \lambda_1^2\mathbf{A}'\mathbf{A}\boldsymbol{\alpha}_1, \quad \mathbf{A}\boldsymbol{\alpha}_1 \neq \mathbf{0}.$$

The vector $\boldsymbol{\beta}_1$ is the proper eigenvector satisfying

$$\mathbf{B}'\mathbf{P}_\mathbf{A}\mathbf{B}\boldsymbol{\beta}_1 = \lambda_1^2\mathbf{B}'\mathbf{B}\boldsymbol{\beta}_1, \quad \mathbf{B}\boldsymbol{\beta}_1 \neq \mathbf{0}.$$

17.6 Consider an n-dimensional random vector $\mathbf{u}$ such that $\operatorname{cov}(\mathbf{u}) = \mathbf{I}_n$ and define $\mathbf{x} = \mathbf{A}'\mathbf{u}$ and $\mathbf{y} = \mathbf{B}'\mathbf{u}$ where $\mathbf{A} \in \mathbb{R}^{n\times a}$ and $\mathbf{B} \in \mathbb{R}^{n\times b}$. Then

$$\operatorname{cov}\begin{pmatrix}\mathbf{x}\\ \mathbf{y}\end{pmatrix} = \operatorname{cov}\begin{pmatrix}\mathbf{A}'\mathbf{u}\\ \mathbf{B}'\mathbf{u}\end{pmatrix} = \begin{pmatrix}\mathbf{A}'\mathbf{A} & \mathbf{A}'\mathbf{B}\\ \mathbf{B}'\mathbf{A} & \mathbf{B}'\mathbf{B}\end{pmatrix} := \begin{pmatrix}\boldsymbol{\Sigma}_{\mathbf{xx}} & \boldsymbol{\Sigma}_{\mathbf{xy}}\\ \boldsymbol{\Sigma}_{\mathbf{yx}} & \boldsymbol{\Sigma}_{\mathbf{yy}}\end{pmatrix} = \boldsymbol{\Sigma}.$$

Let ϱ_i denote the ith largest canonical correlation between the random vectors $\mathbf{A}'\mathbf{u}$ and $\mathbf{B}'\mathbf{u}$ and let $\boldsymbol{\alpha}_1'\mathbf{A}'\mathbf{u}$ and $\boldsymbol{\beta}_1'\mathbf{B}'\mathbf{u}$ be the first canonical variables. In view of 17.5,

$$\varrho_1^2 = \max_{\substack{\mathbf{A}\boldsymbol{\alpha}\neq\mathbf{0}\\ \mathbf{B}\boldsymbol{\beta}\neq\mathbf{0}}} \frac{(\boldsymbol{\alpha}'\mathbf{A}'\mathbf{B}\boldsymbol{\beta})^2}{\boldsymbol{\alpha}'\mathbf{A}'\mathbf{A}\boldsymbol{\alpha}\cdot\boldsymbol{\beta}'\mathbf{B}'\mathbf{B}\boldsymbol{\beta}} = \frac{\boldsymbol{\alpha}_1'\mathbf{A}'\mathbf{P}_\mathbf{B}\mathbf{A}\boldsymbol{\alpha}_1}{\boldsymbol{\alpha}_1'\mathbf{A}'\mathbf{A}\boldsymbol{\alpha}_1} = \operatorname{ch}_1(\mathbf{P}_\mathbf{A}\mathbf{P}_\mathbf{B}).$$

In other words, $(\varrho_1^2, \boldsymbol{\alpha}_1)$ is the first proper eigenpair for $(\mathbf{A}'\mathbf{P}_\mathbf{B}\mathbf{A}, \mathbf{A}'\mathbf{A})$:

$$\mathbf{A}'\mathbf{P}_\mathbf{B}\mathbf{A}\boldsymbol{\alpha}_1 = \varrho_1^2\mathbf{A}'\mathbf{A}\boldsymbol{\alpha}_1, \quad \mathbf{A}\boldsymbol{\alpha}_1 \neq \mathbf{0}.$$

Moreover, $(\varrho_1^2, \mathbf{A}\boldsymbol{\alpha}_1)$ is the first eigenpair of $\mathbf{P}_\mathbf{A}\mathbf{P}_\mathbf{B}$: $\mathbf{P}_\mathbf{A}\mathbf{P}_\mathbf{B}\mathbf{A}\boldsymbol{\alpha}_1 = \varrho_1^2\mathbf{A}\boldsymbol{\alpha}_1$.

17.7 Suppose that $r = \operatorname{r}(\mathbf{A}) \le \operatorname{r}(\mathbf{B})$ and denote

$$\operatorname{cc}(\mathbf{A}'\mathbf{u}, \mathbf{B}'\mathbf{u}) = \{\varrho_1, \dots, \varrho_m, \varrho_{m+1}, \dots, \varrho_h\} = \text{the set of all cc's},$$
$$\operatorname{cc}_+(\mathbf{A}'\mathbf{u}, \mathbf{B}'\mathbf{u}) = \{\varrho_1, \dots, \varrho_m\}$$
$$= \text{the set of nonzero cc's},\ m = \operatorname{r}(\mathbf{A}'\mathbf{B}).$$

Then

(a) there are $h = \mathrm{r}(\mathbf{A})$ pairs of canonical variables $\boldsymbol{\alpha}_i'\mathbf{A}'\mathbf{u}$, $\boldsymbol{\beta}_i'\mathbf{B}'\mathbf{u}$, and h corresponding canonical correlations $\varrho_1 \geq \varrho_2 \geq \cdots \geq \varrho_h \geq 0$,

(b) the vectors $\boldsymbol{\alpha}_i$ are the proper eigenvectors of $\mathbf{A}'\mathbf{P_B}\mathbf{A}$ w.r.t. $\mathbf{A}'\mathbf{A}$, and the ϱ_i^2's are the corresponding proper eigenvalues,

(c) the ϱ_i^2's are the h largest eigenvalues of $\mathbf{P_A}\mathbf{P_B}$,

(d) the nonzero ϱ_i^2's are the nonzero eigenvalues of $\mathbf{P_A}\mathbf{P_B}$, i.e.,

$$\mathrm{cc}_+^2(\mathbf{A}'\mathbf{u}, \mathbf{B}'\mathbf{u}) = \mathrm{nzch}(\mathbf{P_A}\mathbf{P_B}) = \mathrm{nzch}(\boldsymbol{\Sigma}_{\mathbf{yx}}\boldsymbol{\Sigma}_{\mathbf{xx}}^{-}\boldsymbol{\Sigma}_{\mathbf{xy}}\boldsymbol{\Sigma}_{\mathbf{yy}}^{-}),$$

(e) the vectors $\boldsymbol{\beta}_i$ are the proper eigenvectors of $\mathbf{B}'\mathbf{P_A}\mathbf{B}$ w.r.t. $\mathbf{B}'\mathbf{B}$,

(f) the number of nonzero ϱ_i's is $m = \mathrm{r}(\mathbf{A}'\mathbf{B}) = \mathrm{r}(\mathbf{A}) - \dim \mathscr{C}(\mathbf{A}) \cap \mathscr{C}(\mathbf{B})^{\perp}$,

(g) the number of unit ϱ_i's is $u = \dim \mathscr{C}(\mathbf{A}) \cap \mathscr{C}(\mathbf{B})$,

(h) the number of zero ϱ_i's is $s = \mathrm{r}(\mathbf{A}) - \mathrm{r}(\mathbf{A}'\mathbf{B}) = \dim \mathscr{C}(\mathbf{A}) \cap \mathscr{C}(\mathbf{B})^{\perp}$.

17.8 Let $\mathbf{u}$ denote a random vector with covariance matrix $\mathrm{cov}(\mathbf{u}) = \mathbf{I}_n$ and let $\mathbf{K} \in \mathbb{R}^{n\times p}$, $\mathbf{L} \in \mathbb{R}^{n\times q}$, and $\mathbf{F}$ has the property $\mathscr{C}(\mathbf{F}) = \mathscr{C}(\mathbf{K}) \cap \mathscr{C}(\mathbf{L})$. Then

(a) The canonical correlations between $\mathbf{K}'\mathbf{Q_L}\mathbf{u}$ and $\mathbf{L}'\mathbf{Q_K}\mathbf{u}$ are all less than 1, and are precisely those canonical correlations between $\mathbf{K}'\mathbf{u}$ and $\mathbf{L}'\mathbf{u}$ that are not equal to 1.

(b) The nonzero eigenvalues of $\mathbf{P_K}\mathbf{P_L} - \mathbf{P_F}$ are all less than 1, and are precisely those canonical correlations between the vectors $\mathbf{K}'\mathbf{u}$ and $\mathbf{L}'\mathbf{u}$ that are not equal to 1.

18 Column space properties and rank rules

18.1 For conformable matrices $\mathbf{A}$ and $\mathbf{B}$, the following statements hold:

(a) $\mathscr{N}(\mathbf{A}) = \mathscr{N}(\mathbf{A}'\mathbf{A})$,

(b) $\mathscr{C}(\mathbf{A})^{\perp} = \mathscr{C}(\mathbf{A}^{\perp}) = \mathscr{N}(\mathbf{A}')$,

(c) $\mathscr{C}(\mathbf{A}) \subset \mathscr{C}(\mathbf{B}) \iff \mathscr{C}(\mathbf{B})^{\perp} \subset \mathscr{C}(\mathbf{A})^{\perp}$,

(d) $\mathscr{C}(\mathbf{A}) = \mathscr{C}(\mathbf{A}\mathbf{A}')$,

(e) $\mathrm{r}(\mathbf{A}) = \mathrm{r}(\mathbf{A}') = \mathrm{r}(\mathbf{A}\mathbf{A}') = \mathrm{r}(\mathbf{A}'\mathbf{A})$,

(f) $\mathbb{R}^n = \mathscr{C}(\mathbf{A}_{n\times m}) \boxplus \mathscr{C}(\mathbf{A}_{n\times m})^{\perp} = \mathscr{C}(\mathbf{A}) \boxplus \mathscr{N}(\mathbf{A}')$,

(g) $\mathrm{r}(\mathbf{A}_{n\times m}) = n - \mathrm{r}(\mathbf{A}^{\perp}) = n - \dim \mathscr{N}(\mathbf{A}')$,

(h) $\mathscr{C}(\mathbf{A} : \mathbf{B}) = \mathscr{C}(\mathbf{A}) + \mathscr{C}(\mathbf{B})$,

(i) $\mathrm{r}(\mathbf{A} : \mathbf{B}) = \mathrm{r}(\mathbf{A}) + \mathrm{r}(\mathbf{B}) - \dim \mathscr{C}(\mathbf{A}) \cap \mathscr{C}(\mathbf{B})$,

(j) $\mathrm{r}\begin{pmatrix}\mathbf{A}\\ \mathbf{B}\end{pmatrix} = \mathrm{r}(\mathbf{A}) + \mathrm{r}(\mathbf{B}) - \dim \mathscr{C}(\mathbf{A}') \cap \mathscr{C}(\mathbf{B}')$,

(k) $\mathscr{C}(\mathbf{A} + \mathbf{B}) \subset \mathscr{C}(\mathbf{A}) + \mathscr{C}(\mathbf{B}) = \mathscr{C}(\mathbf{A} : \mathbf{B})$,

(l) $\mathrm{r}(\mathbf{A} + \mathbf{B}) \leq \mathrm{r}(\mathbf{A} : \mathbf{B}) \leq \mathrm{r}(\mathbf{A}) + \mathrm{r}(\mathbf{B})$,

(m) $\mathscr{C}(\mathbf{A} : \mathbf{B})^{\perp} = \mathscr{C}(\mathbf{A})^{\perp} \cap \mathscr{C}(\mathbf{B})^{\perp}$.

18.2 (a) $\mathbf{LAY} = \mathbf{MAY}\ \&\ \mathrm{r}(\mathbf{AY}) = \mathrm{r}(\mathbf{A}) \implies \mathbf{LA} = \mathbf{MA}$ rank cancellation rule

(b) $\mathbf{DAM} = \mathbf{DAN}\ \&\ \mathrm{r}(\mathbf{DA}) = \mathrm{r}(\mathbf{A}) \implies \mathbf{AM} = \mathbf{AN}$

18.3 $\mathrm{r}(\mathbf{AB}) = \mathrm{r}(\mathbf{A})\ \&\ \mathbf{FAB} = \mathbf{0} \implies \mathbf{FA} = \mathbf{0}$

18.4 (a) $\mathrm{r}(\mathbf{AB}) = \mathrm{r}(\mathbf{A}) \implies \mathrm{r}(\mathbf{FAB}) = \mathrm{r}(\mathbf{FA})$ for all $\mathbf{F}$

(b) $\mathrm{r}(\mathbf{AB}) = \mathrm{r}(\mathbf{B}) \implies \mathrm{r}(\mathbf{ABG}) = \mathrm{r}(\mathbf{BG})$ for all $\mathbf{G}$

18.5 $\mathrm{r}(\mathbf{AB}) = \mathrm{r}(\mathbf{A}) - \dim \mathscr{C}(\mathbf{A}') \cap \mathscr{C}(\mathbf{B})^{\perp}$ rank of the product

18.6
$$\begin{aligned}\mathrm{r}(\mathbf{A} : \mathbf{B}) &= \mathrm{r}(\mathbf{A}) + \mathrm{r}[(\mathbf{I} - \mathbf{P}_{\mathbf{A}})\mathbf{B}] = \mathrm{r}(\mathbf{A}) + \mathrm{r}[(\mathbf{I} - \mathbf{A}\mathbf{A}^{-})\mathbf{B}] \\ &= \mathrm{r}(\mathbf{A}) + \mathrm{r}(\mathbf{B}) - \dim \mathscr{C}(\mathbf{A}) \cap \mathscr{C}(\mathbf{B})\end{aligned}$$

18.7
$$\begin{aligned}\mathrm{r}\begin{pmatrix}\mathbf{A}\\ \mathbf{B}\end{pmatrix} &= \mathrm{r}(\mathbf{A}) + \mathrm{r}[(\mathbf{I} - \mathbf{P}_{\mathbf{A}'})\mathbf{B}'] = \mathrm{r}(\mathbf{A}) + \mathrm{r}[\mathbf{B}'(\mathbf{I} - \mathbf{A}^{-}\mathbf{A})] \\ &= \mathrm{r}(\mathbf{A}) + \mathrm{r}(\mathbf{B}) - \dim \mathscr{C}(\mathbf{A}') \cap \mathscr{C}(\mathbf{B}')\end{aligned}$$

18.8
$$\begin{aligned}&\mathrm{r}(\mathbf{A}'\mathbf{U}\mathbf{A}) = \mathrm{r}(\mathbf{A}'\mathbf{U}) \\ &\iff \mathscr{C}(\mathbf{A}'\mathbf{U}\mathbf{A}) = \mathscr{C}(\mathbf{A}'\mathbf{U}) \\ &\iff \mathbf{A}'\mathbf{U}\mathbf{A}(\mathbf{A}'\mathbf{U}\mathbf{A})^{-}\mathbf{A}'\mathbf{U} = \mathbf{A}'\mathbf{U}\end{aligned}$$
[holds e.g. if $\mathbf{U} \geq_{\mathsf{L}} \mathbf{0}$]

18.9 $\mathrm{r}(\mathbf{A}'\mathbf{U}\mathbf{A}) = \mathrm{r}(\mathbf{A}) \iff \mathscr{C}(\mathbf{A}'\mathbf{U}\mathbf{A}) = \mathscr{C}(\mathbf{A}') \iff \mathbf{A}'\mathbf{U}\mathbf{A}(\mathbf{A}'\mathbf{U}\mathbf{A})^{-}\mathbf{A}' = \mathbf{A}'$

18.10
$$\begin{aligned}&\mathrm{r}(\mathbf{A} + \mathbf{B}) = \mathrm{r}(\mathbf{A}) + \mathrm{r}(\mathbf{B}) \\ &\iff \dim \mathscr{C}(\mathbf{A}) \cap \mathscr{C}(\mathbf{B}) = 0 = \dim \mathscr{C}(\mathbf{A}') \cap \mathscr{C}(\mathbf{B}')\end{aligned}$$

18.11 $\mathscr{C}(\mathbf{U} + \mathbf{V}) = \mathscr{C}(\mathbf{U} : \mathbf{V})$ if $\mathbf{U}$ and $\mathbf{V}$ are nnd

18.12 $\mathscr{C}(\mathbf{G}) = \mathscr{C}(\mathbf{G}_{*}) \implies \mathscr{C}(\mathbf{AG}) = \mathscr{C}(\mathbf{AG}_{*})$

18.13 $\mathscr{C}(\mathbf{A}_{n\times m}) = \mathscr{C}(\mathbf{B}_{n\times p})$

iff $\exists\, \mathbf{F}\colon \mathbf{A}_{n\times m} = \mathbf{B}_{n\times p}\mathbf{F}_{p\times m}$, where $\mathscr{C}(\mathbf{B}') \cap \mathscr{C}(\mathbf{F}^{\perp}) = \{\mathbf{0}\}$

18.14 $\mathbf{AA}' = \mathbf{BB}' \iff \exists$ orthogonal $\mathbf{Q}$: $\mathbf{A} = \mathbf{BQ}$

18.15 Let $\mathbf{y} \in \mathbb{R}^n$ be a given vector, $\mathbf{y} \neq \mathbf{0}$. Then $\mathbf{y}'\mathbf{Qx} = 0 \; \forall \, \mathbf{Q}_{n\times p} \implies \mathbf{x} = \mathbf{0}$.

18.16 Frobenius inequality: $\mathrm{r}(\mathbf{AZB}) \geq \mathrm{r}(\mathbf{AZ}) + \mathrm{r}(\mathbf{ZB}) - \mathrm{r}(\mathbf{Z})$

18.17 Sylvester's inequality:

$$\mathrm{r}(\mathbf{A}_{m\times n}\mathbf{B}_{n\times p}) \geq \mathrm{r}(\mathbf{A}) + \mathrm{r}(\mathbf{B}) - n, \text{ with equality iff } \mathscr{N}(\mathbf{A}) \subset \mathscr{C}(\mathbf{B})$$

18.18 For $\mathbf{A} \in \mathbb{R}^{n\times m}$ we have

(a) $(\mathbf{I}_n - \mathbf{AA}^-)' \in \{\mathbf{A}^\perp\}$,

(b) $\mathbf{I}_n - (\mathbf{A}^-)'\mathbf{A}' \in \{\mathbf{A}^\perp\}$,

(c) $\mathbf{I}_n - (\mathbf{A}')^-\mathbf{A}' \in \{\mathbf{A}^\perp\}$,

(d) $\mathbf{I}_m - \mathbf{A}^-\mathbf{A} \in \{(\mathbf{A}')^\perp\}$,

(e) $\mathbf{I}_n - \mathbf{AA}^+ = \mathbf{I}_n - \mathbf{P_A} = \mathbf{Q_A} \in \{\mathbf{A}^\perp\}$.

18.19 $\begin{pmatrix} \mathbf{I}_n \\ -\mathbf{B}' \end{pmatrix} \mathbf{Q_A} \in \left\{ \begin{pmatrix} \mathbf{A}_{n\times p} & \mathbf{B}_{n\times q} \\ \mathbf{0} & \mathbf{I}_q \end{pmatrix}^\perp \right\}, \quad \begin{pmatrix} \mathbf{Q_A} & \mathbf{0} \\ \mathbf{0} & \mathbf{0} \end{pmatrix} \in \left\{ \begin{pmatrix} \mathbf{A}_{n\times p} \\ \mathbf{0}_{q\times p} \end{pmatrix}^\perp \right\},$

$\mathbf{Q_A} = \mathbf{I} - \mathbf{P_A}$

18.20 $\begin{pmatrix} \mathbf{I}_n \\ -\mathbf{B}' \end{pmatrix} \in \left\{ \begin{pmatrix} \mathbf{B}_{n\times q} \\ \mathbf{I}_q \end{pmatrix}^\perp \right\}$

18.21 Consider $\mathbf{A}_{n\times p}$, $\mathbf{Z}_{n\times q}$, $\mathbf{D} = \mathrm{diag}(d_1, \dots, d_q) \in \mathrm{PD}_q$, and suppose that $\mathbf{Z}'\mathbf{Z} = \mathbf{I}_q$ and $\mathscr{C}(\mathbf{A}) \subset \mathscr{C}(\mathbf{Z})$. Then $\mathbf{D}^{1/2}\mathbf{Z}'(\mathbf{I}_n - \mathbf{P_A}) \in \{(\mathbf{D}^{-1/2}\mathbf{Z}'\mathbf{A})^\perp\}$.

18.22 $\mathscr{C}(\mathbf{A}) \cap \mathscr{C}(\mathbf{B}) = \mathscr{C}[\mathbf{A}(\mathbf{A}'\mathbf{B}^\perp)^\perp] = \mathscr{C}[\mathbf{A}(\mathbf{A}'\mathbf{Q_B})^\perp] = \mathscr{C}[\mathbf{A}(\mathbf{I} - \mathbf{P}_{\mathbf{A}'\mathbf{Q_B}})]$

18.23 $\mathscr{C}(\mathbf{A}) \cap \mathscr{C}(\mathbf{B})^\perp = \mathscr{C}[\mathbf{A}(\mathbf{A}'\mathbf{B})^\perp] = \mathscr{C}[\mathbf{A}(\mathbf{I} - \mathbf{P}_{\mathbf{A}'\mathbf{B}})] = \mathscr{C}[\mathbf{P_A}(\mathbf{I} - \mathbf{P}_{\mathbf{A}'\mathbf{B}})]$

18.24 $\mathscr{C}(\mathbf{A}) \cap \mathscr{C}(\mathbf{B}) = \mathscr{C}[\mathbf{AA}'(\mathbf{AA}' + \mathbf{BB}')^-\mathbf{BB}']$

18.25 Disjointness: $\mathscr{C}(\mathbf{A}) \cap \mathscr{C}(\mathbf{B}) = \{\mathbf{0}\}$. The following statements are equivalent:

(a) $\mathscr{C}(\mathbf{A}) \cap \mathscr{C}(\mathbf{B}) = \{\mathbf{0}\}$,

(b) $\begin{pmatrix} \mathbf{A}' \\ \mathbf{B}' \end{pmatrix} (\mathbf{AA}' + \mathbf{BB}')^-(\mathbf{AA}' : \mathbf{BB}') = \begin{pmatrix} \mathbf{A}' & \mathbf{0} \\ \mathbf{0} & \mathbf{B}' \end{pmatrix}$,

(c) $\mathbf{A}'(\mathbf{AA}' + \mathbf{BB}')^-\mathbf{AA}' = \mathbf{A}'$,

(d) $\mathbf{A}'(\mathbf{AA}' + \mathbf{BB}')^-\mathbf{B} = \mathbf{0}$,

(e) $(\mathbf{AA}' + \mathbf{BB}')^-$ is a generalized inverse of $\mathbf{AA}'$,

(f) $\mathbf{A}'(\mathbf{AA}' + \mathbf{BB}')^-\mathbf{A} = \mathbf{P}_{\mathbf{A}'}$,

(g) $\mathscr{C}\begin{pmatrix}\mathbf{0}\\ \mathbf{B}'\end{pmatrix} \subset \mathscr{C}\begin{pmatrix}\mathbf{A}'\\ \mathbf{B}'\end{pmatrix}$,

(h) $\mathscr{N}(\mathbf{A} : \mathbf{B}) \subset \mathscr{N}(\mathbf{0} : \mathbf{B})$,

(i) $\mathbf{Y}(\mathbf{A} : \mathbf{B}) = (\mathbf{0} : \mathbf{B})$ has a solution for $\mathbf{Y}$,

(j) $\mathbf{P}_{\binom{\mathbf{A}'}{\mathbf{B}'}} = \begin{pmatrix}\mathbf{P}_{\mathbf{A}'} & \mathbf{0}\\ \mathbf{0} & \mathbf{P}_{\mathbf{B}'}\end{pmatrix}$,

(k) $\mathrm{r}(\mathbf{Q}_\mathbf{B}\mathbf{A}) = \mathrm{r}(\mathbf{A})$,

(l) $\mathscr{C}(\mathbf{A}'\mathbf{Q}_\mathbf{B}) = \mathscr{C}(\mathbf{A}')$,

(m) $\mathbf{P}_{\mathbf{A}'\mathbf{Q}_\mathbf{B}}\mathbf{A}' = \mathbf{A}'$,

(n) $\mathrm{ch}_1(\mathbf{P}_\mathbf{A}\mathbf{P}_\mathbf{B}) < 1$,

(o) $\det(\mathbf{I} - \mathbf{P}_\mathbf{A}\mathbf{P}_\mathbf{B}) \neq 0$.

18.26 Let us denote $\mathbf{P}_{\mathbf{A}\cdot\mathbf{B}} = \mathbf{A}(\mathbf{A}'\mathbf{Q}_\mathbf{B}\mathbf{A})^-\mathbf{A}'\mathbf{Q}_\mathbf{B}$, $\mathbf{Q}_\mathbf{B} = \mathbf{I} - \mathbf{P}_\mathbf{B}$. Then the following statements are equivalent:

(a) $\mathbf{P}_{\mathbf{A}\cdot\mathbf{B}}$ is invariant w.r.t. the choice of $(\mathbf{A}'\mathbf{Q}_\mathbf{B}\mathbf{A})^-$,

(b) $\mathscr{C}[\mathbf{A}(\mathbf{A}'\mathbf{Q}_\mathbf{B}\mathbf{A})^-\mathbf{A}'\mathbf{Q}_\mathbf{B}]$ is invariant w.r.t. the choice of $(\mathbf{A}'\mathbf{Q}_\mathbf{B}\mathbf{A})^-$,

(c) $\mathrm{r}(\mathbf{Q}_\mathbf{B}\mathbf{A}) = \mathrm{r}(\mathbf{A})$,

(d) $\mathbf{P}_{\mathbf{A}\cdot\mathbf{B}}\mathbf{A} = \mathbf{A}$,

(e) $\mathbf{P}_{\mathbf{A}\cdot\mathbf{B}}$ is the projector onto $\mathscr{C}(\mathbf{A})$ along $\mathscr{C}(\mathbf{B}) \boxplus \mathscr{C}(\mathbf{A} : \mathbf{B})^\perp$,

(f) $\mathscr{C}(\mathbf{A}) \cap \mathscr{C}(\mathbf{B}) = \{\mathbf{0}\}$,

(g) $\mathbf{P}_{(\mathbf{A}:\mathbf{B})} = \mathbf{P}_{\mathbf{A}\cdot\mathbf{B}} + \mathbf{P}_{\mathbf{B}\cdot\mathbf{A}}$.

18.27 Let $\mathbf{A} \in \mathbb{R}^{n\times p}$ and $\mathbf{B} \in \mathbb{R}^{n\times q}$. Then

$$\begin{aligned}\mathrm{r}(\mathbf{P}_\mathbf{A}\mathbf{P}_\mathbf{B}\mathbf{Q}_\mathbf{A}) &= \mathrm{r}(\mathbf{P}_\mathbf{A}\mathbf{P}_\mathbf{B}) + \mathrm{r}(\mathbf{P}_\mathbf{A} : \mathbf{P}_\mathbf{B}) - \mathrm{r}(\mathbf{A}) - \mathrm{r}(\mathbf{B})\\ &= \mathrm{r}(\mathbf{P}_\mathbf{A}\mathbf{P}_\mathbf{B}) + \mathrm{r}(\mathbf{Q}_\mathbf{A}\mathbf{P}_\mathbf{B}) - \mathrm{r}(\mathbf{B})\\ &= \mathrm{r}(\mathbf{P}_\mathbf{B}\mathbf{P}_\mathbf{A}) + \mathrm{r}(\mathbf{Q}_\mathbf{B}\mathbf{P}_\mathbf{A}) - \mathrm{r}(\mathbf{A})\\ &= \mathrm{r}(\mathbf{P}_\mathbf{B}\mathbf{P}_\mathbf{A}\mathbf{Q}_\mathbf{B}).\end{aligned}$$

18.28 (a) $\mathscr{C}(\mathbf{A}\mathbf{P}_\mathbf{B}) = \mathscr{C}(\mathbf{AB})$,

(b) $\mathrm{r}(\mathbf{AB}) = \mathrm{r}(\mathbf{A}\mathbf{P}_\mathbf{B}) = \mathrm{r}(\mathbf{P}_\mathbf{B}\mathbf{A}') = \mathrm{r}(\mathbf{B}'\mathbf{A}') = \mathrm{r}(\mathbf{P}_\mathbf{B}\mathbf{P}_{\mathbf{A}'}) = \mathrm{r}(\mathbf{P}_{\mathbf{A}'}\mathbf{P}_\mathbf{B})$.

19 Inverse of a matrix

19.1 (a) $\mathbf{A}[\alpha, \beta]$: submatrix of $\mathbf{A}_{n\times n}$, obtained by choosing the elements of $\mathbf{A}$ which lie in rows α and columns β; α and β are index sets of the rows and the columns of $\mathbf{A}$, respectively.

(b) $\mathbf{A}[\alpha] = \mathbf{A}[\alpha, \alpha]$: principal submatrix; same rows and columns chosen.

(c) $\mathbf{A}_i^{\mathrm{L}} = i$th leading principal submatrix of $\mathbf{A}$: $\mathbf{A}_i^{\mathrm{L}} = \mathbf{A}[1, \dots, i]$.

(d) $\mathbf{A}(\alpha, \beta)$: submatrix of $\mathbf{A}$, obtained by choosing the elements of $\mathbf{A}$ which *do not* lie in rows α and columns β.

(e) $\mathbf{A}(i, j) =$ submatrix of $\mathbf{A}$, obtained by deleting row i and column j.

(f) $\operatorname{minor}(a_{ij}) = \det(\mathbf{A}(i, j) = ij$th minor of $\mathbf{A}$ corresponding to a_{ij}.

(g) $\operatorname{cof}(a_{ij}) = (-1)^{i+j} \operatorname{minor}(a_{ij}) = ij$th cofactor of $\mathbf{A}$ corresponding to a_{ij}.

(h) $\det(\mathbf{A}[\alpha]) =$ principal minor.

(i) $\det(\mathbf{A}_i^{\mathrm{L}}) =$ leading principal minor of order i.

19.2 Determinant. The determinant of matrix $\mathbf{A}_{n\times n}$, denoted as $|\mathbf{A}|$ or $\det(\mathbf{A})$, is $\det(a) = a$ when $a \in \mathbb{R}$; when $n > 1$, we have the Laplace expansion of the determinant by minors along the ith row:

$$\det(\mathbf{A}) = \sum_{j=1}^{n} a_{ij} \operatorname{cof}(a_{ij}), \quad i \in \{1, \dots, n\}.$$

19.3 An alternative equivalent definition of the determinant is the following:

$$\det(\mathbf{A}) = \sum (-1)^{f(i_1,\dots,i_n)} a_{1i_1} a_{2i_2} \cdots a_{ni_n},$$

where the summation is taken over all permutations $\{i_1, \dots, i_n\}$ of the set of integers $\{1, \dots, n\}$, and the function $f(i_1, \dots, i_n)$ equals the number of transpositions necessary to change $\{i_1, \dots, i_n\}$ to $\{1, \dots, n\}$. A transposition is the interchange of two of the integers. The determinant produces all products of n terms of the elements of $\mathbf{A}$ such that exactly one element is selected from each row and each column of $\mathbf{A}$; there are $n!$ of such products.

19.4 If $\mathbf{A}_{n\times n}\mathbf{B}_{n\times n} = \mathbf{I}_n$ then $\mathbf{B} = \mathbf{A}^{-1}$ and $\mathbf{A}$ is said to be nonsingular. $\mathbf{A}_{n\times n}$ is nonsingular iff $\det(\mathbf{A}) \neq 0$ iff $\operatorname{rank}(\mathbf{A}) = n$.

19.5 If $\mathbf{A} = \begin{pmatrix} a & b \\ c & d \end{pmatrix}$, then $\mathbf{A}^{-1} = \dfrac{1}{ad - bc} \begin{pmatrix} d & -b \\ -c & a \end{pmatrix}$, $\det(\mathbf{A}) = ad - bc$.

If $\mathbf{A} = \{a_{ij}\}$, then

$$\mathbf{A}^{-1} = \{a^{ij}\} = \frac{1}{\det(\mathbf{A})}[\operatorname{cof}(\mathbf{A})]' = \frac{1}{\det(\mathbf{A})}\operatorname{adj}(\mathbf{A}),$$

where $\operatorname{cof}(\mathbf{A}) = \{\operatorname{cof}(a_{ij})\}$. The matrix $\operatorname{adj}(\mathbf{A}) = [\operatorname{cof}(\mathbf{A})]'$ is the adjoint matrix of $\mathbf{A}$.

19.6 Let a nonsingular matrix $\mathbf{A}$ be partitioned as $\mathbf{A} = \begin{pmatrix} \mathbf{A}_{11} & \mathbf{A}_{12} \\ \mathbf{A}_{21} & \mathbf{A}_{22} \end{pmatrix}$, where $\mathbf{A}_{11}$ is a square matrix. Then

(a) $\mathbf{A} = \begin{pmatrix} \mathbf{I} & \mathbf{0} \\ \mathbf{A}_{21}\mathbf{A}_{11}^{-1} & \mathbf{I} \end{pmatrix} \begin{pmatrix} \mathbf{A}_{11} & \mathbf{0} \\ \mathbf{0} & \mathbf{A}_{22} - \mathbf{A}_{21}\mathbf{A}_{11}^{-1}\mathbf{A}_{12} \end{pmatrix} \begin{pmatrix} \mathbf{I} & \mathbf{A}_{11}^{-1}\mathbf{A}_{12} \\ \mathbf{0} & \mathbf{I} \end{pmatrix}$

(b) $$\mathbf{A}^{-1} = \begin{pmatrix} \mathbf{A}_{11}^{-1} + \mathbf{A}_{11}^{-1}\mathbf{A}_{12}\mathbf{A}_{22\cdot1}^{-1}\mathbf{A}_{21}\mathbf{A}_{11}^{-1} & -\mathbf{A}_{11}^{-1}\mathbf{A}_{12}\mathbf{A}_{22\cdot1}^{-1} \\ -\mathbf{A}_{22\cdot1}^{-1}\mathbf{A}_{21}\mathbf{A}_{11}^{-1} & \mathbf{A}_{22\cdot1}^{-1} \end{pmatrix}$$
$$= \begin{pmatrix} \mathbf{A}_{11\cdot2}^{-1} & -\mathbf{A}_{11\cdot2}^{-1}\mathbf{A}_{12}\mathbf{A}_{22}^{-1} \\ -\mathbf{A}_{22}^{-1}\mathbf{A}_{21}\mathbf{A}_{11\cdot2}^{-1} & \mathbf{A}_{22}^{-1} + \mathbf{A}_{22}^{-1}\mathbf{A}_{21}\mathbf{A}_{11\cdot2}^{-1}\mathbf{A}_{12}\mathbf{A}_{22}^{-1} \end{pmatrix}$$
$$= \begin{pmatrix} \mathbf{A}_{11}^{-1} & \mathbf{0} \\ \mathbf{0} & \mathbf{0} \end{pmatrix} + \begin{pmatrix} -\mathbf{A}_{11}^{-1}\mathbf{A}_{12} \\ \mathbf{I} \end{pmatrix} \mathbf{A}_{22\cdot1}^{-1}(-\mathbf{A}_{21}\mathbf{A}_{11}^{-1} : \mathbf{I}), \text{ where}$$

(c) $\mathbf{A}_{11\cdot2} = \mathbf{A}_{11} - \mathbf{A}_{12}\mathbf{A}_{22}^{-1}\mathbf{A}_{21} =$ the Schur complement of $\mathbf{A}_{22}$ in $\mathbf{A}$,
$\mathbf{A}_{22\cdot1} = \mathbf{A}_{22} - \mathbf{A}_{21}\mathbf{A}_{11}^{-1}\mathbf{A}_{12} := \mathbf{A}/\mathbf{A}_{11}$.

(d) For possibly singular $\mathbf{A}$ the generalized Schur complements are

$$\mathbf{A}_{11\cdot2} = \mathbf{A}_{11} - \mathbf{A}_{12}\mathbf{A}_{22}^{-}\mathbf{A}_{21}, \quad \mathbf{A}_{22\cdot1} = \mathbf{A}_{22} - \mathbf{A}_{21}\mathbf{A}_{11}^{-}\mathbf{A}_{12}.$$

(e) $$\begin{aligned} \operatorname{r}(\mathbf{A}) &= \operatorname{r}(\mathbf{A}_{11}) + \operatorname{r}(\mathbf{A}_{22\cdot1}) \quad \text{if } \mathscr{C}(\mathbf{A}_{12}) \subset \mathscr{C}(\mathbf{A}_{11}),\ \mathscr{C}(\mathbf{A}_{21}') \subset \mathscr{C}(\mathbf{A}_{11}') \\ &= \operatorname{r}(\mathbf{A}_{22}) + \operatorname{r}(\mathbf{A}_{11\cdot2}) \quad \text{if } \mathscr{C}(\mathbf{A}_{21}) \subset \mathscr{C}(\mathbf{A}_{22}),\ \mathscr{C}(\mathbf{A}_{12}') \subset \mathscr{C}(\mathbf{A}_{22}') \end{aligned}$$

(f) $$\begin{aligned} |\mathbf{A}| &= |\mathbf{A}_{11}||\mathbf{A}_{22} - \mathbf{A}_{21}\mathbf{A}_{11}^{-}\mathbf{A}_{12}| \quad \text{if } \mathscr{C}(\mathbf{A}_{12}) \subset \mathscr{C}(\mathbf{A}_{11}),\ \mathscr{C}(\mathbf{A}_{21}') \subset \mathscr{C}(\mathbf{A}_{11}') \\ &= |\mathbf{A}_{22}||\mathbf{A}_{11} - \mathbf{A}_{12}\mathbf{A}_{22}^{-}\mathbf{A}_{21}| \quad \text{if } \mathscr{C}(\mathbf{A}_{21}) \subset \mathscr{C}(\mathbf{A}_{22}),\ \mathscr{C}(\mathbf{A}_{12}') \subset \mathscr{C}(\mathbf{A}_{22}') \end{aligned}$$

(g) Let $\mathbf{A}_{ij} \in \mathbb{R}^{n\times n}$ and $\mathbf{A}_{11}\mathbf{A}_{21} = \mathbf{A}_{21}\mathbf{A}_{11}$. Then $|\mathbf{A}| = |\mathbf{A}_{11}\mathbf{A}_{22} - \mathbf{A}_{21}\mathbf{A}_{12}|$.

19.7 Wedderburn–Guttman theorem. Consider $\mathbf{A} \in \mathbb{R}^{n\times p}$, $\mathbf{x} \in \mathbb{R}^n$, $\mathbf{y} \in \mathbb{R}^p$ and suppose that $\alpha := \mathbf{x}'\mathbf{A}\mathbf{y} \neq 0$. Then in view of the rank additivity on the Schur complement, see 19.6e, we have

$$\operatorname{r}\begin{pmatrix} \mathbf{A} & \mathbf{A}\mathbf{y} \\ \mathbf{x}'\mathbf{A} & \mathbf{x}'\mathbf{A}\mathbf{y} \end{pmatrix} = \operatorname{r}(\mathbf{A}) = \operatorname{r}(\mathbf{x}'\mathbf{A}\mathbf{y}) + \operatorname{r}(\mathbf{A} - \alpha^{-1}\mathbf{A}\mathbf{y}\mathbf{x}'\mathbf{A}),$$

and hence $\operatorname{rank}(\mathbf{A} - \alpha^{-1}\mathbf{A}\mathbf{y}\mathbf{x}'\mathbf{A}) = \operatorname{rank}(\mathbf{A}) - 1$.

19.8 $$\begin{pmatrix} \mathbf{I} & \mathbf{A} \\ \mathbf{0} & \mathbf{I} \end{pmatrix}^{-1} = \begin{pmatrix} \mathbf{I} & -\mathbf{A} \\ \mathbf{0} & \mathbf{I} \end{pmatrix}, \quad \begin{vmatrix} \mathbf{E}_{n\times n} & \mathbf{F}_{n\times m} \\ \mathbf{0} & \mathbf{G}_{m\times m} \end{vmatrix} = |\mathbf{E}||\mathbf{G}|$$

19.9 Let $\mathbf{A}$ be nnd. Then there exists $\mathbf{L}$ such that $\mathbf{A} = \mathbf{L}'\mathbf{L}$, where

$$\mathbf{A} = \mathbf{L}'\mathbf{L} = \begin{pmatrix} \mathbf{L}_1' \\ \mathbf{L}_2' \end{pmatrix} (\mathbf{L}_1 : \mathbf{L}_2) = \begin{pmatrix} \mathbf{L}_1'\mathbf{L}_1 & \mathbf{L}_1'\mathbf{L}_2 \\ \mathbf{L}_2'\mathbf{L}_1 & \mathbf{L}_2'\mathbf{L}_2 \end{pmatrix} = \begin{pmatrix} \mathbf{A}_{11} & \mathbf{A}_{12} \\ \mathbf{A}_{21} & \mathbf{A}_{22} \end{pmatrix}.$$

If $\mathbf{A}$ is positive definite, then

$$\mathbf{A}^{-1} = \begin{pmatrix} (\mathbf{L}_1'\mathbf{Q}_2\mathbf{L}_1)^{-1} & -(\mathbf{L}_1'\mathbf{Q}_2\mathbf{L}_1)^{-1}\mathbf{L}_1'\mathbf{L}_2(\mathbf{L}_2'\mathbf{L}_2)^{-1} \\ -(\mathbf{L}_2'\mathbf{L}_2)^{-1}\mathbf{L}_2'\mathbf{L}_1(\mathbf{L}_1'\mathbf{Q}_2\mathbf{L}_1)^{-1} & (\mathbf{L}_2'\mathbf{Q}_1\mathbf{L}_2)^{-1} \end{pmatrix}$$
$$= \begin{pmatrix} \mathbf{A}^{11} & \mathbf{A}^{12} \\ \mathbf{A}^{21} & \mathbf{A}^{22} \end{pmatrix}, \quad \mathbf{Q}_j = \mathbf{I} - \mathbf{P}_{\mathbf{L}_j}, \quad \mathbf{L}_i'\mathbf{Q}_j\mathbf{L}_i = \mathbf{A}_{ii\cdot j},$$

where $\mathbf{A}^{11}$ and $\mathbf{A}^{12}$ (and correspondingly $\mathbf{A}^{22}$ and $\mathbf{A}^{21}$) can be expressed also as

$$\mathbf{A}^{11} = (\mathbf{L}_1'\mathbf{L}_1)^{-1} + (\mathbf{L}_1'\mathbf{L}_1)^{-1}\mathbf{L}_1'\mathbf{L}_2(\mathbf{L}_2'\mathbf{Q}_1\mathbf{L}_2)^{-1}\mathbf{L}_2'\mathbf{L}_1(\mathbf{L}_1'\mathbf{L}_1)^{-1},$$
$$\mathbf{A}^{12} = -(\mathbf{L}_1'\mathbf{L}_1)^{-1}\mathbf{L}_1'\mathbf{L}_2(\mathbf{L}_2'\mathbf{Q}_1\mathbf{L}_2)^{-1}.$$

19.10 $$(\mathbf{X}'\mathbf{X})^{-1} = \begin{pmatrix} n & \mathbf{1}'\mathbf{X}_0 \\ \mathbf{X}_0'\mathbf{1} & \mathbf{X}_0'\mathbf{X}_0 \end{pmatrix}^{-1} = \begin{pmatrix} 1/n + \bar{\mathbf{x}}'\mathbf{T}_{\mathbf{xx}}^{-1}\bar{\mathbf{x}} & -\bar{\mathbf{x}}'\mathbf{T}_{\mathbf{xx}}^{-1} \\ -\mathbf{T}_{\mathbf{xx}}^{-1}\bar{\mathbf{x}} & \mathbf{T}_{\mathbf{xx}}^{-1} \end{pmatrix}$$
$$= \begin{pmatrix} t^{00} & t^{01} & \dots & t^{0k} \\ t^{10} & t^{11} & \dots & t^{1k} \\ \vdots & \vdots & \ddots & \vdots \\ t^{k0} & t^{k1} & \dots & t^{kk} \end{pmatrix}$$

19.11 $$[(\mathbf{X} : \mathbf{y})'(\mathbf{X} : \mathbf{y})]^{-1} = \begin{pmatrix} \mathbf{X}'\mathbf{X} & \mathbf{X}'\mathbf{y} \\ \mathbf{y}'\mathbf{X} & \mathbf{y}'\mathbf{y} \end{pmatrix}^{-1} = \begin{pmatrix} \mathbf{G} & -\hat{\boldsymbol\beta}\frac{1}{\mathrm{SSE}} \\ -\hat{\boldsymbol\beta}'\frac{1}{\mathrm{SSE}} & \frac{1}{\mathrm{SSE}} \end{pmatrix},$$ where

19.12 $\mathbf{G} = [\mathbf{X}'(\mathbf{I} - \mathbf{P}_\mathbf{y})\mathbf{X}]^{-1} = (\mathbf{X}'\mathbf{X})^{-1} + \hat{\boldsymbol\beta}\hat{\boldsymbol\beta}'/\mathrm{SSE}$

19.13 $$|\mathbf{X}'\mathbf{X}| = n|\mathbf{X}_0'(\mathbf{I} - \mathbf{J})\mathbf{X}_0| = n|\mathbf{T}_{\mathbf{xx}}| = |\mathbf{X}_0'\mathbf{X}_0|\mathbf{1}'(\mathbf{I} - \mathbf{P}_{\mathbf{X}_0})\mathbf{1}$$
$$= |\mathbf{X}_0'\mathbf{X}_0| \cdot \|(\mathbf{I} - \mathbf{P}_{\mathbf{X}_0})\mathbf{1}\|^2$$

19.14 $|(\mathbf{X} : \mathbf{y})'(\mathbf{X} : \mathbf{y})| = |\mathbf{X}'\mathbf{X}| \cdot \mathrm{SSE}$

19.15 $$\mathrm{r}(\mathbf{X}) = 1 + \mathrm{r}(\mathbf{X}_0) - \dim \mathscr{C}(\mathbf{1}) \cap \mathscr{C}(\mathbf{X}_0)$$
$$= 1 + \mathrm{r}[(\mathbf{I} - \mathbf{J})\mathbf{X}_0] = 1 + \mathrm{r}(\mathbf{T}_{\mathbf{xx}}) = 1 + \mathrm{r}(\mathbf{R}_{\mathbf{xx}}) = 1 + \mathrm{r}(\mathbf{S}_{\mathbf{xx}})$$

19.16 $$|\mathbf{R}_{\mathbf{xx}}| \neq 0 \iff \mathrm{r}(\mathbf{X}) = k + 1 \iff |\mathbf{X}'\mathbf{X}| \neq 0 \iff |\mathbf{T}_{\mathbf{xx}}| \neq 0$$
$$\iff \mathrm{r}(\mathbf{X}_0) = k \ \&\ \mathbf{1} \notin \mathscr{C}(\mathbf{X}_0)$$

19.17 $r_{ij} = 1$ for some $i \neq j$ implies that $|\mathbf{R}_{\mathbf{xx}}| = 0$ (but not vice versa)

19.18 (a) The columns of $\mathbf{X} = (\mathbf{1} : \mathbf{X}_0)$ are orthogonal iff

(b) the columns of $\mathbf{X}_0$ are centered and $\operatorname{cor}_d(\mathbf{X}_0) = \mathbf{R}_{\mathbf{xx}} = \mathbf{I}_k$.

19.19 The statements (a) the columns of $\mathbf{X}_0$ are orthogonal, (b) $\operatorname{cor}_d(\mathbf{X}_0) = \mathbf{I}_k$ (i.e., orthogonality and uncorrelatedness) are equivalent if $\mathbf{X}_0$ is centered.

19.20 It is possible that

(a) $\cos(\mathbf{x}, \mathbf{y})$ is high but $\operatorname{cor}_d(\mathbf{x}, \mathbf{y}) = 0$,

(b) $\cos(\mathbf{x}, \mathbf{y}) = 0$ but $\operatorname{cor}_d(\mathbf{x}, \mathbf{y}) = 1$.

19.21 $\operatorname{cor}_d(\mathbf{x}, \mathbf{y}) = 0 \iff \mathbf{y} \in \mathscr{C}(\mathbf{Cx})^\perp = \mathscr{C}(\mathbf{1} : \mathbf{x})^\perp \boxplus \mathscr{C}(\mathbf{1})$
[exclude the cases when $\mathbf{x} \in \mathscr{C}(\mathbf{1})$ and/or $\mathbf{y} \in \mathscr{C}(\mathbf{1})$]

19.22 $\mathbf{T} = \begin{pmatrix} \mathbf{T}_{\mathbf{xx}} & \mathbf{t}_{\mathbf{x}y} \\ \mathbf{t}'_{\mathbf{x}y} & t_{yy} \end{pmatrix}, \quad \mathbf{S} = \begin{pmatrix} \mathbf{S}_{\mathbf{xx}} & \mathbf{s}_{\mathbf{x}y} \\ \mathbf{s}'_{\mathbf{x}y} & s_{yy} \end{pmatrix},$

$$\mathbf{R} = \begin{pmatrix} \mathbf{R}_{\mathbf{xx}} & \mathbf{r}_{\mathbf{x}y} \\ \mathbf{r}'_{\mathbf{x}y} & 1 \end{pmatrix} = \left(\begin{array}{cc|c} \mathbf{R}_{11} & \mathbf{r}_{12} & \\ \mathbf{r}'_{12} & 1 & \mathbf{r}_{\mathbf{x}y} \\ \hline \multicolumn{2}{c|}{\mathbf{r}'_{\mathbf{x}y}} & 1 \end{array}\right)$$

19.23 $|\mathbf{R}| = |\mathbf{R}_{\mathbf{xx}}|(1 - \mathbf{r}'_{\mathbf{x}y}\mathbf{R}_{\mathbf{xx}}^{-1}\mathbf{r}_{\mathbf{x}y}) = |\mathbf{R}_{\mathbf{xx}}|(1 - R^2_{y\cdot\mathbf{x}}) \le 1$

19.24 $|\mathbf{R}_{\mathbf{xx}}| = |\mathbf{R}_{11}|(1 - \mathbf{r}'_{12}\mathbf{R}_{11}^{-1}\mathbf{r}_{12}) = |\mathbf{R}_{11}|(1 - R^2_{k\cdot 1\dots k-1})$

19.25 $|\mathbf{R}| = |\mathbf{R}_{11}|(1 - R^2_{k\cdot 1\dots k-1})(1 - R^2_{y\cdot\mathbf{x}}), \quad R^2_{y\cdot\mathbf{x}} = R^2(\mathbf{y}; \mathbf{X})$

19.26 $|\mathbf{R}| = (1 - r^2_{12})(1 - R^2_{3\cdot 12})(1 - R^2_{4\cdot 123})\cdots(1 - R^2_{k\cdot 123\dots k-1})(1 - R^2_{y\cdot\mathbf{x}})$

19.27 $1 - R^2_{y\cdot\mathbf{x}} = \dfrac{|\mathbf{R}|}{|\mathbf{R}_{\mathbf{xx}}|} = (1 - r^2_{y1})(1 - r^2_{y2\cdot 1})(1 - r^2_{y3\cdot 12})\cdots(1 - r^2_{yk\cdot 12\dots k-1})$

19.28 $\mathbf{R}^{-1} = \begin{pmatrix} \mathbf{R}^{\mathbf{xx}} & \mathbf{r}^{\mathbf{x}y} \\ (\mathbf{r}^{\mathbf{x}y})' & r^{yy} \end{pmatrix} = \{\underline{r}^{ij}\}, \quad \mathbf{R}_{\mathbf{xx}}^{-1} = \begin{pmatrix} \mathbf{R}^{11} & \mathbf{r}^{12} \\ (\mathbf{r}^{12})' & r^{kk} \end{pmatrix} = \{r^{ij}\}$

19.29 $\mathbf{T}^{-1} = \begin{pmatrix} \mathbf{T}^{\mathbf{xx}} & \mathbf{t}^{\mathbf{x}y} \\ (\mathbf{t}^{\mathbf{x}y})' & t^{yy} \end{pmatrix} = \{\underline{t}^{ij}\}, \quad \mathbf{T}_{\mathbf{xx}}^{-1} = \begin{pmatrix} \mathbf{T}^{11} & \mathbf{t}^{12} \\ (\mathbf{t}^{12})' & t^{kk} \end{pmatrix} = \{t^{ij}\}$

19.30 $\mathbf{R}^{\mathbf{xx}} = (\mathbf{R}_{\mathbf{xx}} - \mathbf{r}_{\mathbf{x}y}\mathbf{r}'_{\mathbf{x}y})^{-1} = \mathbf{R}_{\mathbf{xx}}^{-1} + \mathbf{R}_{\mathbf{xx}}^{-1}\mathbf{r}_{\mathbf{x}y}\mathbf{r}'_{\mathbf{x}y}\mathbf{R}_{\mathbf{xx}}^{-1}/(1 - R^2)$

19.31 $t^{yy} = 1/\mathrm{SSE} =$ the last diagonal element of $\mathbf{T}^{-1}$

19.32 $t^{ii} = 1/\mathrm{SSE}(\mathbf{x}_i; \mathbf{X}_{(-i)}) =$ the ith diagonal element of $\mathbf{T}_{\mathbf{xx}}^{-1}$

19.33 $r^{yy} = \dfrac{1}{1 - \mathbf{r}'_{\mathbf{x}y}\mathbf{R}_{\mathbf{xx}}^{-1}\mathbf{r}_{\mathbf{x}y}} = \dfrac{1}{1 - R^2_{y\cdot\mathbf{x}}} =$ the last diagonal element of $\mathbf{R}^{-1}$

19.34 $$r^{ii} = \frac{1}{1 - R_i^2}$$
$$= \text{VIF}_i = \text{the } i\text{th diagonal element of } \mathbf{R}_{\mathbf{xx}}^{-1}, \quad R_i^2 = R^2(\mathbf{x}_i; \mathbf{X}_{(-i)})$$

19.35 Assuming that appropriate inverses exist:

(a) $(\mathbf{B} - \mathbf{C}\mathbf{D}^{-1}\mathbf{C}')^{-1} = \mathbf{B}^{-1} + \mathbf{B}^{-1}\mathbf{C}\mathbf{S}^{-1}\mathbf{C}'\mathbf{B}^{-1}$, where $\mathbf{S} = \mathbf{D} - \mathbf{C}'\mathbf{B}^{-1}\mathbf{C}$.

(b) $$(\mathbf{B} + \mathbf{u}\mathbf{v}')^{-1} = \mathbf{B}^{-1} - \frac{\mathbf{B}^{-1}\mathbf{u}\mathbf{v}'\mathbf{B}^{-1}}{1 + \mathbf{v}'\mathbf{B}^{-1}\mathbf{u}}.$$

(c) $$(\mathbf{B} + k\mathbf{i}_i\mathbf{i}_j')^{-1} = \mathbf{B}^{-1} - \frac{1}{1 + k\mathbf{i}_j'\mathbf{B}^{-1}\mathbf{i}_i}\mathbf{B}^{-1}\mathbf{i}_i\mathbf{i}_j'\mathbf{B}^{-1}$$
$$= \mathbf{B}^{-1} - \frac{1}{1 + kb^{(ji)}}\mathbf{b}^i(\mathbf{b}^j)'.$$

(d) $$[\mathbf{X}'(\mathbf{I} - \mathbf{i}_i\mathbf{i}_i')\mathbf{X}]^{-1} = (\mathbf{X}_{(i)}'\mathbf{X}_{(i)})^{-1}$$
$$= (\mathbf{X}'\mathbf{X})^{-1} + \frac{1}{1 - h_{ii}}(\mathbf{X}'\mathbf{X})^{-1}\mathbf{x}_{(i)}\mathbf{x}_{(i)}'(\mathbf{X}'\mathbf{X})^{-1}.$$

(e) $(\mathbf{A} + k\mathbf{I})^{-1} = \mathbf{A}^{-1} - k\mathbf{A}^{-1}(\mathbf{I} + k\mathbf{A}^{-1})^{-1}\mathbf{A}^{-1}$.

20 Generalized inverses

In what follows, let $\mathbf{A} \in \mathbb{R}^{n\times m}$.

20.1 (mp1) $\mathbf{AGA} = \mathbf{A}$, (mp1) $\iff \mathbf{G} \in \{\mathbf{A}^-\}$

(mp2) $\mathbf{GAG} = \mathbf{G}$, (mp1) & (mp2): $\mathbf{G} \in \{\mathbf{A}_r^-\}$ reflexive g-inverse

(mp3) $(\mathbf{AG})' = \mathbf{AG}$, (mp1) & (mp3): $\mathbf{G} \in \{\mathbf{A}_\ell^-\}$: least squares g-i

(mp4) $(\mathbf{GA})' = \mathbf{GA}$, (mp1) & (mp4): $\mathbf{G} \in \{\mathbf{A}_m^-\}$: minimum norm g-i

All four conditions $\iff \mathbf{G} = \mathbf{A}^+$: Moore–Penrose inverse (unique)

20.2 The matrix $\mathbf{G}_{m\times n}$ is a generalized inverse of $\mathbf{A}_{n\times m}$ if any of the following equivalent conditions holds:

(a) The vector $\mathbf{Gy}$ is a solution to $\mathbf{Ab} = \mathbf{y}$ always when this equation is consistent, i.e., always when $\mathbf{y} \in \mathscr{C}(\mathbf{A})$.

(b1) $\mathbf{GA}$ is idempotent and $\text{r}(\mathbf{GA}) = \text{r}(\mathbf{A})$, or equivalently

(b2) $\mathbf{AG}$ is idempotent and $\text{r}(\mathbf{AG}) = \text{r}(\mathbf{A})$.

(c) $\mathbf{AGA} = \mathbf{A}$. (mp1)

20.3 $\mathbf{AGA} = \mathbf{A}$ & $\mathrm{r}(\mathbf{G}) = \mathrm{r}(\mathbf{A}) \iff \mathbf{G}$ is a reflexive generalized inverse of $\mathbf{A}$

20.4 A general solution to a consistent (solvable) equation $\mathbf{Ax} = \mathbf{y}$ is

$$\mathbf{A}^{-}\mathbf{y} + (\mathbf{I}_m - \mathbf{A}^{-}\mathbf{A})\mathbf{z}, \text{ where the vector } \mathbf{z} \in \mathbb{R}^m \text{ is free to vary,}$$

and $\mathbf{A}^{-}$ is an arbitrary (but fixed) generalized inverse.

20.5 The class of all solutions to a consistent equation $\mathbf{Ax} = \mathbf{y}$ is $\{\mathbf{Gy}\}$, where $\mathbf{G}$ varies through all generalized inverses of $\mathbf{A}$.

20.6 The equation $\mathbf{Ax} = \mathbf{y}$ is consistent iff $\mathbf{y} \in \mathscr{C}(\mathbf{A})$ iff $[\mathbf{A}'\mathbf{u} = \mathbf{0} \implies \mathbf{y}'\mathbf{u} = 0]$.

20.7 The equation $\mathbf{AX} = \mathbf{Y}$ has a solution (in $\mathbf{X}$) iff $\mathscr{C}(\mathbf{Y}) \subset \mathscr{C}(\mathbf{A})$ in which case the general solution is

$$\mathbf{A}^{-}\mathbf{Y} + (\mathbf{I}_m - \mathbf{A}^{-}\mathbf{A})\mathbf{Z}, \text{ where } \mathbf{Z} \text{ is free to vary.}$$

20.8 A necessary and sufficient condition for the equation $\mathbf{AXB} = \mathbf{C}$ to have a solution is that $\mathbf{AA}^{-}\mathbf{CB}^{-}\mathbf{B} = \mathbf{C}$, in which case the general solution is

$$\mathbf{X} = \mathbf{A}^{-}\mathbf{CB}^{-} + \mathbf{Z} - \mathbf{A}^{-}\mathbf{AZBB}^{-}, \text{ where } \mathbf{Z} \text{ is free to vary.}$$

20.9 Two alternative representations of a general solution to g-inverse of $\mathbf{A}$ are

(a) $\mathbf{G} = \mathbf{A}^{-} + \mathbf{U} - \mathbf{A}^{-}\mathbf{AUAA}^{-}$,

(b) $\mathbf{G} = \mathbf{A}^{-} + \mathbf{V}(\mathbf{I}_n - \mathbf{AA}^{-}) + (\mathbf{I}_m - \mathbf{A}^{-}\mathbf{A})\mathbf{W}$,

where $\mathbf{A}^{-}$ is a particular g-inverse and $\mathbf{U}, \mathbf{V}, \mathbf{W}$ are free to vary. In particular, choosing $\mathbf{A}^{-} = \mathbf{A}^{+}$, the general representations can be expressed as

(c) $\mathbf{G} = \mathbf{A}^{+} + \mathbf{U} - \mathbf{P}_{\mathbf{A}'}\mathbf{UP}_{\mathbf{A}}$, (d) $\mathbf{G} = \mathbf{A}^{+} + \mathbf{V}(\mathbf{I}_n - \mathbf{P}_{\mathbf{A}}) + (\mathbf{I}_m - \mathbf{P}_{\mathbf{A}'})\mathbf{W}$.

20.10 Let $\mathbf{A} \neq \mathbf{0}$, $\mathbf{C} \neq \mathbf{0}$. Then $\mathbf{AB}^{-}\mathbf{C}$ is invariant w.r.t. the choice of $\mathbf{B}^{-}$ iff $\mathscr{C}(\mathbf{C}) \subset \mathscr{C}(\mathbf{B})$ & $\mathscr{C}(\mathbf{A}') \subset \mathscr{C}(\mathbf{B}')$.

20.11 $\mathbf{AA}^{-}\mathbf{C} = \mathbf{C}$ for some $\mathbf{A}^{-}$ iff $\mathbf{AA}^{-}\mathbf{C} = \mathbf{C}$ for all $\mathbf{A}^{-}$ iff $\mathbf{AA}^{-}\mathbf{C}$ is invariant w.r.t. the choice of $\mathbf{A}^{-}$ iff $\mathscr{C}(\mathbf{C}) \subset \mathscr{C}(\mathbf{A})$.

20.12 $\mathbf{AB}^{-}\mathbf{B} = \mathbf{A} \iff \mathscr{C}(\mathbf{A}') \subset \mathscr{C}(\mathbf{B}')$

20.13 Let $\mathbf{C} \neq \mathbf{0}$. Then $\mathscr{C}(\mathbf{AB}^{-}\mathbf{C})$ is invariant w.r.t. the choice of $\mathbf{B}^{-}$ iff $\mathscr{C}(\mathbf{A}') \subset \mathscr{C}(\mathbf{B}')$ holds along with $\mathscr{C}(\mathbf{C}) \subset \mathscr{C}(\mathbf{B})$ or along with $\mathrm{r}(\mathbf{AB}^{+}\mathbf{L}) = \mathrm{r}(\mathbf{A})$, where $\mathbf{L}$ is any matrix such that $\mathscr{C}(\mathbf{L}) = \mathscr{C}(\mathbf{B}) \cap \mathscr{C}(\mathbf{C})$.

20.14 $\operatorname{rank}(\mathbf{AB^-C})$ is invariant w.r.t. the choice of $\mathbf{B}^-$ iff at least one of the column spaces $\mathscr{C}(\mathbf{AB^-C})$ and $\mathscr{C}(\mathbf{C}'(\mathbf{B}')^-\mathbf{A}')$ is invariant w.r.t. the choice of $\mathbf{B}^-$.

20.15 $\mathrm{r}(\mathbf{A}) = \mathrm{r}(\mathbf{AA^-A}) \le \mathrm{r}(\mathbf{AA^-}) \le \mathrm{r}(\mathbf{A^-}), \quad \mathrm{r}(\mathbf{A^+}) = \mathrm{r}(\mathbf{A})$

20.16 $\mathscr{C}(\mathbf{AA^-}) = \mathscr{C}(\mathbf{A})$ but $\mathscr{C}(\mathbf{A^-A}) = \mathscr{C}(\mathbf{A}') \iff \mathbf{A}^- \in \{\mathbf{A}_{14}^-\}$

20.17 $\mathscr{C}(\mathbf{I}_m - \mathbf{A^-A}) = \mathscr{N}(\mathbf{A}), \quad \mathscr{N}(\mathbf{I}_n - \mathbf{AA^-}) = \mathscr{C}(\mathbf{A})$

20.18 $\mathscr{C}(\mathbf{A}^+) = \mathscr{C}(\mathbf{A}')$

20.19 $\mathbf{A}^+ = (\mathbf{A}'\mathbf{A})^+\mathbf{A}' = \mathbf{A}'(\mathbf{AA}')^+$

20.20 $(\mathbf{A}^+)' = (\mathbf{A}')^+, \quad (\mathbf{A}^-)' \in \{(\mathbf{A}')^-\}$

20.21 If $\mathbf{A}$ has a full rank decomposition $\mathbf{A} = \mathbf{UV}'$ then

$$\mathbf{A}^+ = \mathbf{V}(\mathbf{V}'\mathbf{V})^{-1}(\mathbf{U}'\mathbf{U})^{-1}\mathbf{U}'.$$

20.22 $(\mathbf{AB})^+ = \mathbf{B}^+\mathbf{A}^+ \iff \mathscr{C}(\mathbf{BB}'\mathbf{A}') \subset \mathscr{C}(\mathbf{A}')\ \&\ \mathscr{C}(\mathbf{A}'\mathbf{AB}) \subset \mathscr{C}(\mathbf{B})$

20.23 Let $\mathbf{A}_{n\times m}$ have a singular value decomposition $\mathbf{A} = \mathbf{U}\begin{pmatrix}\mathbf{\Delta}_1 & \mathbf{0}\\ \mathbf{0} & \mathbf{0}\end{pmatrix}\mathbf{V}'$. Then:

$$\mathbf{G} \in \{\mathbf{A}^-\} \iff \mathbf{G} = \mathbf{V}\begin{pmatrix}\mathbf{\Delta}_1^{-1} & \mathbf{K}\\ \mathbf{L} & \mathbf{N}\end{pmatrix}\mathbf{U}',$$

$$\mathbf{G} \in \{\mathbf{A}_{12}^-\} \iff \mathbf{G} = \mathbf{V}\begin{pmatrix}\mathbf{\Delta}_1^{-1} & \mathbf{K}\\ \mathbf{L} & \mathbf{L\Delta}_1\mathbf{K}\end{pmatrix}\mathbf{U}',$$

$$\mathbf{G} \in \{\mathbf{A}_{13}^-\} = \{\mathbf{A}_\ell^-\} \iff \mathbf{G} = \mathbf{V}\begin{pmatrix}\mathbf{\Delta}_1^{-1} & \mathbf{0}\\ \mathbf{L} & \mathbf{N}\end{pmatrix}\mathbf{U}',$$

$$\mathbf{G} \in \{\mathbf{A}_{14}^-\} = \{\mathbf{A}_m^-\} \iff \mathbf{G} = \mathbf{V}\begin{pmatrix}\mathbf{\Delta}_1^{-1} & \mathbf{K}\\ \mathbf{0} & \mathbf{N}\end{pmatrix}\mathbf{U}',$$

$$\mathbf{G} = \mathbf{A}^+ \iff \mathbf{G} = \mathbf{V}\begin{pmatrix}\mathbf{\Delta}_1^{-1} & \mathbf{0}\\ \mathbf{0} & \mathbf{0}\end{pmatrix}\mathbf{U}',$$

where $\mathbf{K}$, $\mathbf{L}$, and $\mathbf{N}$ are arbitrary matrices.

20.24 Let $\mathbf{A} = \mathbf{T\Lambda T}' = \mathbf{T}_1\mathbf{\Lambda}_1\mathbf{T}_1'$ be the EVD of $\mathbf{A}$ with $\mathbf{\Lambda}_1$ comprising the nonzero eigenvalues. Then $\mathbf{G}$ is a symmetric and nnd reflexive g-inverse of $\mathbf{A}$ iff

$$\mathbf{G} = \mathbf{T}\begin{pmatrix}\mathbf{\Lambda}_1^{-1} & \mathbf{L}\\ \mathbf{L}' & \mathbf{L}'\mathbf{\Lambda}_1\mathbf{L}\end{pmatrix}\mathbf{T}', \quad \text{where } \mathbf{L} \text{ is an arbitrary matrix.}$$

20.25 If $\mathbf{A}_{n\times m} = \begin{pmatrix}\mathbf{B} & \mathbf{C}\\ \mathbf{D} & \mathbf{E}\end{pmatrix}$ and $\mathrm{r}(\mathbf{A}) = r = \mathrm{r}(\mathbf{B}_{r\times r})$, then

$$\mathbf{G}_{m\times n} = \begin{pmatrix} \mathbf{B}^{-1} & \mathbf{0} \\ \mathbf{0} & \mathbf{0} \end{pmatrix} \in \{\mathbf{A}^-\}.$$

20.26 In (a)–(c) below we consider a nonnegative definite $\mathbf{A}$ partitioned as

$$\mathbf{A} = \mathbf{L}'\mathbf{L} = \begin{pmatrix} \mathbf{L}_1'\mathbf{L}_1 & \mathbf{L}_1'\mathbf{L}_2 \\ \mathbf{L}_2'\mathbf{L}_1 & \mathbf{L}_2'\mathbf{L}_2 \end{pmatrix} = \begin{pmatrix} \mathbf{A}_{11} & \mathbf{A}_{12} \\ \mathbf{A}_{21} & \mathbf{A}_{22} \end{pmatrix}.$$

(a) Block diagonalization of a nonnegative definite matrix:

(i) $$(\mathbf{L}_1 : \mathbf{L}_2)\begin{pmatrix} \mathbf{I} & -(\mathbf{L}_1'\mathbf{L}_1)^-\mathbf{L}_1'\mathbf{L}_2 \\ \mathbf{0} & \mathbf{I} \end{pmatrix} = \mathbf{L}\begin{pmatrix} \mathbf{I} & -\mathbf{A}_{11}^-\mathbf{A}_{12} \\ \mathbf{0} & \mathbf{I} \end{pmatrix} := \mathbf{L}\mathbf{U} = [\mathbf{L}_1 : (\mathbf{I} - \mathbf{P}_{\mathbf{L}_1})\mathbf{L}_2],$$

(ii) $$\mathbf{U}'\mathbf{L}'\mathbf{L}\mathbf{U} = \mathbf{U}'\mathbf{A}\mathbf{U} = \begin{pmatrix} \mathbf{L}_1'\mathbf{L}_1 & \mathbf{0} \\ \mathbf{0} & \mathbf{L}_2'(\mathbf{I} - \mathbf{P}_{\mathbf{L}_1})\mathbf{L}_2 \end{pmatrix},$$

(iii) $$\begin{pmatrix} \mathbf{I} & \mathbf{0} \\ -\mathbf{A}_{21}\mathbf{A}_{11}^{=} & \mathbf{I} \end{pmatrix}\cdot\mathbf{A}\cdot\begin{pmatrix} \mathbf{I} & -\mathbf{A}_{11}^{\sim}\mathbf{A}_{12} \\ \mathbf{0} & \mathbf{I} \end{pmatrix} = \begin{pmatrix} \mathbf{A}_{11} & \mathbf{0} \\ \mathbf{0} & \mathbf{A}_{22} - \mathbf{A}_{21}\mathbf{A}_{11}^-\mathbf{A}_{12} \end{pmatrix},$$

(iv) $$\mathbf{A} = \begin{pmatrix} \mathbf{I} & \mathbf{0} \\ \mathbf{A}_{21}\mathbf{A}_{11}^{=} & \mathbf{I} \end{pmatrix}\begin{pmatrix} \mathbf{A}_{11} & \mathbf{0} \\ \mathbf{0} & \mathbf{A}_{22} - \mathbf{A}_{21}\mathbf{A}_{11}^-\mathbf{A}_{12} \end{pmatrix}\begin{pmatrix} \mathbf{I} & \mathbf{A}_{11}^{\sim}\mathbf{A}_{12} \\ \mathbf{0} & \mathbf{I} \end{pmatrix},$$

where $\mathbf{A}_{11}^{=}$, $\mathbf{A}_{11}^-$, and $\mathbf{A}_{11}^{\sim}$ are arbitrary generalized inverses of $\mathbf{A}_{11}$, and $\mathbf{A}_{22\cdot1} = \mathbf{A}_{22} - \mathbf{A}_{21}\mathbf{A}_{11}^-\mathbf{A}_{12} =$ the Schur complement of $\mathbf{A}_{11}$ in $\mathbf{A}$.

(b) The matrix $\mathbf{A}^\#$ is one generalized inverse of $\mathbf{A}$:

$$\mathbf{A}^\# = \begin{pmatrix} \mathbf{A}_{11\cdot2}^{\sim} & -\mathbf{A}_{11\cdot2}^{\sim}\mathbf{A}_{12}\mathbf{A}_{22}^{\sim} \\ -\mathbf{A}_{22}^{\sim}\mathbf{A}_{21}\mathbf{A}_{11\cdot2}^{\sim} & \mathbf{A}_{22}^{\sim} + \mathbf{A}_{22}^{\sim}\mathbf{A}_{21}\mathbf{A}_{11\cdot2}^{\sim}\mathbf{A}_{12}\mathbf{A}_{22}^{\sim} \end{pmatrix},$$

where $\mathbf{A}_{11\cdot2} = \mathbf{L}_1'\mathbf{Q}_2\mathbf{L}_1$, with $\mathbf{B}^{\sim}$ denoting a g-inverse of $\mathbf{B}$ and $\mathbf{Q}_2 = \mathbf{I} - \mathbf{P}_{\mathbf{L}_2}$. In particular, the matrix $\mathbf{A}^\#$ is a symmetric reflexive g-inverse of $\mathbf{A}$ for any choices of symmetric reflexive g-inverses $(\mathbf{L}_2'\mathbf{L}_2)^{\sim}$ and $(\mathbf{L}_1'\mathbf{Q}_2\mathbf{L}_1)^{\sim}$. We say that $\mathbf{A}^\#$ is in Banachiewicz–Schur form.

(c) If any of the following conditions hold, the all three hold:

(i) $\mathrm{r}(\mathbf{A}) = \mathrm{r}(\mathbf{A}_{11}) + \mathrm{r}(\mathbf{A}_{22})$, i.e., $\mathscr{C}(\mathbf{L}_1) \cap \mathscr{C}(\mathbf{L}_2) = \{\mathbf{0}\}$,

(ii) $$\mathbf{A}^+ = \begin{pmatrix} \mathbf{A}_{11\cdot2}^+ & -\mathbf{A}_{11\cdot2}^+\mathbf{A}_{12}\mathbf{A}_{22}^+ \\ -\mathbf{A}_{22}^+\mathbf{A}_{21}\mathbf{A}_{11\cdot2}^+ & \mathbf{A}_{22}^+ + \mathbf{A}_{22}^+\mathbf{A}_{21}\mathbf{A}_{11\cdot2}^+\mathbf{A}_{12}\mathbf{A}_{22}^+ \end{pmatrix},$$

(iii) $$\mathbf{A}^+ = \begin{pmatrix} \mathbf{A}_{11}^+ + \mathbf{A}_{11}^+\mathbf{A}_{12}\mathbf{A}_{22\cdot1}^+\mathbf{A}_{21}\mathbf{A}_{11}^+ & -\mathbf{A}_{11}^+\mathbf{A}_{12}\mathbf{A}_{22\cdot1}^+ \\ -\mathbf{A}_{22\cdot1}^+\mathbf{A}_{21}\mathbf{A}_{11}^+ & \mathbf{A}_{22\cdot1}^+ \end{pmatrix}.$$

20.27 Consider $\mathbf{A}_{n\times m}$ and $\mathbf{B}_{n\times m}$ and let the pd inner product matrix be $\mathbf{V}$. Then

$$\|\mathbf{Ax}\|_\mathbf{V} \le \|\mathbf{Ax} + \mathbf{By}\|_\mathbf{V} \ \forall\, \mathbf{x}, \mathbf{y} \in \mathbb{R}^m \iff \mathbf{A}'\mathbf{V}\mathbf{B} = \mathbf{0}.$$

The statement above holds also for nnd $\mathbf{V}$, i.e., $\mathbf{t}'\mathbf{V}\mathbf{u}$ is a semi-inner product.

20.28 Let $\mathbf{G}$ be a g-inverse of $\mathbf{A}$ such that $\mathbf{Gy}$ is a minimum norm solution (w.r.t. standard norm) of $\mathbf{Ax} = \mathbf{y}$ for any $\mathbf{y} \in \mathscr{C}(\mathbf{A})$. Then it is necessary and sufficient that $\mathbf{AGA} = \mathbf{A}$ and $(\mathbf{GA})' = \mathbf{GA}$, i.e., $\mathbf{G} \in \{\mathbf{A}_{14}^-\}$. Such a $\mathbf{G}$ is called a minimum norm g-inverse and denoted as $\mathbf{A}_m^-$.

20.29 Let $\mathbf{A} \in \mathbb{R}^{n\times m}$ and let the inner product matrix be $\mathbf{N} \in \mathrm{PD}_n$ and denote a minimum norm g-inverse as $\mathbf{A}_{m(\mathbf{N})}^- \in \mathbb{R}^{m\times n}$. Then the following statements are equivalent:

(a) $\mathbf{G} = \mathbf{A}_{m(\mathbf{N})}^-$,

(b) $\mathbf{AGA} = \mathbf{A}$, $(\mathbf{GA})'\mathbf{N} = \mathbf{NGA}$ (here $\mathbf{N}$ can be nnd),

(c) $\mathbf{GAN}^{-1}\mathbf{A}' = \mathbf{N}^{-1}\mathbf{A}'$,

(d) $\mathbf{GA} = \mathbf{P}_{\mathbf{N}^{-1}\mathbf{A}';\mathbf{N}}$, i.e., $(\mathbf{GA})' = \mathbf{P}_{\mathbf{A}';\mathbf{N}^{-1}}$.

20.30 (a) $(\mathbf{N} + \mathbf{A}'\mathbf{A})^-\mathbf{A}'[\mathbf{A}'(\mathbf{N} + \mathbf{A}'\mathbf{A})^-\mathbf{A}]^- \in \{\mathbf{A}_{m(\mathbf{N})}^-\}$,

(b) $\mathscr{C}(\mathbf{A}') \subset \mathscr{C}(\mathbf{N}) \implies \mathbf{N}^-\mathbf{A}'(\mathbf{A}'\mathbf{N}^-\mathbf{A})^- \in \{\mathbf{A}_{m(\mathbf{N})}^-\}$.

20.31 Let $\mathbf{G}$ be a matrix (not necessarily a g-inverse) such that $\mathbf{Gy}$ is a least-squares solution (w.r.t. standard norm) of $\mathbf{Ax} = \mathbf{y}$ for any $\mathbf{y}$, that is

$$\|\mathbf{y} - \mathbf{AGy}\| \leq \|\mathbf{y} - \mathbf{Ax}\| \quad \text{for all } \mathbf{x} \in \mathbb{R}^m,\ \mathbf{y} \in \mathbb{R}^n.$$

Then it is necessary and sufficient that $\mathbf{AGA} = \mathbf{A}$ and $(\mathbf{AG})' = \mathbf{AG}$, that is, $\mathbf{G} \in \{\mathbf{A}_{13}^-\}$. Such a $\mathbf{G}$ is called a least-squares g-inverse and denoted as $\mathbf{A}_\ell^-$.

20.32 Let the inner product matrix be $\mathbf{V} \in \mathrm{PD}_n$ and denote a least-squares g-inverse as $\mathbf{A}_{\ell(\mathbf{V})}^-$. Then the following statements are equivalent:

(a) $\mathbf{G} = \mathbf{A}_{\ell(\mathbf{V})}^-$,

(b) $\mathbf{AGA} = \mathbf{A}$, $(\mathbf{AG})'\mathbf{V} = \mathbf{VAG}$,

(c) $\mathbf{A}'\mathbf{VAG} = \mathbf{A}'\mathbf{V}$ (here $\mathbf{V}$ can be nnd),

(d) $\mathbf{AG} = \mathbf{P}_{\mathbf{A};\mathbf{V}}$.

20.33 $(\mathbf{A}'\mathbf{VA})^-\mathbf{A}'\mathbf{V} \in \{\mathbf{A}_{\ell(\mathbf{V})}^-\}$.

20.34 Let the inner product matrix $\mathbf{V}_{n\times n}$ be pd. Then $\{(\mathbf{A}')_{m(\mathbf{V})}^-\} = \{(\mathbf{A}_{\ell(\mathbf{V}^{-1})}^-)'\}$.

20.35 The minimum norm solution for $\mathbf{X}'\mathbf{a} = \mathbf{k}$ is $\tilde{\mathbf{a}} = (\mathbf{X}')_{m(\mathbf{V})}^-\mathbf{k} := \mathbf{Gk}$, where

$$(\mathbf{X}')_{m(\mathbf{V})}^- := \mathbf{G} = \mathbf{W}^-\mathbf{X}(\mathbf{X}'\mathbf{W}^-\mathbf{X})^-; \quad \mathbf{W} = \mathbf{V} + \mathbf{XX}'.$$

Furthermore, $\mathrm{BLUE}(\mathbf{k}'\boldsymbol{\beta}) = \tilde{\mathbf{a}}'\mathbf{y} = \mathbf{k}'\mathbf{G}'\mathbf{y} = \mathbf{k}'(\mathbf{X}'\mathbf{W}^-\mathbf{X})^-\mathbf{X}'\mathbf{W}^-\mathbf{y}$.

21 Projectors

21.1 Orthogonal projector. Let $\mathbf{A} \in \mathbb{R}_r^{n\times m}$ and let the inner product and the corresponding norm be defined as $\langle \mathbf{t}, \mathbf{u}\rangle = \mathbf{t}'\mathbf{u}$, and $\|\mathbf{t}\| = \sqrt{\langle \mathbf{t}, \mathbf{t}\rangle}$, respectively. Further, let $\mathbf{A}^\perp \in \mathbb{R}^{n\times q}$ be a matrix spanning $\mathscr{C}(\mathbf{A})^\perp = \mathscr{N}(\mathbf{A}')$, and the columns of the matrices $\mathbf{A}_b \in \mathbb{R}^{n\times r}$ and $\mathbf{A}_b^\perp \in \mathbb{R}^{n\times(n-r)}$ form bases for $\mathscr{C}(\mathbf{A})$ and $\mathscr{C}(\mathbf{A})^\perp$, respectively. Then the following conditions are equivalent ways to define the unique matrix $\mathbf{P}$:

(a) The matrix $\mathbf{P}$ transforms every $\mathbf{y} \in \mathbb{R}^n$,

$$\mathbf{y} = \mathbf{y}_A + \mathbf{y}_{A^\perp}, \quad \mathbf{y}_A \in \mathscr{C}(\mathbf{A}), \quad \mathbf{y}_{A^\perp} \in \mathscr{C}(\mathbf{A})^\perp,$$

into its projection onto $\mathscr{C}(\mathbf{A})$ along $\mathscr{C}(\mathbf{A})^\perp$; that is, for each $\mathbf{y}$ above, the multiplication $\mathbf{Py}$ gives the projection $\mathbf{y}_A$: $\mathbf{Py} = \mathbf{y}_A$.

(b) $\mathbf{P}(\mathbf{Ab} + \mathbf{A}^\perp\mathbf{c}) = \mathbf{Ab}$ for all $\mathbf{b} \in \mathbb{R}^m$, $\mathbf{c} \in \mathbb{R}^q$.

(c) $\mathbf{P}(\mathbf{A} : \mathbf{A}^\perp) = (\mathbf{A} : \mathbf{0})$.

(d) $\mathbf{P}(\mathbf{A}_b : \mathbf{A}_b^\perp) = (\mathbf{A}_b : \mathbf{0})$.

(e) $\mathscr{C}(\mathbf{P}) \subset \mathscr{C}(\mathbf{A})$, $\min_{\mathbf{b}} \|\mathbf{y} - \mathbf{Ab}\|^2 = \|\mathbf{y} - \mathbf{Py}\|^2$ for all $\mathbf{y} \in \mathbb{R}^n$.

(f) $\mathscr{C}(\mathbf{P}) \subset \mathscr{C}(\mathbf{A})$, $\mathbf{P}'\mathbf{A} = \mathbf{A}$.

(g) $\mathscr{C}(\mathbf{P}) = \mathscr{C}(\mathbf{A})$, $\mathbf{P}'\mathbf{P} = \mathbf{P}$.

(h) $\mathscr{C}(\mathbf{P}) = \mathscr{C}(\mathbf{A})$, $\mathbf{P}^2 = \mathbf{P}$, $\mathbf{P}' = \mathbf{P}$.

(i) $\mathscr{C}(\mathbf{P}) = \mathscr{C}(\mathbf{A})$, $\mathbb{R}^n = \mathscr{C}(\mathbf{P}) \boxplus \mathscr{C}(\mathbf{I}_n - \mathbf{P})$.

(j) $\mathbf{P} = \mathbf{A}(\mathbf{A}'\mathbf{A})^-\mathbf{A}' = \mathbf{AA}^+$.

(k) $\mathbf{P} = \mathbf{A}_b(\mathbf{A}_b'\mathbf{A}_b)^{-1}\mathbf{A}_b'$.

(l) $\mathbf{P} = \mathbf{A}_o\mathbf{A}_o' = \mathbf{U}\left(\begin{smallmatrix}\mathbf{I}_r & \mathbf{0}\\ \mathbf{0} & \mathbf{0}\end{smallmatrix}\right)\mathbf{U}'$, where $\mathbf{U} = (\mathbf{A}_o : \mathbf{A}_o^\perp)$, the columns of $\mathbf{A}_o$ and $\mathbf{A}_o^\perp$ forming orthonormal bases for $\mathscr{C}(\mathbf{A})$ and $\mathscr{C}(\mathbf{A})^\perp$, respectively.

The matrix $\mathbf{P} = \mathbf{P_A}$ is the orthogonal projector onto the column space $\mathscr{C}(\mathbf{A})$ w.r.t. the inner product $\langle \mathbf{t}, \mathbf{u}\rangle = \mathbf{t}'\mathbf{u}$. Correspondingly, the matrix $\mathbf{I}_n - \mathbf{P_A}$ is the orthogonal projector onto $\mathscr{C}(\mathbf{A}^\perp)$: $\mathbf{I}_n - \mathbf{P_A} = \mathbf{P}_{\mathbf{A}^\perp}$.

21.2 In 21.3–21.10 we consider orthogonal projectors defined w.r.t. the standard inner product in $\mathbb{R}^n$ so that they are symmetric idempotent matrices. If $\mathbf{P}$ is idempotent but not necessarily symmetric, it is called an oblique projector or simply a projector.

21.3 $\mathbf{P_A} + \mathbf{P_B}$ is an orthogonal projector $\iff$ $\mathbf{A}'\mathbf{B} = \mathbf{0}$, in which case $\mathbf{P_A} + \mathbf{P_B}$ is the orthogonal projector onto $\mathscr{C}(\mathbf{A} : \mathbf{B})$.

21.4 $\mathbf{P}_{(\mathbf{A}:\mathbf{B})} = \mathbf{P}_\mathbf{A} + \mathbf{P}_{(\mathbf{I}-\mathbf{P}_\mathbf{A})\mathbf{B}}$

21.5 The following statements are equivalent:

(a) $\mathbf{P}_\mathbf{A} - \mathbf{P}_\mathbf{B}$ is an orthogonal projector, (b) $\mathbf{P}_\mathbf{A}\mathbf{P}_\mathbf{B} = \mathbf{P}_\mathbf{B}\mathbf{P}_\mathbf{A} = \mathbf{P}_\mathbf{B}$,

(c) $\|\mathbf{P}_\mathbf{A}\mathbf{x}\| \geq \|\mathbf{P}_\mathbf{B}\mathbf{x}\|$ for all $\mathbf{x} \in \mathbb{R}^n$, (d) $\mathbf{P}_\mathbf{A} - \mathbf{P}_\mathbf{B} \geq_\mathsf{L} \mathbf{0}$,

(e) $\mathscr{C}(\mathbf{B}) \subset \mathscr{C}(\mathbf{A})$.

If any of the above conditions hold, then $\mathbf{P}_\mathbf{A} - \mathbf{P}_\mathbf{B} = \mathbf{P}_{(\mathbf{I}-\mathbf{P}_\mathbf{B})\mathbf{A}} = \mathbf{P}_{\mathscr{C}(\mathbf{A})\cap\mathscr{C}(\mathbf{B})^\perp}$.

21.6 Let $\mathbf{L}$ be a matrix with property $\mathscr{C}(\mathbf{L}) = \mathscr{C}(\mathbf{A}) \cap \mathscr{C}(\mathbf{B})$. Then

(a) $\mathscr{C}(\mathbf{A}) = \mathscr{C}[\mathbf{L} : (\mathbf{I} - \mathbf{P}_\mathbf{L})\mathbf{A}] = \mathscr{C}(\mathbf{A}) \cap \mathscr{C}(\mathbf{B}) \boxplus \mathscr{C}[(\mathbf{I} - \mathbf{P}_\mathbf{L})\mathbf{A}]$,

(b) $\mathbf{P}_\mathbf{A} = \mathbf{P}_\mathbf{L} + \mathbf{P}_{(\mathbf{I}-\mathbf{P}_\mathbf{L})\mathbf{A}} = \mathbf{P}_\mathbf{L} + \mathbf{P}_{\mathscr{C}(\mathbf{A})\cap\mathscr{C}(\mathbf{L})^\perp}$,

(c) $\mathbf{P}_\mathbf{A}\mathbf{P}_\mathbf{B} = \mathbf{P}_\mathbf{L} + \mathbf{P}_{(\mathbf{I}-\mathbf{P}_\mathbf{L})\mathbf{A}}\mathbf{P}_{(\mathbf{I}-\mathbf{P}_\mathbf{L})\mathbf{B}}$,

(d) $(\mathbf{I} - \mathbf{P}_\mathbf{L})\mathbf{P}_\mathbf{A} = \mathbf{P}_\mathbf{A}(\mathbf{I} - \mathbf{P}_\mathbf{L}) = \mathbf{P}_\mathbf{A} - \mathbf{P}_\mathbf{L} = \mathbf{P}_{\mathscr{C}(\mathbf{A})\cap\mathscr{C}(\mathbf{L})^\perp}$,

(e) $\mathscr{C}[(\mathbf{I} - \mathbf{P}_\mathbf{L})\mathbf{A}] = \mathscr{C}[(\mathbf{I} - \mathbf{P}_\mathbf{L})\mathbf{P}_\mathbf{A}] = \mathscr{C}(\mathbf{A}) \cap \mathscr{C}(\mathbf{L})^\perp$,

(f) $\mathrm{r}(\mathbf{A}) = \dim \mathscr{C}(\mathbf{A}) \cap \mathscr{C}(\mathbf{B}) + \dim \mathscr{C}(\mathbf{A}) \cap \mathscr{C}(\mathbf{L})^\perp$.

21.7 Commuting projectors. Denote $\mathscr{C}(\mathbf{L}) = \mathscr{C}(\mathbf{A}) \cap \mathscr{C}(\mathbf{B})$. Then the following statements are equivalent;

(a) $\mathbf{P}_\mathbf{A}\mathbf{P}_\mathbf{B} = \mathbf{P}_\mathbf{B}\mathbf{P}_\mathbf{A}$,

(b) $\mathbf{P}_\mathbf{A}\mathbf{P}_\mathbf{B} = \mathbf{P}_{\mathscr{C}(\mathbf{A})\cap\mathscr{C}(\mathbf{B})} = \mathbf{P}_\mathbf{L}$,

(c) $\mathbf{P}_{(\mathbf{A}:\mathbf{B})} = \mathbf{P}_\mathbf{A} + \mathbf{P}_\mathbf{B} - \mathbf{P}_\mathbf{A}\mathbf{P}_\mathbf{B}$,

(d) $\mathscr{C}(\mathbf{A} : \mathbf{B}) \cap \mathscr{C}(\mathbf{B})^\perp = \mathscr{C}(\mathbf{A}) \cap \mathscr{C}(\mathbf{B})^\perp$,

(e) $\mathscr{C}(\mathbf{A}) = \mathscr{C}(\mathbf{A}) \cap \mathscr{C}(\mathbf{B}) \boxplus \mathscr{C}(\mathbf{A}) \cap \mathscr{C}(\mathbf{B})^\perp$,

(f) $\mathrm{r}(\mathbf{A}) = \dim \mathscr{C}(\mathbf{A}) \cap \mathscr{C}(\mathbf{B}) + \dim \mathscr{C}(\mathbf{A}) \cap \mathscr{C}(\mathbf{B})^\perp$,

(g) $\mathrm{r}(\mathbf{A}'\mathbf{B}) = \dim \mathscr{C}(\mathbf{A}) \cap \mathscr{C}(\mathbf{B})$,

(h) $\mathbf{P}_{(\mathbf{I}-\mathbf{P}_\mathbf{A})\mathbf{B}} = \mathbf{P}_\mathbf{B} - \mathbf{P}_\mathbf{A}\mathbf{P}_\mathbf{B}$,

(i) $\mathscr{C}(\mathbf{P}_\mathbf{A}\mathbf{B}) = \mathscr{C}(\mathbf{A}) \cap \mathscr{C}(\mathbf{B})$,

(j) $\mathscr{C}(\mathbf{P}_\mathbf{A}\mathbf{B}) \subset \mathscr{C}(\mathbf{B})$,

(k) $\mathbf{P}_{(\mathbf{I}-\mathbf{P}_\mathbf{L})\mathbf{A}}\mathbf{P}_{(\mathbf{I}-\mathbf{P}_\mathbf{L})\mathbf{B}} = \mathbf{0}$.

21.8 Let $\mathbf{P}_\mathbf{A}$ and $\mathbf{P}_\mathbf{B}$ be orthogonal projectors of order $n \times n$. Then

(a) $-1 \leq \mathrm{ch}_i(\mathbf{P}_\mathbf{A} - \mathbf{P}_\mathbf{B}) \leq 1, \quad i = 1, \dots, n$,

(b) $0 \leq \mathrm{ch}_i(\mathbf{P}_\mathbf{A}\mathbf{P}_\mathbf{B}) \leq 1, \quad i = 1, \dots, n$,

(c) $\operatorname{trace}(\mathbf{P_A P_B}) \leq \operatorname{r}(\mathbf{P_A P_B})$,

(d) $\#\{\operatorname{ch}_i(\mathbf{P_A P_B}) = 1\} = \dim \mathscr{C}(\mathbf{A}) \cap \mathscr{C}(\mathbf{B})$, where $\#\{\operatorname{ch}_i(\mathbf{Z}) = 1\}$ = the number of unit eigenvalues of $\mathbf{Z}$.

21.9 The following statements are equivalent:

$$\operatorname{ch}_1(\mathbf{P_A P_B}) < 1, \quad \mathscr{C}(\mathbf{A}) \cap \mathscr{C}(\mathbf{B}) = \{\mathbf{0}\}, \quad \det(\mathbf{I} - \mathbf{P_A P_B}) \neq 0.$$

21.10 Let $\mathbf{A} \in \mathbb{R}^{n\times a}$, $\mathbf{B} \in \mathbb{R}^{n\times b}$, and $\mathbf{T} \in \mathbb{R}^{n\times t}$, and assume that $\mathscr{C}(\mathbf{T}) \subset \mathscr{C}(\mathbf{A})$. Moreover, let $\mathbf{U}$ be any matrix satisfying $\mathscr{C}(\mathbf{U}) = \mathscr{C}(\mathbf{A}) \cap \mathscr{C}(\mathbf{B})$. Then

$$\mathscr{C}(\mathbf{Q_T U}) = \mathscr{C}(\mathbf{Q_T A}) \cap \mathscr{C}(\mathbf{Q_T B}), \quad \text{where } \mathbf{Q_T} = \mathbf{I}_n - \mathbf{P_T}.$$

21.11 As a generalization to $\mathbf{y}'(\mathbf{I}_n - \mathbf{P_X})\mathbf{y} \leq (\mathbf{y} - \mathbf{Xb})'(\mathbf{y} - \mathbf{Xb})$ for all $\mathbf{b}$, we have, for given $\mathbf{Y}_{n\times q}$ and $\mathbf{X}_{n\times p}$, the Löwner ordering

$$\mathbf{Y}'(\mathbf{I}_n - \mathbf{P_X})\mathbf{Y} \leq_{\mathsf{L}} (\mathbf{Y} - \mathbf{XB})'(\mathbf{Y} - \mathbf{XB}) \quad \text{for all } \mathbf{B}_{p\times q}.$$

21.12 Let $\mathbf{P} = \begin{pmatrix} \mathbf{P}_{11} & \mathbf{P}_{12} \\ \mathbf{P}_{21} & \mathbf{P}_{22} \end{pmatrix}$ be an orthogonal projector where $\mathbf{P}_{11}$ is a square matrix. Then $\mathbf{P}_{22\cdot1} = \mathbf{P}_{22} - \mathbf{P}_{21}\mathbf{P}_{11}^{-}\mathbf{P}_{12}$ is also an orthogonal projector.

21.13 The matrix $\mathbf{P} \in \mathbb{R}_r^{n\times n}$ is idempotent iff any of the following conditions holds:

(a) $\mathbf{P} = \mathbf{AA}^{-}$ for some $\mathbf{A}$,

(b) $\mathbf{I}_n - \mathbf{P}$ is idempotent,

(c) $\operatorname{r}(\mathbf{P}) + \operatorname{r}(\mathbf{I}_n - \mathbf{P}) = n$,

(d) $\mathbb{R}^n = \mathscr{C}(\mathbf{P}) \oplus \mathscr{C}(\mathbf{I}_n - \mathbf{P})$,

(e) $\mathbf{P}$ has a full rank decomposition $\mathbf{P} = \mathbf{UV}'$, where $\mathbf{V}'\mathbf{U} = \mathbf{I}_r$,

(f) $\mathbf{P} = \mathbf{B}\begin{pmatrix} \mathbf{I}_r & \mathbf{0} \\ \mathbf{0} & \mathbf{0} \end{pmatrix}\mathbf{B}^{-1}$, where $\mathbf{B}$ is a nonsingular matrix,

(g) $\mathbf{P} = \mathbf{D}\begin{pmatrix} \mathbf{I}_r & \mathbf{C} \\ \mathbf{0} & \mathbf{0} \end{pmatrix}\mathbf{D}'$, where $\mathbf{D}$ is an orthogonal matrix and $\mathbf{C} \in \mathbb{R}^{r\times(n-r)}$,

(h) $\operatorname{r}(\mathbf{P}) = \operatorname{tr}(\mathbf{P})$ and $\operatorname{r}(\mathbf{I}_n - \mathbf{P}) = \operatorname{tr}(\mathbf{I}_n - \mathbf{P})$.

21.14 If $\mathbf{P}^2 = \mathbf{P}$ then $\operatorname{rank}(\mathbf{P}) = \operatorname{tr}(\mathbf{P})$ and

$$\operatorname{ch}(\mathbf{P}) = \{0, \dots, 0, 1, \dots, 1\}, \quad \#\{\operatorname{ch}(\mathbf{P}) = 1\} = \operatorname{rank}(\mathbf{P}),$$

but $\operatorname{ch}(\mathbf{P}) = \{0, \dots, 0, 1, \dots, 1\}$ does not imply $\mathbf{P}^2 = \mathbf{P}$ (unless $\mathbf{P}' = \mathbf{P}$).

21.15 Let $\mathbf{P}$ be symmetric. Then $\mathbf{P}^2 = \mathbf{P} \iff \operatorname{ch}(\mathbf{P}) = \{0, \dots, 0, 1, \dots, 1\}$; here $\#\{\operatorname{ch}(\mathbf{P}) = 1\} = \operatorname{rank}(\mathbf{P}) = \operatorname{tr}(\mathbf{P})$.

21.16 $\mathbf{P}^2 = \mathbf{P} \implies \mathbf{P}$ is the oblique projector onto $\mathscr{C}(\mathbf{P})$ along $\mathscr{N}(\mathbf{P})$, where the direction space can be also written as $\mathscr{N}(\mathbf{P}) = \mathscr{C}(\mathbf{I} - \mathbf{P})$.

21.17 $\mathbf{AA}^-$ is the oblique projector onto $\mathscr{C}(\mathbf{A})$ along $\mathscr{N}(\mathbf{AA}^-)$, where the direction space can be written as $\mathscr{N}(\mathbf{AA}^-) = \mathscr{C}(\mathbf{I}_n - \mathbf{AA}^-)$. Correspondingly, $\mathbf{A}^-\mathbf{A}$ is the oblique projector onto $\mathscr{C}(\mathbf{A}^-\mathbf{A})$ along $\mathscr{N}(\mathbf{A}^-\mathbf{A}) = \mathscr{N}(\mathbf{A})$.

21.18 $\mathbf{AA}^+ = \mathbf{P}_{\mathbf{A}}, \quad \mathbf{A}^+\mathbf{A} = \mathbf{P}_{\mathbf{A}'}$

21.19 Generalized projector $\mathbf{P}_{\mathbf{A}|\mathbf{B}}$. By $\mathbf{P}_{\mathbf{A}|\mathbf{B}}$ we mean any matrix $\mathbf{G}$, say, satisfying

(a) $\mathbf{G}(\mathbf{A} : \mathbf{B}) = (\mathbf{A} : \mathbf{0})$,

where it is assumed that $\mathscr{C}(\mathbf{A}) \cap \mathscr{C}(\mathbf{B}) = \{\mathbf{0}\}$, which is a necessary and sufficient condition for the solvability of (a). Matrix $\mathbf{G}$ is a generalized projector onto $\mathscr{C}(\mathbf{A})$ along $\mathscr{C}(\mathbf{B})$ but it need not be unique and idempotent as is the case when $\mathscr{C}(\mathbf{A} : \mathbf{B}) = \mathbb{R}^n$. We denote the set of matrices $\mathbf{G}$ satisfying (a) as $\{\mathbf{P}_{\mathbf{A}|\mathbf{B}}\}$ and the general expression for $\mathbf{G}$ is, for example,

$$\mathbf{G} = (\mathbf{A} : \mathbf{0})(\mathbf{A} : \mathbf{B})^- + \mathbf{F}(\mathbf{I} - \mathbf{P}_{(\mathbf{A}:\mathbf{B})}), \quad \text{where } \mathbf{F} \text{ is free to vary.}$$

21.20 Suppose that $\mathscr{C}(\mathbf{A}) \cap \mathscr{C}(\mathbf{B}) = \{\mathbf{0}_n\} = \mathscr{C}(\mathbf{C}) \cap \mathscr{C}(\mathbf{D})$. Then

$$\{\mathbf{P}_{\mathbf{C}|\mathbf{D}}\} \subset \{\mathbf{P}_{\mathbf{A}|\mathbf{B}}\} \iff \mathscr{C}(\mathbf{A}) \subset \mathscr{C}(\mathbf{C}) \text{ and } \mathscr{C}(\mathbf{B}) \subset \mathscr{C}(\mathbf{D}).$$

21.21 Orthogonal projector w.r.t. the inner product matrix $\mathbf{V}$. Let $\mathbf{A} \in \mathbb{R}^{n\times m}_r$, and let the inner product (and the corresponding norm) be defined as $\langle \mathbf{t}, \mathbf{u} \rangle_{\mathbf{V}} = \mathbf{t}'\mathbf{Vu}$ where $\mathbf{V}$ is pd. Further, let $\mathbf{A}^{\perp}_{\mathbf{V}}$ be an $n \times q$ matrix spanning $\mathscr{C}(\mathbf{A})^{\perp}_{\mathbf{V}} = \mathscr{N}(\mathbf{A}'\mathbf{V}) = \mathscr{C}(\mathbf{VA})^{\perp} = \mathscr{C}(\mathbf{V}^{-1}\mathbf{A}^{\perp})$. Then the following conditions are equivalent ways to define the unique matrix $\mathbf{P}_*$:

(a) $\mathbf{P}_*(\mathbf{Ab} + \mathbf{A}^{\perp}_{\mathbf{V}}\mathbf{c}) = \mathbf{Ab} \quad$ for all $\mathbf{b} \in \mathbb{R}^m$, $\mathbf{c} \in \mathbb{R}^q$.

(b) $\mathbf{P}_*(\mathbf{A} : \mathbf{A}^{\perp}_{\mathbf{V}}) = \mathbf{P}_*(\mathbf{A} : \mathbf{V}^{-1}\mathbf{A}^{\perp}) = (\mathbf{A} : \mathbf{0})$.

(c) $\mathscr{C}(\mathbf{P}_*) \subset \mathscr{C}(\mathbf{A}), \quad \min_{\mathbf{b}} \|\mathbf{y} - \mathbf{Ab}\|^2_{\mathbf{V}} = \|\mathbf{y} - \mathbf{P}_*\mathbf{y}\|^2_{\mathbf{V}}$ for all $\mathbf{y} \in \mathbb{R}^n$.

(d) $\mathscr{C}(\mathbf{P}_*) \subset \mathscr{C}(\mathbf{A}), \quad \mathbf{P}'_*\mathbf{VA} = \mathbf{VA}$.

(e) $\mathbf{P}'_*(\mathbf{VA} : \mathbf{A}^{\perp}) = (\mathbf{VA} : \mathbf{0})$.

(f) $\mathscr{C}(\mathbf{P}_*) = \mathscr{C}(\mathbf{A}), \quad \mathbf{P}'_*\mathbf{V}(\mathbf{I}_n - \mathbf{P}_*) = \mathbf{0}$.

(g) $\mathscr{C}(\mathbf{P}_*) = \mathscr{C}(\mathbf{A}), \quad \mathbf{P}^2_* = \mathbf{P}_*, \quad (\mathbf{VP}_*)' = \mathbf{VP}_*$.

(h) $\mathscr{C}(\mathbf{P}_*) = \mathscr{C}(\mathbf{A}), \quad \mathbb{R}^n = \mathscr{C}(\mathbf{P}_*) \boxplus \mathscr{C}(\mathbf{I}_n - \mathbf{P}_*)$; here $\boxplus$ refers to the orthogonality with respect to the given inner product.

(i) $\mathbf{P}_* = \mathbf{A}(\mathbf{A}'\mathbf{VA})^-\mathbf{A}'\mathbf{V}$, which is invariant for any choice of $(\mathbf{A}'\mathbf{VA})^-$.

The matrix $\mathbf{P}_* = \mathbf{P}_{\mathbf{A};\mathbf{V}}$ is the orthogonal projector onto the column space $\mathscr{C}(\mathbf{A})$ w.r.t. the inner product $\langle \mathbf{t}, \mathbf{u} \rangle_{\mathbf{V}} = \mathbf{t}'\mathbf{Vu}$. Correspondingly, the matrix

$\mathbf{I}_n - \mathbf{P}_{\mathbf{A};\mathbf{V}}$ is the orthogonal projector onto $\mathscr{C}(\mathbf{A}_{\mathbf{V}}^{\perp})$:

$$\mathbf{I}_n - \mathbf{P}_{\mathbf{A};\mathbf{V}} = \mathbf{P}_{\mathbf{A}_{\mathbf{V}}^{\perp};\mathbf{V}} = \mathbf{V}^{-1}\mathbf{Z}(\mathbf{Z}'\mathbf{V}^{-1}\mathbf{Z})^{-}\mathbf{Z}' = \mathbf{P}'_{\mathbf{Z};\mathbf{V}^{-1}} = \mathbf{P}_{\mathbf{V}^{-1}\mathbf{Z};\mathbf{V}},$$

where $\mathbf{Z} \in \{\mathbf{A}^{\perp}\}$, i.e., $\mathbf{P}_{\mathbf{A};\mathbf{V}} = \mathbf{I}_n - \mathbf{P}'_{\mathbf{A}^{\perp};\mathbf{V}^{-1}} = \mathbf{I}_n - \mathbf{P}_{\mathbf{V}^{-1}\mathbf{A}^{\perp};\mathbf{V}}$.

21.22 Consider the linear model $\{\mathbf{y}, \mathbf{X}\boldsymbol{\beta}, \mathbf{V}\}$ and let $\mathscr{C}(\mathbf{X})_{\mathbf{V}^{-1}}^{\perp}$ denote the set of vectors which are orthogonal to every vector in $\mathscr{C}(\mathbf{X})$ with respect to the inner product matrix $\mathbf{V}^{-1}$. Then the following sets are identical:

(a) $\mathscr{C}(\mathbf{X})_{\mathbf{V}^{-1}}^{\perp}$, (b) $\mathscr{C}(\mathbf{V}\mathbf{X}^{\perp})$, (c) $\mathscr{N}(\mathbf{X}'\mathbf{V}^{-1})$,

(d) $\mathscr{C}(\mathbf{V}^{-1}\mathbf{X})^{\perp}$, (e) $\mathscr{N}(\mathbf{P}_{\mathbf{X};\mathbf{V}^{-1}})$, (f) $\mathscr{C}(\mathbf{I}_n - \mathbf{P}_{\mathbf{X};\mathbf{V}^{-1}})$.

Denote $\mathbf{W} = \mathbf{V} + \mathbf{X}\mathbf{U}\mathbf{X}'$, where $\mathscr{C}(\mathbf{W}) = \mathscr{C}(\mathbf{X} : \mathbf{V})$. Then

$$\mathscr{C}(\mathbf{V}\mathbf{X}^{\perp}) = \mathscr{C}(\mathbf{W}^{-}\mathbf{X} : \mathbf{I}_n - \mathbf{W}^{-}\mathbf{W})^{\perp},$$

where $\mathbf{W}^{-}$ is an arbitrary (but fixed) generalized inverse of $\mathbf{W}$. The column space $\mathscr{C}(\mathbf{V}\mathbf{X}^{\perp})$ can be expressed also as

$$\mathscr{C}(\mathbf{V}\mathbf{X}^{\perp}) = \mathscr{C}[(\mathbf{W}^{-})'\mathbf{X} : \mathbf{I}_n - (\mathbf{W}^{-})'\mathbf{W}']^{\perp}.$$

Moreover, let $\mathbf{V}$ be possibly singular and assume that $\mathscr{C}(\mathbf{X}) \subset \mathscr{C}(\mathbf{V})$. Then

$$\mathscr{C}(\mathbf{V}\mathbf{X}^{\perp}) = \mathscr{C}(\mathbf{V}^{-}\mathbf{X} : \mathbf{I}_n - \mathbf{V}^{-}\mathbf{V})^{\perp} \subset \mathscr{C}(\mathbf{V}^{-}\mathbf{X})^{\perp},$$

where the inclusion becomes equality iff $\mathbf{V}$ is positive definite.

21.23 Let $\mathbf{V}$ be pd and let $\mathbf{X}$ be partitioned as $\mathbf{X} = (\mathbf{X}_1 : \mathbf{X}_2)$. Then

(a) $\mathbf{P}_{\mathbf{X}_1;\mathbf{V}^{-1}} + \mathbf{P}_{\mathbf{X}_2;\mathbf{V}}$ is an orthogonal projector iff $\mathbf{X}_1'\mathbf{V}^{-1}\mathbf{X}_2 = \mathbf{0}$, in which case $\mathbf{P}_{\mathbf{X}_1;\mathbf{V}^{-1}} + \mathbf{P}_{\mathbf{X}_2;\mathbf{V}^{-1}} = \mathbf{P}_{(\mathbf{X}_1:\mathbf{X}_2);\mathbf{V}^{-1}}$,

(b) $\mathbf{P}_{(\mathbf{X}_1:\mathbf{X}_2);\mathbf{V}^{-1}} = \mathbf{P}_{\mathbf{X}_1;\mathbf{V}^{-1}} + \mathbf{P}_{(\mathbf{I}-\mathbf{P}_{\mathbf{X}_1;\mathbf{V}^{-1}})\mathbf{X}_2;\mathbf{V}^{-1}}$,

(c) $\mathbf{P}_{(\mathbf{I}-\mathbf{P}_{\mathbf{X}_1;\mathbf{V}^{-1}})\mathbf{X}_2;\mathbf{V}^{-1}} = \mathbf{P}_{\mathbf{V}\dot{\mathbf{M}}_1\mathbf{X}_2;\mathbf{V}^{-1}} = \mathbf{V}\dot{\mathbf{M}}_1\mathbf{X}_2(\mathbf{X}_2'\dot{\mathbf{M}}_1\mathbf{X}_2)^{-1}\mathbf{X}_2'\dot{\mathbf{M}}_1$,

(d) $\dot{\mathbf{M}}_1 = \mathbf{V}^{-1} - \mathbf{V}^{-1}\mathbf{X}_1(\mathbf{X}_1'\mathbf{V}^{-1}\mathbf{X}_1)^{-1}\mathbf{X}_1'\mathbf{V}^{-1} = \mathbf{M}_1(\mathbf{M}_1\mathbf{V}\mathbf{M}_1)^{-}\mathbf{M}_1$.

21.24 Consider a weakly singular partitioned linear model and denote $\mathbf{P}_{\mathbf{A};\mathbf{V}^+} = \mathbf{A}(\mathbf{A}'\mathbf{V}^+\mathbf{A})^{-}\mathbf{A}'\mathbf{V}^+$ and $\dot{\mathbf{M}}_1 = \mathbf{M}_1(\mathbf{M}_1\mathbf{V}\mathbf{M}_1)^{-}\mathbf{M}_1$. Then

$$\begin{aligned}
\mathbf{P}_{\mathbf{X};\mathbf{V}^+} &= \mathbf{X}(\mathbf{X}'\mathbf{V}^+\mathbf{X})^{-}\mathbf{X}'\mathbf{V}^+ = \mathbf{V}^{1/2}\mathbf{P}_{\mathbf{V}^{+1/2}\mathbf{X}}\mathbf{V}^{+1/2} \\
&= \mathbf{V}^{1/2}(\mathbf{P}_{\mathbf{V}^{+1/2}\mathbf{X}_1} + \mathbf{P}_{(\mathbf{I}-\mathbf{P}_{\mathbf{V}^{+1/2}\mathbf{X}_1})\mathbf{V}^{+1/2}\mathbf{X}_2})\mathbf{V}^{+1/2} \\
&= \mathbf{P}_{\mathbf{X}_1;\mathbf{V}^+} + \mathbf{V}^{1/2}\mathbf{P}_{(\mathbf{I}-\mathbf{P}_{\mathbf{V}^{+1/2}\mathbf{X}_1})\mathbf{V}^{+1/2}\mathbf{X}_2}\mathbf{V}^{+1/2} \\
&= \mathbf{P}_{\mathbf{X}_1;\mathbf{V}^+} + \mathbf{V}^{1/2}\mathbf{P}_{\mathbf{V}^{1/2}\dot{\mathbf{M}}_1\mathbf{X}_2}\mathbf{V}^{+1/2} \\
&= \mathbf{P}_{\mathbf{X}_1;\mathbf{V}^+} + \mathbf{V}\dot{\mathbf{M}}_1\mathbf{X}_2(\mathbf{X}_2'\dot{\mathbf{M}}_1\mathbf{X}_2)^{-}\mathbf{X}_2'\dot{\mathbf{M}}_1\mathbf{P}_{\mathbf{V}}.
\end{aligned}$$

22 Eigenvalues

22.1 Definition: The scalar λ is an eigenvalue of $\mathbf{A}_{n\times n}$ if

$$\mathbf{At} = \lambda\mathbf{t} \text{ for some nonzero vector } \mathbf{t}, \text{ i.e., } (\mathbf{A} - \lambda\mathbf{I}_n)\mathbf{t} = \mathbf{0},$$

in which case $\mathbf{t}$ is an eigenvector of $\mathbf{A}$ corresponding to λ, and $(\lambda, \mathbf{t})$ is an eigenpair for $\mathbf{A}$. If $\mathbf{A}$ is symmetric, then all eigenvalues are real. The eigenvalues are the n roots of the characteristic equation

$$p_{\mathbf{A}}(\lambda) = \det(\mathbf{A} - \lambda\mathbf{I}_n) = 0,$$

where $p_{\mathbf{A}}(\lambda) = \det(\mathbf{A} - \lambda\mathbf{I}_n)$ is the characteristic polynomial of $\mathbf{A}$. We denote (when eigenvalues are real)

$$\begin{aligned} &\lambda_1 \geq \cdots \geq \lambda_n, \quad \mathrm{ch}_i(\mathbf{A}) = \lambda_i, \\ &\mathrm{ch}(\mathbf{A}) = \{\lambda_1, \dots, \lambda_n\} = \text{ spectrum of } \mathbf{A}, \\ &\mathrm{nzch}(\mathbf{A}) = \{\, \lambda_i : \lambda_i \neq 0 \,\}. \end{aligned}$$

22.2 The characteristic equation can be written as

$$\begin{aligned} p_{\mathbf{A}}(\lambda) &= (-\lambda)^n + S_1(-\lambda)^{n-1} + \cdots + S_{n-1}(-\lambda) + S_n = 0, \\ p_{\mathbf{A}}(\lambda) &= (\lambda_1 - \lambda)\cdots(\lambda_n - \lambda) = 0, \end{aligned}$$

for appropriate real coefficients $S_1, \dots, S_n$: S_i is the sum of all $i \times i$ principal minors and hence $S_n = p_{\mathbf{A}}(0) = \det(\mathbf{A}) = \lambda_1 \cdots \lambda_n$, $S_1 = \mathrm{tr}(\mathbf{A}) = \lambda_1 + \cdots + \lambda_n$, and thereby always

$$\det(\mathbf{A}) = \lambda_1 \cdots \lambda_n, \quad \mathrm{tr}(\mathbf{A}) = \lambda_1 + \cdots + \lambda_n.$$

For $n = 3$, we have

$$\begin{aligned} &\det(\mathbf{A} - \lambda\mathbf{I}_3) \\ &= (-\lambda)^3 + \mathrm{tr}(\mathbf{A})(-\lambda)^2 \\ &\quad + \left(\begin{vmatrix} a_{11} & a_{12} \\ a_{21} & a_{22} \end{vmatrix} + \begin{vmatrix} a_{11} & a_{13} \\ a_{31} & a_{33} \end{vmatrix} + \begin{vmatrix} a_{22} & a_{23} \\ a_{32} & a_{33} \end{vmatrix} \right)(-\lambda) + \det(\mathbf{A}). \end{aligned}$$

22.3 The characteristic polynomial of $\mathbf{A} = \left(\begin{smallmatrix} a & b \\ b & c \end{smallmatrix}\right)$ is

$$p_{\mathbf{A}}(\lambda) = \lambda^2 - (a + c)\lambda + (ac - b^2) = \lambda^2 - \mathrm{tr}(\mathbf{A}) \cdot \lambda + \det(\mathbf{A}).$$

The eigenvalues of $\mathbf{A}$ are $\frac{1}{2}\big(a + c \pm \sqrt{(a - c)^2 + 4b^2}\,\big)$.

22.4 The following statements are equivalent (for a nonnull $\mathbf{t}$ and pd $\mathbf{A}$):

(a) $\mathbf{t}$ is an eigenvector of $\mathbf{A}$, (b) $\cos(\mathbf{t}, \mathbf{At}) = 1$,

(c) $\mathbf{t}'\mathbf{At} - (\mathbf{t}'\mathbf{A}^{-1}\mathbf{t})^{-1} = 0$ with $\|\mathbf{t}\| = 1$,

(d) $\cos(\mathbf{A}^{1/2}\mathbf{t}, \mathbf{A}^{-1/2}\mathbf{t}) = 1$,

(e) $\cos(\mathbf{y}, \mathbf{At}) = 0$ for all $\mathbf{y} \in \mathscr{C}(\mathbf{t})^{\perp}$.

22.5 The spectral radius of $\mathbf{A} \in \mathbb{R}^{n\times n}$ is defined as $\rho(\mathbf{A}) = \max\{|\mathrm{ch}_i(\mathbf{A})|\}$; the eigenvalue corresponding to $\rho(\mathbf{A})$ is called the dominant eigenvalue.

EVD Eigenvalue decomposition. A symmetric $\mathbf{A}_{n\times n}$ can be written as

$$\mathbf{A} = \mathbf{T\Lambda T}' = \lambda_1 \mathbf{t}_1\mathbf{t}_1' + \cdots + \lambda_n \mathbf{t}_n\mathbf{t}_n',$$

and thereby

$$(\mathbf{At}_1 : \mathbf{At}_2 : \ldots : \mathbf{At}_n) = (\lambda_1\mathbf{t}_1 : \lambda_2\mathbf{t}_2 : \ldots : \lambda_n\mathbf{t}_n), \quad \mathbf{AT} = \mathbf{T\Lambda},$$

where $\mathbf{T}_{n\times n}$ is orthogonal, $\mathbf{\Lambda} = \mathrm{diag}(\lambda_1, \ldots, \lambda_n)$, and $\lambda_1 \geq \cdots \geq \lambda_n$ are the ordered eigenvalues of $\mathbf{A}$; $\mathrm{ch}_i(\mathbf{A}) = \lambda_i$. The columns $\mathbf{t}_i$ of $\mathbf{T}$ are the orthonormal eigenvectors of $\mathbf{A}$.

Consider the distinct eigenvalues of $\mathbf{A}$, $\lambda_{\{1\}} > \cdots > \lambda_{\{s\}}$, and let $\mathbf{T}_{\{i\}}$ be an $n \times m_i$ matrix consisting of the orthonormal eigenvectors corresponding to $\lambda_{\{i\}}$; m_i is the multiplicity of $\lambda_{\{i\}}$. Then

$$\mathbf{A} = \mathbf{T\Lambda T}' = \lambda_{\{1\}}\mathbf{T}_{\{1\}}\mathbf{T}_{\{1\}}' + \cdots + \lambda_{\{s\}}\mathbf{T}_{\{s\}}\mathbf{T}_{\{s\}}'.$$

With this ordering, $\mathbf{\Lambda}$ is unique and $\mathbf{T}$ is unique up to postmultiplying by a blockdiagonal matrix $\mathbf{U} = \mathrm{blockdiag}(\mathbf{U}_1, \ldots, \mathbf{U}_s)$, where $\mathbf{U}_i$ is an orthogonal $m_i \times m_i$ matrix. If all the eigenvalues are distinct, then $\mathbf{U}$ is a diagonal matrix with diagonal elements equal to ± 1.

22.6 For a nonnegative definite $n \times n$ matrix $\mathbf{A}$ with rank $r > 0$ we have

$$\begin{aligned}\mathbf{A} = \mathbf{T\Lambda T}' &= (\mathbf{T}_1 : \mathbf{T}_0)\begin{pmatrix}\mathbf{\Lambda}_1 & \mathbf{0}\\ \mathbf{0} & \mathbf{0}\end{pmatrix}\begin{pmatrix}\mathbf{T}_1'\\ \mathbf{T}_0'\end{pmatrix} = \mathbf{T}_1\mathbf{\Lambda}_1\mathbf{T}_1'\\ &= \lambda_1\mathbf{t}_1\mathbf{t}_1' + \cdots + \lambda_r\mathbf{t}_r\mathbf{t}_r',\end{aligned}$$

where $\lambda_1 \geq \cdots \geq \lambda_r > 0$, $\mathbf{\Lambda}_1 = \mathrm{diag}(\lambda_1, \ldots, \lambda_r)$, and $\mathbf{T}_1 = (\mathbf{t}_1 : \ldots : \mathbf{t}_r)$, $\mathbf{T}_0 = (\mathbf{t}_{r+1} : \ldots : \mathbf{t}_n)$.

22.7 The nnd square root of the nnd matrix $\mathbf{A} = \mathbf{T\Lambda T}'$ is defined as

$$\mathbf{A}^{1/2} = \mathbf{T\Lambda}^{1/2}\mathbf{T}' = \mathbf{T}_1\mathbf{\Lambda}_1^{1/2}\mathbf{T}_1'; \quad (\mathbf{A}^{1/2})^+ = \mathbf{T}_1\mathbf{\Lambda}_1^{-1/2}\mathbf{T}_1' := \mathbf{A}^{+1/2}.$$

22.8 If $\mathbf{A}_{p\times p} = (a-b)\mathbf{I}_p + b\mathbf{1}_p\mathbf{1}_p'$ for some $a, b \in \mathbb{R}$, then $\mathbf{A}$ is called a completely symmetric matrix. If it is a covariance matrix (i.e., nnd) then it is said to have an intraclass correlation structure. Consider the matrices

$$\mathbf{A}_{p\times p} = (a-b)\mathbf{I}_p + b\mathbf{1}_p\mathbf{1}_p', \quad \mathbf{\Sigma}_{p\times p} = (1-\varrho)\mathbf{I}_p + \varrho\mathbf{1}_p\mathbf{1}_p'.$$

(a) $\det(\mathbf{A}) = (a-b)^{n-1}[a + (p-1)b]$,

(b) the eigenvalues of $\mathbf{A}$ are $a + (p-1)b$ (with multiplicity 1), and $a - b$ (with multiplicity $n - 1$),

(c) $\mathbf{A}$ is nonsingular iff $a \neq b$ and $a \neq -(p-1)b$, in which case

$$\mathbf{A}^{-1} = \frac{1}{a-b}\left(\mathbf{I}_p - \frac{b}{a+(p-1)b}\mathbf{1}_p\mathbf{1}_p'\right),$$

(d) $\operatorname{ch}(\boldsymbol{\Sigma}) = \begin{cases} 1+(p-1)\varrho & \text{with multiplicity } 1, \\ 1-\varrho & \text{with multiplicity } p-1, \end{cases}$

(e) $\boldsymbol{\Sigma}$ is nonnegative definite $\iff -\frac{1}{p-1} \leq \varrho \leq 1$,

(f) $\mathbf{t}_1 = \alpha\mathbf{1}_p =$ eigenvector w.r.t. $\lambda_1 = 1+(p-1)\varrho$, $0 \neq \alpha \in \mathbb{R}$,

$\mathbf{t}_2, \dots, \mathbf{t}_p$ are orthonormal eigenvectors w.r.t. $\lambda_i = 1-\varrho$, $i = 2, \dots, p$,

$\mathbf{t}_2, \dots, \mathbf{t}_p$ form an orthonormal basis for $\mathscr{C}(\mathbf{1}_p)^{\perp}$,

(g) $\det(\boldsymbol{\Sigma}) = (1-\varrho)^{p-1}[1+(p-1)\varrho]$,

(h) $\boldsymbol{\Sigma}\mathbf{1}_p = [1+(p-1)\varrho]\mathbf{1}_p := \lambda_1\mathbf{1}_p$; if $\varrho \neq -\frac{1}{p-1}$, then $\mathbf{1}_p = \lambda_1^{-1}\boldsymbol{\Sigma}\mathbf{1}_p$ in which case

$$\mathbf{1}_p'\boldsymbol{\Sigma}^{-}\mathbf{1}_p = \lambda_1^{-2}\mathbf{1}_p'\boldsymbol{\Sigma}\boldsymbol{\Sigma}^{-}\boldsymbol{\Sigma}\mathbf{1}_p = \lambda_1^{-2}\mathbf{1}_p'\boldsymbol{\Sigma}\mathbf{1}_p = \frac{p}{1+(p-1)\varrho},$$

(i) $\boldsymbol{\Sigma}^{-1} = \dfrac{1}{1-\varrho}\left(\mathbf{I}_p - \dfrac{\varrho}{1+(p-1)\varrho}\mathbf{1}_p\mathbf{1}_p'\right)$, for $\varrho \neq 1$, $\varrho \neq -\dfrac{1}{p-1}$.

(j) Suppose that

$$\operatorname{cov}\begin{pmatrix}\mathbf{x}\\ y\end{pmatrix} = \begin{pmatrix}\boldsymbol{\Sigma}_{\mathbf{xx}} & \boldsymbol{\sigma}_{\mathbf{x}y}\\ \boldsymbol{\sigma}_{\mathbf{x}y}' & 1\end{pmatrix} = (1-\varrho)\mathbf{I}_{p+1} + \varrho\mathbf{1}_{p+1}\mathbf{1}_{p+1}',$$

where (necessarily) $-\frac{1}{p} \leq \varrho \leq 1$. Then

$$\varrho^2_{y\cdot\mathbf{x}} = \boldsymbol{\sigma}_{\mathbf{x}y}'\boldsymbol{\Sigma}_{\mathbf{xx}}^{-}\boldsymbol{\sigma}_{\mathbf{x}y} = \frac{p\varrho^2}{1+(p-1)\varrho}.$$

(k) Assume that $\operatorname{cov}\begin{pmatrix}\mathbf{x}\\ \mathbf{y}\end{pmatrix} = \begin{pmatrix}\boldsymbol{\Sigma} & \varrho\mathbf{1}_p\mathbf{1}_p'\\ \varrho\mathbf{1}_p\mathbf{1}_p' & \boldsymbol{\Sigma}\end{pmatrix}$, where

$\boldsymbol{\Sigma} = (1-\varrho)\mathbf{I}_p + \varrho\mathbf{1}_p\mathbf{1}_p'$. Then $\operatorname{cor}(\mathbf{1}_p'\mathbf{x}, \mathbf{1}_p'\mathbf{y}) = \dfrac{p\varrho}{1+(p-1)\varrho}$.

22.9 Consider a completely symmetric matrix $\mathbf{A}_{p\times p} = (a-b)\mathbf{I}_p + b\mathbf{1}_p\mathbf{1}_p'$, Then

$$\mathbf{L}^{-1}\mathbf{A}\mathbf{L} = \begin{pmatrix} a-b & 0 & 0 & \dots & 0 & b\\ 0 & a-b & 0 & \dots & 0 & b\\ \vdots & \vdots & \vdots & & \vdots & \vdots\\ 0 & 0 & 0 & \dots & a-b & b\\ 0 & 0 & 0 & \dots & 0 & a+(p-1)b \end{pmatrix} := \mathbf{F},$$

where $\mathbf{L}$ carries out elementary column operations:

$$\mathbf{L} = \begin{pmatrix} \mathbf{I}_{p-1} & \mathbf{0}_{p-1} \\ -\mathbf{1}'_{p-1} & 1 \end{pmatrix}, \quad \mathbf{L}^{-1} = \begin{pmatrix} \mathbf{I}_{p-1} & \mathbf{0}_{p-1} \\ \mathbf{1}'_{p-1} & 1 \end{pmatrix},$$

and hence $\det(\mathbf{A}) = (a-b)^{p-1}[a+(p-1)b]$. Moreover, $\det(\mathbf{A} - \lambda\mathbf{I}_p) = 0$ iff $\det(\mathbf{I}_p - \lambda\mathbf{F}) = 0$.

22.10 The algebraic multiplicity of λ, denoted as $\operatorname{alg\,mult}_{\mathbf{A}}(\lambda)$ is its multiplicity as a root of $\det(\mathbf{A} - \lambda\mathbf{I}_n) = 0$. The set $\{\mathbf{t} \neq \mathbf{0} : \mathbf{At} = \lambda\mathbf{t}\}$ is the set of all eigenvectors associated with λ. The eigenspace of $\mathbf{A}$ corresponding to λ is

$$\{\mathbf{t} : (\mathbf{A} - \lambda\mathbf{I}_n)\mathbf{t} = \mathbf{0}\} = \mathscr{N}(\mathbf{A} - \lambda\mathbf{I}_n).$$

If $\operatorname{alg\,mult}_{\mathbf{A}}(\lambda) = 1$, λ is called a simple eigenvalue; otherwise it is a multiple eigenvalue. The geometric multiplicity of λ is the dimension of the eigenspace of $\mathbf{A}$ corresponding to λ:

$$\operatorname{geo\,mult}_{\mathbf{A}}(\lambda) = \dim \mathscr{N}(\mathbf{A} - \lambda\mathbf{I}_n) = n - \operatorname{r}(\mathbf{A} - \lambda\mathbf{I}_n).$$

Moreover, $\operatorname{geo\,mult}_{\mathbf{A}}(\lambda) \leq \operatorname{alg\,mult}_{\mathbf{A}}(\lambda)$ for each $\lambda \in \operatorname{ch}(\mathbf{A})$; here the equality holds e.g. when $\mathbf{A}$ is symmetric.

22.11 Similarity. Two $n \times n$ matrices $\mathbf{A}$ and $\mathbf{B}$ are said to be similar whenever there exists a nonsingular matrix $\mathbf{F}$ such that $\mathbf{F}^{-1}\mathbf{AF} = \mathbf{B}$. The product $\mathbf{F}^{-1}\mathbf{AF}$ is called a similarity transformation of $\mathbf{A}$.

22.12 Diagonalizability. A matrix $\mathbf{A}_{n\times n}$ is said to be diagonalizable whenever there exists a nonsingular matrix $\mathbf{F}_{n\times n}$ such that $\mathbf{F}^{-1}\mathbf{AF} = \mathbf{D}$ for some diagonal matrix $\mathbf{D}_{n\times n}$, i.e., $\mathbf{A}$ is similar to a diagonal matrix. In particular, any symmetric $\mathbf{A}$ is diagonalizable.

22.13 The following statements concerning the matrix $\mathbf{A}_{n\times n}$ are equivalent:

(a) $\mathbf{A}$ is diagonalizable,

(b) $\operatorname{geo\,mult}_{\mathbf{A}}(\lambda) = \operatorname{alg\,mult}_{\mathbf{A}}(\lambda)$ for all $\lambda \in \operatorname{ch}(\mathbf{A})$,

(c) $\mathbf{A}$ has n linearly independent eigenvectors.

22.14 Let $\mathbf{A} \in \mathbb{R}^{n\times m}$, $\mathbf{B} \in \mathbb{R}^{m\times n}$. Then $\mathbf{AB}$ and $\mathbf{BA}$ have the same nonzero eigenvalues: $\operatorname{nzch}(\mathbf{AB}) = \operatorname{nzch}(\mathbf{BA})$. Moreover, $\det(\mathbf{I}_n - \mathbf{AB}) = \det(\mathbf{I}_m - \mathbf{BA})$.

22.15 Let $\mathbf{A}, \mathbf{B} \in \operatorname{NND}_n$. Then

(a) $\operatorname{tr}(\mathbf{AB}) \leq \operatorname{ch}_1(\mathbf{A}) \cdot \operatorname{tr}(\mathbf{B})$,

(b) $\mathbf{A} \geq \mathbf{B} \implies \operatorname{ch}_i(\mathbf{A}) \geq \operatorname{ch}_i(\mathbf{B})$, $i = 1, \dots, n$; here we must have at least one strict inequality if $\mathbf{A} \neq \mathbf{B}$.

22.16 Interlacing theorem. Let $\mathbf{A}_{n\times n}$ be symmetric and let $\mathbf{B}$ be a principal submatrix of $\mathbf{A}$ of order $(n-1)\times(n-1)$. Then

$$\mathrm{ch}_{i+1}(\mathbf{A}) \leq \mathrm{ch}_i(\mathbf{B}) \leq \mathrm{ch}_i(\mathbf{A}), \quad i = 1,\dots,n-1.$$

22.17 Let $\mathbf{A} \in \mathbb{R}^{n\times a}_r$ and denote $\mathbf{B} = \begin{pmatrix} \mathbf{0} & \mathbf{A} \\ \mathbf{A}' & \mathbf{0} \end{pmatrix}$, $\mathrm{sg}(\mathbf{A}) = \{\delta_i,\dots,\delta_r\}$. Then the nonzero eigenvalues of $\mathbf{B}$ are $\delta_1,\dots,\delta_r,-\delta_1,\dots,-\delta_r$.

In 22.18–EY1 we consider $\mathbf{A}_{n\times n}$ whose EVD is $\mathbf{A} = \mathbf{T\Lambda T}'$ and denote $\mathbf{T}_{(k)} = (\mathbf{t}_1 : \dots : \mathbf{t}_k)$, $\mathbf{T}_{[k]} = (\mathbf{t}_{n-k+1} : \dots : \mathbf{t}_n)$.

22.18 (a) $\mathrm{ch}_1(\mathbf{A}) = \lambda_1 = \max\limits_{\mathbf{x}\neq\mathbf{0}} \dfrac{\mathbf{x}'\mathbf{A}\mathbf{x}}{\mathbf{x}'\mathbf{x}} = \max\limits_{\mathbf{x}'\mathbf{x}=1} \mathbf{x}'\mathbf{A}\mathbf{x}$,

(b) $\mathrm{ch}_n(\mathbf{A}) = \lambda_n = \min\limits_{\mathbf{x}\neq\mathbf{0}} \dfrac{\mathbf{x}'\mathbf{A}\mathbf{x}}{\mathbf{x}'\mathbf{x}} = \min\limits_{\mathbf{x}'\mathbf{x}=1} \mathbf{x}'\mathbf{A}\mathbf{x}$,

(c) $\mathrm{ch}_n(\mathbf{A}) = \lambda_n \leq \dfrac{\mathbf{x}'\mathbf{A}\mathbf{x}}{\mathbf{x}'\mathbf{x}} \leq \lambda_1 = \mathrm{ch}_1(\mathbf{A})$, $\dfrac{\mathbf{x}'\mathbf{A}\mathbf{x}}{\mathbf{x}'\mathbf{x}}$ = Rayleigh quotient,

(d) $\mathrm{ch}_2(\mathbf{A}) = \lambda_2 = \max\limits_{\substack{\mathbf{x}'\mathbf{x}=1 \\ \mathbf{t}_1'\mathbf{x}=0}} \mathbf{x}'\mathbf{A}\mathbf{x}$,

(e) $\mathrm{ch}_{k+1}(\mathbf{A}) = \lambda_{k+1} = \max\limits_{\substack{\mathbf{x}'\mathbf{x}=1 \\ \mathbf{T}_{(k)}'\mathbf{x}=\mathbf{0}}} \mathbf{x}'\mathbf{A}\mathbf{x} = \lambda_{k+1}$, $k = 1,\dots,n-1$.

22.19 Let $\mathbf{A}_{n\times n}$ be symmetric and let k be a given integer, $k \leq n$. Then

$$\max_{\mathbf{G}'\mathbf{G}=\mathbf{I}_k} \mathrm{tr}(\mathbf{G}'\mathbf{A}\mathbf{G}) = \max_{\mathbf{G}'\mathbf{G}=\mathbf{I}_k} \mathrm{tr}(\mathbf{P}_\mathbf{G}\mathbf{A}) = \lambda_1 + \dots + \lambda_k,$$

where the upper bound is obtained when $\mathbf{G} = (\mathbf{t}_1 : \dots : \mathbf{t}_k) = \mathbf{T}_{(k)}$.

PST Poincaré separation theorem. Let $\mathbf{A}_{n\times n}$ be a symmetric matrix, and let $\mathbf{G}_{n\times k}$ be such that $\mathbf{G}'\mathbf{G} = \mathbf{I}_k$, $k \leq n$. Then, for $i = 1,\dots,k$,

$$\mathrm{ch}_{n-k+i}(\mathbf{A}) \leq \mathrm{ch}_i(\mathbf{G}'\mathbf{A}\mathbf{G}) \leq \mathrm{ch}_i(\mathbf{A}).$$

Equality holds on the right simultaneously for all $i = 1,\dots,k$ when $\mathbf{G} = \mathbf{T}_{(k)}\mathbf{K}$, and on the left if $\mathbf{G} = \mathbf{T}_{[k]}\mathbf{L}$; $\mathbf{K}$ and $\mathbf{L}$ are arbitrary orthogonal matrices.

CFT Courant–Fischer theorem. Let $\mathbf{A}_{n\times n}$ be symmetric and let k be a given integer with $2 \leq k \leq n$ and let $\mathbf{B} \in \mathbb{R}^{n\times(k-1)}$. Then

(a) $\min\limits_{\mathbf{B}} \max\limits_{\mathbf{B}'\mathbf{x}=\mathbf{0}} \dfrac{\mathbf{x}'\mathbf{A}\mathbf{x}}{\mathbf{x}'\mathbf{x}} = \lambda_k$,

(b) $\max\limits_{\mathbf{B}} \min\limits_{\mathbf{B}'\mathbf{x}=\mathbf{0}} \dfrac{\mathbf{x}'\mathbf{A}\mathbf{x}}{\mathbf{x}'\mathbf{x}} = \lambda_{n-k+1}$.

The result (a) is obtained when $\mathbf{B} = \mathbf{T}_{(k-1)}$ and $\mathbf{x} = \mathbf{t}_k$.

EY1 Eckart–Young theorem. Let $\mathbf{A}_{n\times n}$ be a symmetric matrix of rank r. Then

$$\min_{\mathrm{r}(\mathbf{B})=k} \|\mathbf{A}-\mathbf{B}\|_F^2 = \min_{\mathrm{r}(\mathbf{B})=k} \mathrm{tr}(\mathbf{A}-\mathbf{B})(\mathbf{A}-\mathbf{B})' = \lambda_{k+1}^2 + \cdots + \lambda_r^2,$$

and the minimum is attained when

$$\mathbf{B} = \mathbf{T}_{(k)}\mathbf{\Lambda}_{(k)}\mathbf{T}_{(k)}' = \lambda_1\mathbf{t}_1\mathbf{t}_1' + \cdots + \lambda_k\mathbf{t}_k\mathbf{t}_k'.$$

22.20 Let $\mathbf{A}$ be a given $n \times m$ matrix of rank r, and let $\mathbf{B} \in \mathbb{R}_k^{n\times m}$, $k < r$. Then

$$(\mathbf{A}-\mathbf{B})(\mathbf{A}-\mathbf{B})' \geq_{\mathsf{L}} (\mathbf{A}-\mathbf{A}\mathbf{P}_{\mathbf{B}'})(\mathbf{A}-\mathbf{A}\mathbf{P}_{\mathbf{B}'})' = \mathbf{A}(\mathbf{I}_m - \mathbf{P}_{\mathbf{B}'})\mathbf{A}',$$

and hence $\|\mathbf{A}-\mathbf{B}\|_F^2 \geq \|\mathbf{A}-\mathbf{A}\mathbf{P}_{\mathbf{B}'}\|_F^2$. Moreover, if $\mathbf{B}$ has the full rank decomposition $\mathbf{B} = \mathbf{F}\mathbf{G}'$, where $\mathbf{G}'\mathbf{G} = \mathbf{I}_k$, then $\mathbf{P}_{\mathbf{B}'} = \mathbf{P}_{\mathbf{G}}$ and

$$(\mathbf{A}-\mathbf{B})(\mathbf{A}-\mathbf{B})' \geq_{\mathsf{L}} (\mathbf{A}-\mathbf{F}\mathbf{G}')(\mathbf{A}-\mathbf{F}\mathbf{G}')' = \mathbf{A}(\mathbf{I}_m - \mathbf{G}\mathbf{G}')\mathbf{A}'.$$

22.21 Suppose $\mathbf{V} \in \mathrm{PD}_n$ and $\mathbf{C}$ denotes the centering matrix. Then the following statements are equivalent:

(a) $\mathbf{CVC} = c^2\mathbf{C}$ for some $c \neq 0$,

(b) $\mathbf{V} = \alpha^2\mathbf{I} + \mathbf{a}\mathbf{1}' + \mathbf{1}\mathbf{a}'$, where $\mathbf{a}$ is an arbitrary vector and α is any scalar ensuring the positive definiteness of $\mathbf{V}$.

The eigenvalues of $\mathbf{V}$ in (b) are $\alpha^2 + \mathbf{1}'\mathbf{a} \pm \sqrt{n\mathbf{a}'\mathbf{a}}$, each with multiplicity one, and α^2 with multiplicity $n-2$.

22.22 Eigenvalues of $\mathbf{A}$ w.r.t. pd $\mathbf{B}$. Let $\mathbf{A}_{n\times n}$ and $\mathbf{B}_{n\times n}$ be symmetric of which $\mathbf{B}$ is nonnegative definite. Let λ be a scalar and $\mathbf{w}$ a vector such that

(a) $\mathbf{Aw} = \lambda\mathbf{Bw}, \quad \mathbf{Bw} \neq \mathbf{0}$.

Then we call λ a proper eigenvalue and $\mathbf{w}$ a proper eigenvector of $\mathbf{A}$ with respect to $\mathbf{B}$, or shortly, $(\lambda, \mathbf{w})$ is a proper eigenpair for $(\mathbf{A}, \mathbf{B})$. There may exist a vector $\mathbf{w} \neq \mathbf{0}$ such that $\mathbf{Aw} = \mathbf{Bw} = \mathbf{0}$, in which case (a) is satisfied with arbitrary λ. We call such a vector $\mathbf{w}$ an improper eigenvector of $\mathbf{A}$ with respect to $\mathbf{B}$. The space of improper eigenvectors is $\mathscr{C}(\mathbf{A} : \mathbf{B})^{\perp}$. Consider next the situation when $\mathbf{B}$ is positive definite (in which case the word "proper" can be dropped off). Then (a) becomes

(b) $\mathbf{Aw} = \lambda\mathbf{Bw}, \quad \mathbf{w} \neq \mathbf{0}$.

Premultiplying (b) by $\mathbf{B}^{-1}$ yields the usual eigenvalue equation $\mathbf{B}^{-1}\mathbf{Aw} = \lambda\mathbf{w}$, $\mathbf{w} \neq \mathbf{0}$. We denote

(c) $\mathrm{ch}(\mathbf{A}, \mathbf{B}) = \mathrm{ch}(\mathbf{B}^{-1}\mathbf{A}) = \mathrm{ch}(\mathbf{A}\mathbf{B}^{-1}) = \{\lambda_1, \ldots, \lambda_n\}$.

The matrix $\mathbf{B}^{-1}\mathbf{A}$ is not necessarily symmetric but in view of

(d) $\mathrm{nzch}(\mathbf{B}^{-1}\mathbf{A}) = \mathrm{nzch}(\mathbf{B}^{-1/2}\mathbf{B}^{-1/2}\mathbf{A}) = \mathrm{nzch}(\mathbf{B}^{-1/2}\mathbf{A}\mathbf{B}^{-1/2})$,

and the symmetry of $\mathbf{B}^{-1/2}\mathbf{A}\mathbf{B}^{-1/2}$, the eigenvalues of $\mathbf{B}^{-1}\mathbf{A}$ are all real.

Premultiplying (a) by $\mathbf{B}^{-1/2}$ yields $\mathbf{B}^{-1/2}\mathbf{A}\mathbf{w} = \lambda\mathbf{B}^{1/2}\mathbf{w}$, i.e., $\mathbf{B}^{-1/2}\mathbf{A}\mathbf{B}^{-1/2} \cdot \mathbf{B}^{1/2}\mathbf{w} = \lambda\mathbf{B}^{1/2}\mathbf{w}$, which shows the equivalence of the following statements:

(e) $(\lambda, \mathbf{w})$ is an eigenpair for $(\mathbf{A}, \mathbf{B})$,

(f) $(\lambda, \mathbf{B}^{1/2}\mathbf{w})$ is an eigenpair for $\mathbf{B}^{-1/2}\mathbf{A}\mathbf{B}^{-1/2}$,

(g) $(\lambda, \mathbf{w})$ is an eigenpair for $\mathbf{B}^{-1}\mathbf{A}$.

22.23 Rewriting (b) above as $(\mathbf{A}-\lambda\mathbf{B})\mathbf{w} = \mathbf{0}$, we observe that nontrivial solutions $\mathbf{w}$ for (b) exist iff $\det(\mathbf{A} - \lambda\mathbf{B}) = 0$. The expression $\mathbf{A} - \lambda\mathbf{B}$, with indeterminate λ, is called a matrix pencil or simply a pencil.

22.24 Let $\mathbf{A}_{n\times n}$ be symmetric and $\mathbf{B}_{n\times n}$ positive definite. Then

(a)
$$\begin{aligned}\max_{\mathbf{x}\neq\mathbf{0}} \frac{\mathbf{x}'\mathbf{A}\mathbf{x}}{\mathbf{x}'\mathbf{B}\mathbf{x}} &= \max_{\mathbf{x}\neq\mathbf{0}} \frac{\mathbf{x}'\mathbf{B}^{1/2} \cdot \mathbf{B}^{-1/2}\mathbf{A}\mathbf{B}^{-1/2} \cdot \mathbf{B}^{1/2}\mathbf{x}}{\mathbf{x}'\mathbf{B}^{1/2} \cdot \mathbf{B}^{1/2}\mathbf{x}} \\ &= \max_{\mathbf{z}\neq\mathbf{0}} \frac{\mathbf{z}'\mathbf{B}^{-1/2}\mathbf{A}\mathbf{B}^{-1/2}\mathbf{z}}{\mathbf{z}'\mathbf{z}} \\ &= \mathrm{ch}_1(\mathbf{B}^{-1/2}\mathbf{A}\mathbf{B}^{-1/2}) = \mathrm{ch}_1(\mathbf{B}^{-1}\mathbf{A}) \\ &:= \lambda_1 = \text{the largest root of } \det(\mathbf{A} - \lambda\mathbf{B}) = 0.\end{aligned}$$

(b) Denote $\mathbf{W}_i = (\mathbf{w}_1 : \ldots : \mathbf{w}_i)$. The vectors $\mathbf{w}_1, \ldots, \mathbf{w}_n$ satisfy

$$\max_{\mathbf{x}\neq\mathbf{0}} \frac{\mathbf{x}'\mathbf{A}\mathbf{x}}{\mathbf{x}'\mathbf{B}\mathbf{x}} = \frac{\mathbf{w}_1'\mathbf{A}\mathbf{w}_1}{\mathbf{w}_1'\mathbf{B}\mathbf{w}_1} = \mathrm{ch}_1(\mathbf{B}^{-1}\mathbf{A}) = \lambda_1,$$

$$\max_{\substack{\mathbf{W}_{i-1}'\mathbf{B}\mathbf{x}=\mathbf{0}\\ \mathbf{x}\neq\mathbf{0}}} \frac{\mathbf{x}'\mathbf{A}\mathbf{x}}{\mathbf{x}'\mathbf{B}\mathbf{x}} = \frac{\mathbf{w}_i'\mathbf{A}\mathbf{w}_i}{\mathbf{w}_i'\mathbf{B}\mathbf{w}_i} = \mathrm{ch}_i(\mathbf{B}^{-1}\mathbf{A}) = \lambda_i, \quad i > 1,$$

iff $\mathbf{w}_i$ is an eigenvector of $\mathbf{B}^{-1}\mathbf{A}$ corresponding to the eigenvalue $\mathrm{ch}_i(\mathbf{B}^{-1}\mathbf{A}) = \lambda_i$, i.e., λ_i is the ith largest root of $\det(\mathbf{A} - \lambda\mathbf{B}) = 0$.

(c)
$$\max_{\mathbf{x}\neq\mathbf{0}} \frac{(\mathbf{a}'\mathbf{x})^2}{\mathbf{x}'\mathbf{B}\mathbf{x}} = \max_{\mathbf{x}\neq\mathbf{0}} \frac{\mathbf{x}' \cdot \mathbf{a}\mathbf{a}' \cdot \mathbf{x}}{\mathbf{x}'\mathbf{B}\mathbf{x}} = \mathrm{ch}_1(\mathbf{a}\mathbf{a}'\mathbf{B}^{-1}) = \mathbf{a}'\mathbf{B}^{-1}\mathbf{a}.$$

22.25 Let $\mathbf{B}_{n\times n}$ be nonnegative definite and $\mathbf{a} \in \mathscr{C}(\mathbf{B})$. Then

$$\max_{\mathbf{B}\mathbf{x}\neq\mathbf{0}} \frac{(\mathbf{a}'\mathbf{x})^2}{\mathbf{x}'\mathbf{B}\mathbf{x}} = \mathbf{a}'\mathbf{B}^{-}\mathbf{a},$$

where the equality is obtained iff $\mathbf{B}\mathbf{x} = \alpha\mathbf{a}$ for some $\alpha \in \mathbb{R}$.

22.26 Let $\mathbf{V}_{p\times p}$ be nnd with $\mathbf{V}_\delta = \mathrm{diag}(\mathbf{V})$ being pd. Then

$$\max_{\mathbf{a}\neq\mathbf{0}} \frac{\mathbf{a}'\mathbf{V}\mathbf{a}}{\mathbf{a}'\mathbf{V}_\delta\mathbf{a}} = \mathrm{ch}_1(\mathbf{V}_\delta^{-1/2}\mathbf{V}\mathbf{V}_\delta^{-1/2}) = \mathrm{ch}_1(\mathbf{R}_V),$$

where $\mathbf{R}_V = \mathbf{V}_\delta^{-1/2}\mathbf{V}\mathbf{V}_\delta^{-1/2}$ can be considered as a correlation matrix. Moreover,

$$\frac{\mathbf{a}'\mathbf{V}\mathbf{a}}{\mathbf{a}'\mathbf{V}_\delta\mathbf{a}} \leq p \quad \text{for all } \mathbf{a} \in \mathbb{R}^p,$$

where the equality is obtained iff $\mathbf{V} = \gamma^2\mathbf{q}\mathbf{q}'$ for some $\gamma \in \mathbb{R}$ and some $\mathbf{q} = (q_1, \dots, q_p)'$, and $\mathbf{a}$ is a multiple of $\mathbf{a}_* = \mathbf{V}_\delta^{-1/2}\mathbf{1} = \frac{1}{\gamma}(1/q_1, \dots, 1/q_p)'$.

22.27 Simultaneous diagonalization. Let $\mathbf{A}_{n\times n}$ and $\mathbf{B}_{n\times n}$ be symmetric. Then there exists an orthogonal matrix $\mathbf{Q}$ such that $\mathbf{Q}'\mathbf{A}\mathbf{Q}$ and $\mathbf{Q}'\mathbf{B}\mathbf{Q}$ are both diagonal iff $\mathbf{AB} = \mathbf{BA}$.

22.28 Consider the symmetric matrices $\mathbf{A}_{n\times n}$ and $\mathbf{B}_{n\times n}$.

(a) If $\mathbf{B}$ is pd, then there exists a nonsingular matrix $\mathbf{Q}_{n\times n}$ such that

$$\mathbf{Q}'\mathbf{A}\mathbf{Q} = \mathbf{\Lambda} = \operatorname{diag}(\lambda_1, \dots, \lambda_n), \quad \mathbf{Q}'\mathbf{B}\mathbf{Q} = \mathbf{I}_n,$$

where $\operatorname{ch}(\mathbf{B}^{-1}\mathbf{A}) = \{\lambda_1, \dots, \lambda_n\}$. The columns of $\mathbf{Q}$ are the eigenvectors of $\mathbf{B}^{-1}\mathbf{A}$; $\mathbf{Q}$ is not necessarily orthogonal.

(b) Let $\mathbf{B}$ be nnd with $\operatorname{r}(\mathbf{B}) = b$. Then there exists a matrix $\mathbf{L}_{n\times b}$ such that

$$\mathbf{L}'\mathbf{A}\mathbf{L} = \operatorname{diag}(\lambda_1, \dots, \lambda_b), \quad \mathbf{L}'\mathbf{B}\mathbf{L} = \mathbf{I}_b.$$

(c) Let $\mathbf{B}$ be nnd with $\operatorname{r}(\mathbf{B}) = b$, and assume that $\operatorname{r}(\mathbf{N}'\mathbf{A}\mathbf{N}) = \operatorname{r}(\mathbf{N}'\mathbf{A})$, where $\mathbf{N} = \mathbf{B}^\perp$. Then there exists a nonsingular matrix $\mathbf{Q}_{n\times n}$ such that

$$\mathbf{Q}'\mathbf{A}\mathbf{Q} = \begin{pmatrix} \mathbf{\Lambda}_1 & \mathbf{0} \\ \mathbf{0} & \mathbf{\Lambda}_2 \end{pmatrix}, \quad \mathbf{Q}'\mathbf{B}\mathbf{Q} = \begin{pmatrix} \mathbf{I}_b & \mathbf{0} \\ \mathbf{0} & \mathbf{0} \end{pmatrix},$$

where $\mathbf{\Lambda}_1 \in \mathbb{R}^{b\times b}$ and $\mathbf{\Lambda}_2 \in \mathbb{R}^{(n-b)\times(n-b)}$ are diagonal matrices.

22.29 As in 22.22 (p. 101), consider the eigenvalues of $\mathbf{A}$ w.r.t. nnd $\mathbf{B}$ but allow now $\mathbf{B}$ to be singular. Let λ be a scalar and $\mathbf{w}$ a vector such that

(a) $\mathbf{A}\mathbf{w} = \lambda\mathbf{B}\mathbf{w}$, i.e., $(\mathbf{A} - \lambda\mathbf{B})\mathbf{w} = \mathbf{0}$, $\quad \mathbf{B}\mathbf{w} \neq \mathbf{0}$.

Scalar λ is a proper eigenvalue and $\mathbf{w}$ a proper eigenvector of $\mathbf{A}$ with respect to $\mathbf{B}$. The nontrivial solutions $\mathbf{w}$ for (a) above exist iff $\det(\mathbf{A} - \lambda\mathbf{B}) = 0$. The matrix pencil $\mathbf{A} - \lambda\mathbf{B}$, with indeterminate λ, is is said to be singular if $\det(\mathbf{A} - \lambda\mathbf{B}) = 0$ is satisfied for any λ; otherwise the pencil is regular.

22.30 If λ_i is the ith largest proper eigenvalue of $\mathbf{A}$ with respect to $\mathbf{B}$, then we write

$$\operatorname{ch}_i(\mathbf{A}, \mathbf{B}) = \lambda_i, \quad \lambda_1 \geq \lambda_2 \geq \dots \geq \lambda_b, \quad b = \operatorname{r}(\mathbf{B}),$$
$$\operatorname{ch}(\mathbf{A}, \mathbf{B}) = \{\lambda_1, \dots, \lambda_b\} = \text{set of proper eigenvalues of } \mathbf{A} \text{ w.r.t. } \mathbf{B},$$
$$\operatorname{nzch}(\mathbf{A}, \mathbf{B}) = \text{set of nonzero proper eigenvalues of } \mathbf{A} \text{ w.r.t. } \mathbf{B}.$$

22.31 Let $\mathbf{A}_{n\times n}$ be symmetric, $\mathbf{B}_{n\times n}$ nnd, and $\mathrm{r}(\mathbf{B}) = b$, and assume that $\mathrm{r}(\mathbf{N}'\mathbf{A}\mathbf{N}) = \mathrm{r}(\mathbf{N}'\mathbf{A})$, where $\mathbf{N} = \mathbf{B}^{\perp}$. Then there are precisely b proper eigenvalues of $\mathbf{A}$ with respect to $\mathbf{B}$, $\mathrm{ch}(\mathbf{A},\mathbf{B}) = \{\lambda_1,\dots,\lambda_b\}$, some of which may be repeated or null. Also $\mathbf{w}_1,\dots,\mathbf{w}_b$, the corresponding eigenvectors, can be so chosen that if $\mathbf{w}_i$ is the ith column of $\mathbf{W}_{n\times b}$, then

$$\mathbf{W}'\mathbf{A}\mathbf{W} = \mathbf{\Lambda}_1 = \mathrm{diag}(\lambda_1,\dots,\lambda_b), \quad \mathbf{W}'\mathbf{B}\mathbf{W} = \mathbf{I}_b, \quad \mathbf{W}'\mathbf{A}\mathbf{N} = \mathbf{0}.$$

22.32 Suppose that $\mathrm{r}(\mathbf{Q_B}\mathbf{A}\mathbf{Q_B}) = \mathrm{r}(\mathbf{Q_B}\mathbf{A})$ holds; here $\mathbf{Q_B} = \mathbf{I}_n - \mathbf{P_B}$. Then

(a) $\mathrm{nzch}(\mathbf{A},\mathbf{B}) = \mathrm{nzch}[(\mathbf{A} - \mathbf{A}\mathbf{Q_B}(\mathbf{Q_B}\mathbf{A}\mathbf{Q_B})^{-}\mathbf{Q_B}\mathbf{A})\mathbf{B}^{-}]$,

(b) $\mathscr{C}(\mathbf{A}) \subset \mathscr{C}(\mathbf{B}) \implies \mathrm{nzch}(\mathbf{A},\mathbf{B}) = \mathrm{nzch}(\mathbf{A}\mathbf{B}^{-})$,

where the set $\mathrm{nzch}(\mathbf{A}\mathbf{B}^{-})$ is invariant with respect to the choice of the $\mathbf{B}^{-}$.

22.33 Consider the linear model $\{\mathbf{y}, \mathbf{X}\boldsymbol{\beta}, \mathbf{V}\}$. The nonzero proper eigenvalues of $\mathbf{V}$ with respect to $\mathbf{H}$ are the same as the nonzero eigenvalues of the covariance matrix of the $\mathrm{BLUE}(\mathbf{X}\boldsymbol{\beta})$.

22.34 If $\mathbf{A}_{n\times n}$ is symmetric and $\mathbf{B}_{n\times n}$ is nnd satisfying $\mathscr{C}(\mathbf{A}) \subset \mathscr{C}(\mathbf{B})$, then

$$\max_{\mathbf{Bx}\neq\mathbf{0}} \frac{\mathbf{x}'\mathbf{A}\mathbf{x}}{\mathbf{x}'\mathbf{B}\mathbf{x}} = \frac{\mathbf{w}_1'\mathbf{A}\mathbf{w}_1}{\mathbf{w}_1'\mathbf{B}\mathbf{w}_1} = \lambda_1 = \mathrm{ch}_1(\mathbf{B}^{+}\mathbf{A}) = \mathrm{ch}_1(\mathbf{A},\mathbf{B}),$$

where λ_1 is the largest proper eigenvalue of $\mathbf{A}$ with respect to $\mathbf{B}$ and $\mathbf{w}_1$ the corresponding proper eigenvector satisfying $\mathbf{A}\mathbf{w}_1 = \lambda_1\mathbf{B}\mathbf{w}_1$, $\mathbf{B}\mathbf{w}_1 \neq \mathbf{0}$. If $\mathscr{C}(\mathbf{A}) \subset \mathscr{C}(\mathbf{B})$ does not hold, then $\mathbf{x}'\mathbf{A}\mathbf{x}/\mathbf{x}'\mathbf{B}\mathbf{x}$ has no upper bound.

22.35 Under a weakly singular linear model where $\mathrm{r}(\mathbf{X}) = r$ we have

$$\max_{\mathbf{Vx}\neq\mathbf{0}} \frac{\mathbf{x}'\mathbf{H}\mathbf{x}}{\mathbf{x}'\mathbf{V}\mathbf{x}} = \lambda_1 = \mathrm{ch}_1(\mathbf{V}^{+}\mathbf{H}) = \mathrm{ch}_1(\mathbf{H}\mathbf{V}^{+}\mathbf{H}) = 1/\,\mathrm{ch}_r(\mathbf{H}\mathbf{V}^{+}\mathbf{H})^{+},$$

and hence $1/\lambda_1$ is the smallest nonzero eigenvalue of $\mathrm{cov}(\mathbf{X}\tilde{\boldsymbol{\beta}})$.

22.36 Consider symmetric $n \times n$ matrices $\mathbf{A}$ and $\mathbf{B}$. Then

$$\sum_{i=1}^{n} \mathrm{ch}_i(\mathbf{A})\,\mathrm{ch}_{n-i-1}(\mathbf{B}) \leq \mathrm{tr}(\mathbf{A}\mathbf{B}) \leq \sum_{i=1}^{n} \mathrm{ch}_i(\mathbf{A})\,\mathrm{ch}_i(\mathbf{B}).$$

Suppose that $\mathbf{A} \in \mathbb{R}^{n\times p}$, $\mathbf{B} \in \mathbb{R}^{p\times n}$, and $k = \min(\mathrm{r}(\mathbf{A}), \mathrm{r}(\mathbf{B}))$. Then

$$-\sum_{i=1}^{k} \mathrm{sg}_i(\mathbf{A})\,\mathrm{sg}_i(\mathbf{B}) \leq \mathrm{tr}(\mathbf{A}\mathbf{B}) \leq \sum_{i=1}^{k} \mathrm{sg}_i(\mathbf{A})\,\mathrm{sg}_i(\mathbf{B}).$$

22.37 Suppose that $\mathbf{A}$, $\mathbf{B}$ and $\mathbf{A} - \mathbf{B}$ are nnd $n \times n$ matrices. Then

$$\mathrm{ch}_i(\mathbf{A} - \mathbf{B}) \geq \mathrm{ch}_{i+k}(\mathbf{A}), \quad i = 1,\dots,n-k,$$

and with equality for all i iff $\mathbf{B} = \mathbf{T}_{(k)}\mathbf{\Lambda}_{(k)}\mathbf{T}'_{(k)}$, where $\mathbf{\Lambda}_{(k)}$ comprises the first k eigenvalues of $\mathbf{A}$ and $\mathbf{T}_{(k)} = (\mathbf{t}_1 : \ldots : \mathbf{t}_k)$.

22.38 For $\mathbf{A}, \mathbf{B} \in \mathbb{R}^{n\times m}$, where $\mathrm{r}(\mathbf{A}) = r$ and $\mathrm{r}(\mathbf{B}) = k$, we have

$$\mathrm{sg}_i(\mathbf{A} - \mathbf{B}) \geq \mathrm{sg}_{i+k}(\mathbf{A}), \quad i + k \leq r,$$

and $\mathrm{sg}_i(\mathbf{A} - \mathbf{B}) \geq 0$ for $i + k > r$. The equality is attained above iff $k \leq r$ and $\mathbf{B} = \sum_{i=1}^{k} \mathrm{sg}_i(\mathbf{A})\mathbf{t}_i\mathbf{u}_i'$, where $\mathbf{A} = \sum_{i=1}^{r} \mathrm{sg}_i(\mathbf{A})\mathbf{t}_i\mathbf{u}_i'$ is the SVD of $\mathbf{A}$.

23 Singular value decomposition & other matrix decompositions

SVD Singular value decomposition. Matrix $\mathbf{A} \in \mathbb{R}_r^{n\times m}$, $(m \leq n)$, can be written as

$$\begin{aligned}\mathbf{A} &= (\mathbf{U}_1 : \mathbf{U}_0)\begin{pmatrix}\mathbf{\Delta}_1 & \mathbf{0}\\ \mathbf{0} & \mathbf{0}\end{pmatrix}\begin{pmatrix}\mathbf{V}_1'\\ \mathbf{V}_0'\end{pmatrix} = \mathbf{U}\mathbf{\Delta}\mathbf{V}' = \mathbf{U}_1\mathbf{\Delta}_1\mathbf{V}_1' = \mathbf{U}_*\mathbf{\Delta}_*\mathbf{V}'\\ &= \delta_1\mathbf{u}_1\mathbf{v}_1' + \cdots + \delta_r\mathbf{u}_r\mathbf{v}_r',\end{aligned}$$

where $\mathbf{\Delta}_1 = \mathrm{diag}(\delta_1, \ldots, \delta_r)$, $\delta_1 \geq \cdots \geq \delta_r > 0$, $\mathbf{\Delta} \in \mathbb{R}^{n\times m}$, and

$$\mathbf{\Delta} = \begin{pmatrix}\mathbf{\Delta}_1 & \mathbf{0}\\ \mathbf{0} & \mathbf{0}\end{pmatrix} = \begin{pmatrix}\mathbf{\Delta}_*\\ \mathbf{0}\end{pmatrix} \in \mathbb{R}^{n\times m}, \quad \mathbf{\Delta}_1 \in \mathbb{R}^{r\times r}, \quad \mathbf{\Delta}_* \in \mathbb{R}^{m\times m},$$

$$\mathbf{\Delta}_* = \mathrm{diag}(\delta_1, \ldots, \delta_r, \delta_{r+1}, \ldots, \delta_m) = \text{the first } m \text{ rows of } \mathbf{\Delta},$$

$$\delta_{r+1} = \delta_{r+2} = \cdots = \delta_m = 0,$$

$$\begin{aligned}\delta_i &= \mathrm{sg}_i(\mathbf{A}) = \sqrt{\mathrm{ch}_i(\mathbf{A}'\mathbf{A})}\\ &= i\text{th singular value of } \mathbf{A}, \quad i = 1, \ldots, m,\end{aligned}$$

$$\mathbf{U}_{n\times n} = (\mathbf{U}_1 : \mathbf{U}_0), \quad \mathbf{U}_1 \in \mathbb{R}^{n\times r}, \quad \mathbf{U}'\mathbf{U} = \mathbf{U}\mathbf{U}' = \mathbf{I}_n,$$

$$\mathbf{V}_{m\times m} = (\mathbf{V}_1 : \mathbf{V}_0), \quad \mathbf{V}_1 \in \mathbb{R}^{m\times r}, \quad \mathbf{V}'\mathbf{V} = \mathbf{V}\mathbf{V}' = \mathbf{I}_m,$$

$$\mathbf{U}_* = (\mathbf{u}_1 : \ldots : \mathbf{u}_m) = \text{the first } m \text{ columns of } \mathbf{U}, \quad \mathbf{U}_* \in \mathbb{R}^{n\times m},$$

$$\mathbf{V}'\mathbf{A}'\mathbf{A}\mathbf{V} = \mathbf{\Delta}'\mathbf{\Delta} = \mathbf{\Delta}_*^2 = \begin{pmatrix}\mathbf{\Delta}_1^2 & \mathbf{0}\\ \mathbf{0} & \mathbf{0}\end{pmatrix} \in \mathbb{R}^{m\times m},$$

$$\mathbf{U}'\mathbf{A}\mathbf{A}'\mathbf{U} = \mathbf{\Delta}\mathbf{\Delta}' = \mathbf{\Delta}_\#^2 = \begin{pmatrix}\mathbf{\Delta}_1^2 & \mathbf{0}\\ \mathbf{0} & \mathbf{0}\end{pmatrix} \in \mathbb{R}^{n\times n},$$

$$\mathbf{u}_i = i\text{th left singular vector of } \mathbf{A};\ i\text{th eigenvector of } \mathbf{A}\mathbf{A}',$$

$$\mathbf{v}_i = i\text{th right singular vector of } \mathbf{A};\ i\text{th eigenvector of } \mathbf{A}'\mathbf{A}.$$

23.1 With the above notation,

(a) $\{\delta_1^2, \ldots, \delta_r^2\} = \mathrm{nzch}(\mathbf{A}'\mathbf{A}) = \mathrm{nzch}(\mathbf{A}\mathbf{A}')$,

(b) $\mathscr{C}(\mathbf{V}_1) = \mathscr{C}(\mathbf{A})$, $\quad \mathscr{C}(\mathbf{U}_1) = \mathscr{C}(\mathbf{A}')$,

(c) $\mathbf{A}\mathbf{v}_i = \delta_i\mathbf{u}_i$, $\quad \mathbf{A}'\mathbf{u}_i = \delta_i\mathbf{v}_i$, $\quad i = 1, \ldots, m$,

(d) $\mathbf{A}'\mathbf{u}_i = \mathbf{0}, \quad i = m+1, \dots, n,$

(e) $\mathbf{u}_i'\mathbf{A}\mathbf{v}_i = \delta_i, \quad i = 1, \dots, m, \quad \mathbf{u}_i'\mathbf{A}\mathbf{v}_j = 0 \text{ for } i \neq j.$

23.2 Not all pairs of orthogonal $\mathbf{U}$ and $\mathbf{V}$ satisfying $\mathbf{U}'\mathbf{A}\mathbf{A}'\mathbf{U} = \mathbf{\Delta}_\#^2$ and $\mathbf{V}'\mathbf{A}'\mathbf{A}\mathbf{V} = \mathbf{\Delta}_*^2$ yield $\mathbf{A} = \mathbf{U}\mathbf{\Delta}\mathbf{V}'$.

23.3 Let $\mathbf{A}$ have an SVD $\mathbf{A} = \mathbf{U}\mathbf{\Delta}\mathbf{V}' = \delta_1\mathbf{u}_1\mathbf{v}_1' + \cdots + \delta_r\mathbf{u}_r\mathbf{v}_r'$, $\mathrm{r}(\mathbf{A}) = r$. Then

$$\mathrm{sg}_1(\mathbf{A}) = \delta_1 = \max \mathbf{x}'\mathbf{A}\mathbf{y} \quad \text{subject to} \quad \mathbf{x}'\mathbf{x} = \mathbf{y}'\mathbf{y} = 1,$$

and the maximum is obtained when $\mathbf{x} = \mathbf{u}_1$ and $\mathbf{y} = \mathbf{v}_1$,

$$\mathrm{sg}_1^2(\mathbf{A}) = \delta_1^2 = \max_{\mathbf{x}\neq\mathbf{0},\, \mathbf{y}\neq\mathbf{0}} \frac{(\mathbf{x}'\mathbf{A}\mathbf{y})^2}{\mathbf{x}'\mathbf{x}\cdot\mathbf{y}'\mathbf{y}} = \mathrm{ch}_1(\mathbf{A}'\mathbf{A}).$$

The second largest singular value δ_2 can be obtained as $\delta_2 = \max \mathbf{x}'\mathbf{A}\mathbf{y}$, where the maximum is taken over the set

$$\{\, \mathbf{x} \in \mathbb{R}^n,\ \mathbf{y} \in \mathbb{R}^m : \mathbf{x}'\mathbf{x} = \mathbf{y}'\mathbf{y} = 1,\ \mathbf{x}'\mathbf{u}_1 = \mathbf{y}'\mathbf{v}_1 = 0 \,\}.$$

The ith largest singular value can be defined correspondingly as

$$\max_{\substack{\mathbf{x}\neq\mathbf{0},\, \mathbf{y}\neq\mathbf{0} \\ \mathbf{U}_{(k)}\mathbf{x}=\mathbf{0},\, \mathbf{V}_{(k)}\mathbf{y}=\mathbf{0}}} \frac{\mathbf{x}'\mathbf{A}\mathbf{y}}{\sqrt{\mathbf{x}'\mathbf{x}\cdot\mathbf{y}'\mathbf{y}}} = \delta_{k+1}, \quad k = 1, \dots, r-1,$$

where $\mathbf{U}_{(k)} = (\mathbf{u}_1 : \ldots : \mathbf{u}_k)$ and $\mathbf{V}_{(k)} = (\mathbf{v}_1 : \ldots : \mathbf{v}_k)$; the maximum occurs when $\mathbf{x} = \mathbf{u}_{k+1}$ and $\mathbf{y} = \mathbf{v}_{k+1}$.

23.4 Let $\mathbf{A} \in \mathbb{R}^{n\times m}$, $\mathbf{B} \in \mathbb{R}^{n\times n}$, $\mathbf{C} \in \mathbb{R}^{m\times m}$ and $\mathbf{B}$ and $\mathbf{C}$ are pd. Then

$$\begin{aligned}\max_{\mathbf{x}\neq\mathbf{0},\, \mathbf{y}\neq\mathbf{0}} \frac{(\mathbf{x}'\mathbf{A}\mathbf{y})^2}{\mathbf{x}'\mathbf{B}\mathbf{x}\cdot\mathbf{y}'\mathbf{C}\mathbf{y}} &= \max_{\mathbf{x}\neq\mathbf{0},\, \mathbf{y}\neq\mathbf{0}} \frac{(\mathbf{x}'\mathbf{B}^{1/2}\mathbf{B}^{-1/2}\mathbf{A}\mathbf{C}^{-1/2}\mathbf{C}^{1/2}\mathbf{y})^2}{\mathbf{x}'\mathbf{B}^{1/2}\mathbf{B}^{1/2}\mathbf{x}\cdot\mathbf{y}'\mathbf{C}^{1/2}\mathbf{C}^{1/2}\mathbf{y}} \\ &= \max_{\mathbf{t}\neq\mathbf{0},\, \mathbf{u}\neq\mathbf{0}} \frac{(\mathbf{t}'\mathbf{B}^{-1/2}\mathbf{A}\mathbf{C}^{-1/2}\mathbf{u})^2}{\mathbf{t}'\mathbf{t}\cdot\mathbf{u}'\mathbf{u}} = \mathrm{sg}_1^2(\mathbf{B}^{-1/2}\mathbf{A}\mathbf{C}^{-1/2}).\end{aligned}$$

With minor changes the above holds for possibly singular $\mathbf{B}$ and $\mathbf{C}$.

23.5 The matrix 2-norm (or the spectral norm) is defined as

$$\begin{aligned}\|\mathbf{A}\|_2 &= \max_{\|\mathbf{x}\|_2=1} \|\mathbf{A}\mathbf{x}\|_2 = \max_{\mathbf{x}\neq\mathbf{0}} \left(\frac{\mathbf{x}'\mathbf{A}'\mathbf{A}\mathbf{x}}{\mathbf{x}'\mathbf{x}}\right)^{1/2} \\ &= \sqrt{\mathrm{ch}_1(\mathbf{A}'\mathbf{A})} = \mathrm{sg}_1(\mathbf{A}),\end{aligned}$$

where $\|\mathbf{x}\|_2$ refers to the standard Euclidean vector norm, and $\mathrm{sg}_i(\mathbf{A}) = \sqrt{\mathrm{ch}_i(\mathbf{A}'\mathbf{A})} = \delta_i =$ the ith largest singular value of $\mathbf{A}$. Obviously we have $\|\mathbf{A}\mathbf{x}\|_2 \leq \|\mathbf{A}\|_2\|\mathbf{x}\|_2$. Recall that

$$\|\mathbf{A}\|_F = \sqrt{\delta_1^2 + \cdots + \delta_r^2}, \quad \|\mathbf{A}\|_2 = \delta_1, \quad \text{where } r = \mathrm{rank}(\mathbf{A}).$$

EY2 Eckart–Young theorem. Let $\mathbf{A}_{n\times m}$ be a given matrix of rank r, with the singular value decomposition $\mathbf{A} = \mathbf{U}_1\mathbf{\Delta}_1\mathbf{V}_1' = \delta_1\mathbf{u}_1\mathbf{v}_1' + \cdots + \delta_r\mathbf{u}_r\mathbf{v}_r'$. Let $\mathbf{B}$ be an $n \times m$ matrix of rank k $(< r)$. Then

$$\min_{\mathbf{B}} \|\mathbf{A} - \mathbf{B}\|_F^2 = \delta_{k+1}^2 + \cdots + \delta_r^2,$$

and the minimum is attained taking $\mathbf{B} = \dot{\mathbf{B}}_k = \delta_1\mathbf{u}_1\mathbf{v}_1' + \cdots + \delta_k\mathbf{u}_k\mathbf{v}_k'$.

23.6 Let $\mathbf{A}_{n\times m}$ and $\mathbf{B}_{m\times n}$ have the SVDs $\mathbf{A} = \mathbf{U}\mathbf{\Delta}_{\mathbf{A}}\mathbf{V}'$, $\mathbf{B} = \mathbf{R}\mathbf{\Delta}_{\mathbf{B}}\mathbf{S}'$, where $\mathbf{\Delta}_{\mathbf{A}} = \operatorname{diag}(\alpha_1, \dots, \alpha_r)$, $\mathbf{\Delta}_{\mathbf{B}} = \operatorname{diag}(\beta_1, \dots, \beta_r)$, and $r = \min(n, m)$. Then

$$|\operatorname{tr}(\mathbf{AXBY})| \leq \alpha_1\beta_1 + \cdots + \alpha_r\beta_r$$

for all orthogonal $\mathbf{X}_{m\times m}$ and $\mathbf{Y}_{n\times n}$. The upper bound is attained when $\mathbf{X} = \mathbf{V}\mathbf{R}'$ and $\mathbf{Y} = \mathbf{S}\mathbf{U}'$.

23.7 Let $\mathbf{A}$ and $\mathbf{B}$ be $n \times m$ matrices with SVDs corresponding to 23.6. Then

(a) the minimum of $\|\mathbf{XA} - \mathbf{BY}\|_F$ when $\mathbf{X}$ and $\mathbf{Y}$ run through orthogonal matrices is attained at $\mathbf{X} = \mathbf{R}\mathbf{U}'$ and $\mathbf{Y} = \mathbf{S}\mathbf{V}'$,

(b) the minimum of $\|\mathbf{A} - \mathbf{BZ}\|_F$ when $\mathbf{Z}$ varies over orthogonal matrices is attained when $\mathbf{Z} = \mathbf{L}\mathbf{K}'$, where $\mathbf{L}\mathbf{\Delta}\mathbf{K}'$ is the SVD of $\mathbf{B}'\mathbf{A}$.

FRD Full rank decomposition of $\mathbf{A}_{n\times m}$, $\operatorname{r}(\mathbf{A}) = r$:

$$\mathbf{A} = \mathbf{B}\mathbf{C}', \quad \text{where } \operatorname{r}(\mathbf{B}_{n\times r}) = \operatorname{r}(\mathbf{C}_{m\times r}) = r.$$

CHO Cholesky decomposition. Let $\mathbf{A}_{n\times n}$ be nnd. Then there exists a lower triangular matrix $\mathbf{U}_{n\times n}$ having all $u_{ii} \geq 0$ such that $\mathbf{A} = \mathbf{U}\mathbf{U}'$.

QRD QR-decomposition. Matrix $\mathbf{A}_{n\times m}$ can be expressed as $\mathbf{A} = \mathbf{Q}_{n\times n}\mathbf{R}_{n\times m}$, where $\mathbf{Q}$ is orthogonal and $\mathbf{R}$ is upper triangular.

POL Polar decomposition. Matrix $\mathbf{A}_{n\times m}$ $(n \leq m)$ can be expressed as $\mathbf{A} = \mathbf{P}_{n\times n}\mathbf{U}_{n\times m}$, where $\mathbf{P}$ is nonnegative definite with $\operatorname{r}(\mathbf{P}) = \operatorname{r}(\mathbf{A})$ and $\mathbf{U}\mathbf{U}' = \mathbf{I}_n$.

SCH Schur's triangularization theorem. Let $\mathbf{A}_{n\times n}$ have real eigenvalues $\lambda_1, \dots, \lambda_n$. Then there exists an orthogonal $\mathbf{U}_{n\times n}$ such that $\mathbf{U}'\mathbf{A}\mathbf{U} = \mathbf{T}$, where $\mathbf{T}$ is an upper-triangular matrix with λ_i's as its diagonal elements.

ROT Orthogonal rotation. Denote

$$\mathbf{A}_\theta = \begin{pmatrix} \cos\theta & -\sin\theta \\ \sin\theta & \cos\theta \end{pmatrix}, \quad \mathbf{B}_\theta = \begin{pmatrix} \cos\theta & \sin\theta \\ \sin\theta & -\cos\theta \end{pmatrix}.$$

Then any 2×2 orthogonal matrix $\mathbf{Q}$ is $\mathbf{A}_\theta$ or $\mathbf{B}_\theta$ for some θ. Transformation $\mathbf{A}_\theta\mathbf{u}_{(i)}$ rotates the observation $\mathbf{u}_{(i)}$ by the angle θ in the counter-clockwise direction, and $\mathbf{B}_\theta \mathbf{u}_{(i)}$ makes the reflection of the observation $\mathbf{u}_{(i)}$ w.r.t. the line $y = \tan(\frac{\theta}{2})x$. Matrix

$$\mathbf{C}_\theta = \begin{pmatrix} \cos\theta & \sin\theta \\ -\sin\theta & \cos\theta \end{pmatrix}$$

carries out an orthogonal rotation clockwise.

HSD Hartwig–Spindelböck decomposition. Let $\mathbf{A} \in \mathbb{R}^{n\times n}$ be of rank r. Then there exists an orthogonal $\mathbf{U} \in \mathbb{R}^{n\times n}$ such that

$$\mathbf{A} = \mathbf{U}\begin{pmatrix} \mathbf{\Delta K} & \mathbf{\Delta L} \\ \mathbf{0} & \mathbf{0} \end{pmatrix}\mathbf{U}', \quad \text{where } \mathbf{\Delta} = \operatorname{diag}(\delta_1 \mathbf{I}_{r_1}, \dots, \delta_t \mathbf{I}_{r_t}),$$

with $\mathbf{\Delta}$ being the diagonal matrix of singular values of $\mathbf{A}$, $\delta_1 > \dots > \delta_t > 0$, $r_1 + \dots + r_t = r$, while $\mathbf{K} \in \mathbb{R}^{r\times r}$, $\mathbf{L} \in \mathbb{R}^{r\times(n-r)}$ satisfy $\mathbf{KK}' + \mathbf{LL}' = \mathbf{I}_r$.

23.8 Consider the matrix $\mathbf{A} \in \mathbb{R}^{n\times n}$ with representation HSD. Then:

(a) $\mathbf{A}'\mathbf{A} = \mathbf{U}\begin{pmatrix} \mathbf{K}'\mathbf{\Delta}^2\mathbf{K} & \mathbf{K}'\mathbf{\Delta}^2\mathbf{L} \\ \mathbf{L}'\mathbf{\Delta}^2\mathbf{K} & \mathbf{L}'\mathbf{\Delta}^2\mathbf{L} \end{pmatrix}\mathbf{U}', \quad \mathbf{AA}' = \mathbf{U}\begin{pmatrix} \mathbf{\Delta}^2 & \mathbf{0} \\ \mathbf{0} & \mathbf{0} \end{pmatrix}\mathbf{U}',$

$\mathbf{A}^+\mathbf{A} = \mathbf{U}\begin{pmatrix} \mathbf{K}'\mathbf{K} & \mathbf{K}'\mathbf{L} \\ \mathbf{L}'\mathbf{K} & \mathbf{L}'\mathbf{L} \end{pmatrix}\mathbf{U}', \quad \mathbf{AA}^+ = \mathbf{U}\begin{pmatrix} \mathbf{I}_r & \mathbf{0} \\ \mathbf{0} & \mathbf{0} \end{pmatrix}\mathbf{U}',$

(b) $\mathbf{A}$ is an oblique projector, i.e., $\mathbf{A}^2 = \mathbf{A}$, iff $\mathbf{\Delta K} = \mathbf{I}_r$,

(c) $\mathbf{A}$ is an orthogonal projector iff $\mathbf{L} = \mathbf{0}$, $\mathbf{\Delta} = \mathbf{I}_r$, $\mathbf{K} = \mathbf{I}_r$.

23.9 A matrix $\mathbf{E} \in \mathbb{R}^{n\times n}$ is a general permutation matrix if it is a product of elementary permutation matrices $\mathbf{E}_{ij}$; $\mathbf{E}_{ij}$ is the identity matrix $\mathbf{I}_n$ with the ith and jth rows (or equivalently columns) interchanged.

23.10 (a) $\mathbf{A}_{n\times n}$ is nonnegative if all $a_{ij} \geq 0$,

(b) $\mathbf{A}_{n\times n}$ is reducible if there is a permutation matrix $\mathbf{Q}$, such that

$$\mathbf{Q}'\mathbf{AQ} = \begin{pmatrix} \mathbf{A}_{11} & \mathbf{A}_{12} \\ \mathbf{0} & \mathbf{A}_{22} \end{pmatrix},$$

where $\mathbf{A}_{11}$ and $\mathbf{A}_{22}$ are square, and it is otherwise irreducible.

PFT Perron–Frobenius theorem. If $\mathbf{A}_{n\times n}$ is nonnegative and irreducible, then

(a) $\mathbf{A}$ has a positive eigenvalue, ϱ, equal to the spectral radius of $\mathbf{A}$, $\rho(\mathbf{A}) = \max\{|\operatorname{ch}_i(\mathbf{A})|\}$; the eigenvalue corresponding to $\rho(\mathbf{A})$ is called the dominant eigenvalue,

(b) ϱ has multiplicity 1,

(c) there is a positive eigenvector (all elements ≥ 0) corresponding to ϱ.

23.11 Stochastic matrix. If $\mathbf{A} \in \mathbb{R}^{n \times n}$ is nonnegative and $\mathbf{A1}_n = \mathbf{1}_n$, then $\mathbf{A}$ is a stochastic matrix. If also $\mathbf{1}_n'\mathbf{A} = \mathbf{1}_n'$, then $\mathbf{A}$ is a doubly stochastic matrix. If, in addition, both diagonal sums are 1, then $\mathbf{A}$ is a superstochastic matrix.

23.12 Magic square. The matrix

$$\mathbf{A} = \begin{pmatrix} 16 & 3 & 2 & 13 \\ 5 & 10 & 11 & 8 \\ 9 & 6 & 7 & 12 \\ 4 & 15 & 14 & 1 \end{pmatrix},$$

appearing in Albrecht Dürer's copper-plate engraving *Melencolia I* is a magic square, i.e., a $k \times k$ array such that the numbers in every row, column and in each of the two main diagonals add up to the same magic sum, 34 in this case. The matrix $\mathbf{A}$ here defines a classic magic square since the entries in $\mathbf{A}$ are the consecutive integers $1, 2, \dots, k^2$. The Moore–Penrose inverse $\mathbf{A}^+$ also a magic square (though not a classic one) and its magic sum is $1/34$.

24 Löwner ordering

24.1 Let $\mathbf{A}_{n \times n}$ be symmetric.

(a) The following statements are equivalent:

(i) $\mathbf{A}$ is positive definite; shortly $\mathbf{A} \in \mathrm{PD}_n$, $\mathbf{A} >_{\mathsf{L}} \mathbf{0}$

(ii) $\mathbf{x}'\mathbf{A}\mathbf{x} > 0$ for all vectors $\mathbf{x} \neq \mathbf{0}$

(iii) $\mathrm{ch}_i(\mathbf{A}) > 0$ for $i = 1, \dots, n$

(iv) $\mathbf{A} = \mathbf{FF}'$ for some $\mathbf{F}_{n \times n}$, $\mathrm{r}(\mathbf{F}) = n$

(v) all leading principal minors > 0

(vi) $\mathbf{A}^{-1}$ is positive definite

(b) The following statements are equivalent:

(i) $\mathbf{A}$ is nonnegative definite; shortly $\mathbf{A} \in \mathrm{NND}_n$, $\mathbf{A} \geq_{\mathsf{L}} \mathbf{0}$

(ii) $\mathbf{x}'\mathbf{A}\mathbf{x} \geq 0$ for all vectors $\mathbf{x}$

(iii) $\mathrm{ch}_i(\mathbf{A}) \geq 0$ for $i = 1, \dots, n$

(iv) $\mathbf{A} = \mathbf{FF}'$ for some matrix $\mathbf{F}$

(v) all principal minors ≥ 0

24.2 $\mathbf{A} \leq_{\mathsf{L}} \mathbf{B} \iff \mathbf{B} - \mathbf{A} \geq_{\mathsf{L}} \mathbf{0} \iff \mathbf{B} - \mathbf{A} = \mathbf{FF}'$ for some matrix $\mathbf{F}$

(definition of Löwner ordering)

24.3 Let $\mathbf{A}, \mathbf{B} \in \text{NND}_n$. Then

(a) $\mathbf{A} \leq_{\mathsf{L}} \mathbf{B} \iff \mathscr{C}(\mathbf{A}) \subset \mathscr{C}(\mathbf{B})$ & $\text{ch}_1(\mathbf{AB}^-) \leq 1$
in which case $\text{ch}_1(\mathbf{AB}^-)$ is invariant with respect to $\mathbf{B}^-$,

(b) $\mathbf{A} \leq_{\mathsf{L}} \mathbf{B} \iff \mathscr{C}(\mathbf{A}) \subset \mathscr{C}(\mathbf{B})$ & $\mathbf{AB}^-\mathbf{A} \leq_{\mathsf{L}} \mathbf{A}$.

24.4 Let $\mathbf{A} \geq_{\mathsf{L}} \mathbf{B}$ where $\mathbf{A} >_{\mathsf{L}} \mathbf{0}$ and $\mathbf{B} \geq_{\mathsf{L}} \mathbf{0}$. Then $\det(\mathbf{A}) = \det(\mathbf{B}) \implies \mathbf{A} = \mathbf{B}$.

24.5 Albert's theorem. Let $\mathbf{A} = \begin{pmatrix} \mathbf{A}_{11} & \mathbf{A}_{12} \\ \mathbf{A}_{21} & \mathbf{A}_{22} \end{pmatrix}$ be a symmetric matrix where $\mathbf{A}_{11}$ is a square matrix. Then the following three statements are equivalent:

(a) $\mathbf{A} \geq_{\mathsf{L}} \mathbf{0}$,

(b$_1$) $\mathbf{A}_{11} \geq_{\mathsf{L}} \mathbf{0}$, (b$_2$) $\mathscr{C}(\mathbf{A}_{12}) \subset \mathscr{C}(\mathbf{A}_{11})$, (b$_3$) $\mathbf{A}_{22} - \mathbf{A}_{21}\mathbf{A}_{11}^-\mathbf{A}_{12} \geq_{\mathsf{L}} \mathbf{0}$,

(c$_1$) $\mathbf{A}_{22} \geq_{\mathsf{L}} \mathbf{0}$, (c$_2$) $\mathscr{C}(\mathbf{A}_{21}) \subset \mathscr{C}(\mathbf{A}_{22})$, (c$_3$) $\mathbf{A}_{11} - \mathbf{A}_{12}\mathbf{A}_{22}^-\mathbf{A}_{21} \geq_{\mathsf{L}} \mathbf{0}$.

24.6 (Continued ...) The following three statements are equivalent:

(a) $\mathbf{A} >_{\mathsf{L}} \mathbf{0}$,

(b$_1$) $\mathbf{A}_{11} >_{\mathsf{L}} \mathbf{0}$, (b$_2$) $\mathbf{A}_{22} - \mathbf{A}_{21}\mathbf{A}_{11}^{-1}\mathbf{A}_{12} >_{\mathsf{L}} \mathbf{0}$,

(c$_1$) $\mathbf{A}_{22} >_{\mathsf{L}} \mathbf{0}$, (c$_2$) $\mathbf{A}_{11} - \mathbf{A}_{12}\mathbf{A}_{22}^{-1}\mathbf{A}_{21} >_{\mathsf{L}} \mathbf{0}$.

24.7 The following three statements are equivalent:

(a) $\mathbf{A} = \begin{pmatrix} \mathbf{B} & \mathbf{b} \\ \mathbf{b}' & \alpha \end{pmatrix} \geq_{\mathsf{L}} \mathbf{0}$,

(b) $\mathbf{B} - \mathbf{bb}'/\alpha \geq_{\mathsf{L}} \mathbf{0}$,

(c) $\mathbf{B} \geq_{\mathsf{L}} \mathbf{0}$, $\mathbf{b} \in \mathscr{C}(\mathbf{B})$, $\mathbf{b}'\mathbf{B}^-\mathbf{b} \leq \alpha$.

24.8 Let $\mathbf{U}$ and $\mathbf{V}$ be pd. Then:

(a) $\mathbf{U} - \mathbf{B}'\mathbf{V}^{-1}\mathbf{B} >_{\mathsf{L}} \mathbf{0} \iff \mathbf{V} - \mathbf{BU}^{-1}\mathbf{B}' >_{\mathsf{L}} \mathbf{0}$,

(b) $\mathbf{U} - \mathbf{B}'\mathbf{V}^{-1}\mathbf{B} \geq_{\mathsf{L}} \mathbf{0} \iff \mathbf{V} - \mathbf{BU}^{-1}\mathbf{B}' \geq_{\mathsf{L}} \mathbf{0}$.

24.9 Let $\mathbf{A} >_{\mathsf{L}} \mathbf{0}$ and $\mathbf{B} >_{\mathsf{L}} \mathbf{0}$. Then: $\mathbf{A} <_{\mathsf{L}} \mathbf{B} \iff \mathbf{A}^{-1} >_{\mathsf{L}} \mathbf{B}^{-1}$.

24.10 Let $\mathbf{A} \geq_{\mathsf{L}} \mathbf{0}$ and $\mathbf{B} \geq_{\mathsf{L}} \mathbf{0}$. Then any two of the following conditions imply the third: $\mathbf{A} \leq_{\mathsf{L}} \mathbf{B}$, $\mathbf{A}^+ \geq_{\mathsf{L}} \mathbf{B}^+$, $\text{r}(\mathbf{A}) = \text{r}(\mathbf{B})$.

24.11 Let $\mathbf{A}$ be symmetric. Then $\mathbf{A} - \mathbf{BB}' \geq_{\mathsf{L}} \mathbf{0}$ iff

$$\mathbf{A} \geq_{\mathsf{L}} \mathbf{0}, \quad \mathscr{C}(\mathbf{B}) \subset \mathscr{C}(\mathbf{A}), \quad \text{and} \quad \mathbf{I} - \mathbf{B}'\mathbf{A}^-\mathbf{B} \geq_{\mathsf{L}} \mathbf{0}.$$

24.12 Consider a symmetric matrix

$$\mathbf{R} = \begin{pmatrix} 1 & r_{12} & r_{13} \\ r_{21} & 1 & r_{23} \\ r_{31} & r_{32} & 1 \end{pmatrix}, \quad \text{where all } r_{ij}^2 \leq 1.$$

Then $\mathbf{R}$ is a correlation matrix iff $\mathbf{R}$ is nod which holds off $\det(\mathbf{R}) = 1 - r_{12}^2 - r_{13}^2 - r_{23}^2 + 2r_{12}r_{13}r_{23} \geq 0$, or, equivalently, $(r_{12} - r_{13}r_{23})^2 \leq (1 - r_{13}^2)(1 - r_{23}^2)$.

24.13 The inertia $\operatorname{In}(\mathbf{A})$ of a symmetric $\mathbf{A}_{n\times n}$ is defined as the triple $\{\pi, \nu, \zeta\}$, where π is the number of positive eigenvalues of $\mathbf{A}$, ν is the number that are negative, and ζ is the number that are zero. Thus $\pi + \nu = \mathrm{r}(\mathbf{A})$ and $\pi + \nu + \zeta = n$. Matrix $\mathbf{A}$ is nnd if $\nu = 0$, and pd if $\nu = \zeta = 0$.

24.14 The inertia of the symmetric partitioned matrix $\mathbf{A} = \begin{pmatrix} \mathbf{A}_{11} & \mathbf{A}_{12} \\ \mathbf{A}_{21} & \mathbf{A}_{22} \end{pmatrix}$ satisfies

(a) $\operatorname{In}(\mathbf{A}) = \operatorname{In}(\mathbf{A}_{11}) + \operatorname{In}(\mathbf{A}_{22} - \mathbf{A}_{21}\mathbf{A}_{11}^{-}\mathbf{A}_{12}) \quad \text{if } \mathscr{C}(\mathbf{A}_{12}) \subset \mathscr{C}(\mathbf{A}_{11})$,

(b) $\operatorname{In}(\mathbf{A}) = \operatorname{In}(\mathbf{A}_{22}) + \operatorname{In}(\mathbf{A}_{11} - \mathbf{A}_{12}\mathbf{A}_{22}^{-}\mathbf{A}_{21}) \quad \text{if } \mathscr{C}(\mathbf{A}_{21}) \subset \mathscr{C}(\mathbf{A}_{22})$.

24.15 The minus (or rank-subtractivity) partial ordering for $\mathbf{A}_{n\times m}$ and $\mathbf{B}_{n\times m}$:

$$\mathbf{A} \leq^{-} \mathbf{B} \iff \mathrm{r}(\mathbf{B} - \mathbf{A}) = \mathrm{r}(\mathbf{B}) - \mathrm{r}(\mathbf{A}),$$

or equivalently, $\mathbf{A} \leq^{-} \mathbf{B}$ holds iff any of the following conditions holds:

(a) $\mathbf{A}^{-}\mathbf{A} = \mathbf{A}^{-}\mathbf{B}$ and $\mathbf{A}\mathbf{A}^{-} = \mathbf{B}\mathbf{A}^{-}$ for some $\mathbf{A}^{-} \in \{\mathbf{A}^{-}\}$,

(b) $\mathbf{A}^{-}\mathbf{A} = \mathbf{A}^{-}\mathbf{B}$ and $\mathbf{A}\mathbf{A}^{\sim} = \mathbf{B}\mathbf{A}^{\sim}$ for some $\mathbf{A}^{-}, \mathbf{A}^{\sim} \in \{\mathbf{A}^{-}\}$,

(c) $\{\mathbf{B}^{-}\} \subset \{\mathbf{A}^{-}\}$,

(d) $\mathscr{C}(\mathbf{A}) \cap \mathscr{C}(\mathbf{B} - \mathbf{A}) = \{\mathbf{0}\}$ and $\mathscr{C}(\mathbf{A}') \cap \mathscr{C}(\mathbf{B}' - \mathbf{A}') = \{\mathbf{0}\}$,

(e) $\mathscr{C}(\mathbf{A}) \subset \mathscr{C}(\mathbf{B})$, $\mathscr{C}(\mathbf{A}') \subset \mathscr{C}(\mathbf{B}')$ & $\mathbf{A}\mathbf{B}^{-}\mathbf{A} = \mathbf{A}$ for some (and hence for all) $\mathbf{B}^{-}$.

24.16 Let $\mathbf{V} \in \mathrm{NND}_n$, $\mathbf{X} \in \mathbb{R}^{n\times p}$, and denote

$$\mathcal{U} = \{\,\mathbf{U} : \mathbf{0} \leq_{\mathsf{L}} \mathbf{U} \leq_{\mathsf{L}} \mathbf{V},\ \mathscr{C}(\mathbf{U}) \subset \mathscr{C}(\mathbf{X})\,\}.$$

The maximal element (in the Löwner partial ordering) $\mathbf{U}$ in $\mathcal{U}$ is the shorted matrix of $\mathbf{V}$ with respect to $\mathbf{X}$, and denoted as $\operatorname{Sh}(\mathbf{V} \mid \mathbf{X})$.

24.17 The shorted matrix $\mathbf{S} = \operatorname{Sh}(\mathbf{V} \mid \mathbf{X})$ has the following properties:

(a) $\mathbf{S} = \operatorname{cov}(\mathbf{X}\tilde{\boldsymbol{\beta}})$ under $\{\mathbf{y}, \mathbf{X}\boldsymbol{\beta}, \mathbf{V}\}$, (b) $\mathscr{C}(\mathbf{S}) = \mathscr{C}(\mathbf{X}) \cap \mathscr{C}(\mathbf{V})$,

(c) $\mathscr{C}(\mathbf{V}) = \mathscr{C}(\mathbf{S}) \oplus \mathscr{C}(\mathbf{V} - \mathbf{S})$, (d) $\mathrm{r}(\mathbf{V}) = \mathrm{r}(\mathbf{S}) + \mathrm{r}(\mathbf{V} - \mathbf{S})$,

(e) $\mathbf{S}\mathbf{V}^{+}(\mathbf{V} - \mathbf{S}) = \mathbf{0}$,

(f) $\mathbf{V}^{-} \in \{\mathbf{S}^{-}\}$ for some (and hence for all) $\mathbf{V}^{-} \in \{\mathbf{V}^{-}\}$, i.e., $\{\mathbf{V}^{-}\} \subset \{\mathbf{S}^{-}\}$.

25 Inequalities

CSI $(\mathbf{x}'\mathbf{y})^2 \leq \mathbf{x}'\mathbf{x} \cdot \mathbf{y}'\mathbf{y}$ Cauchy–Schwarz inequality with equality holding iff $\mathbf{x}$ and $\mathbf{y}$ are linearly dependent

25.1 Recall: (a) Equality holds in CSI if $\mathbf{x} = \mathbf{0}$ or $\mathbf{y} = \mathbf{0}$. (b) The nonnull vectors $\mathbf{x}$ and $\mathbf{y}$ are linearly dependent (l.d.) iff $\mathbf{x} = \lambda\mathbf{y}$ for some $\lambda \in \mathbb{R}$.

In what follows, the matrix $\mathbf{V}_{n\times n}$ is nnd or pd, depending on the case. The ordered eigenvalues are $\lambda_1, \dots, \lambda_n$ and the corresponding orthonormal eigenvectors are $\mathbf{t}_1, \dots, \mathbf{t}_n$.

25.2 $(\mathbf{x}'\mathbf{V}\mathbf{y})^2 \leq \mathbf{x}'\mathbf{V}\mathbf{x} \cdot \mathbf{y}'\mathbf{V}\mathbf{y}$ equality iff $\mathbf{V}\mathbf{x}$ and $\mathbf{V}\mathbf{y}$ are l.d.

25.3 $(\mathbf{x}'\mathbf{y})^2 \leq \mathbf{x}'\mathbf{V}\mathbf{x} \cdot \mathbf{y}'\mathbf{V}^{-1}\mathbf{y}$ equality iff $\mathbf{x}$ and $\mathbf{V}\mathbf{y}$ are l.d.; $\mathbf{V}$ is pd

25.4 $(\mathbf{x}'\mathbf{P}_{\mathbf{V}}\mathbf{y})^2 \leq \mathbf{x}'\mathbf{V}\mathbf{x} \cdot \mathbf{y}'\mathbf{V}^{+}\mathbf{y}$ equality iff $\mathbf{V}^{1/2}\mathbf{x}$ and $\mathbf{V}^{+1/2}\mathbf{y}$ are l.d.

25.5 $(\mathbf{x}'\mathbf{y})^2 \leq \mathbf{x}'\mathbf{V}\mathbf{x} \cdot \mathbf{y}'\mathbf{V}^{-}\mathbf{y}$ for all $\mathbf{y} \in \mathscr{C}(\mathbf{V})$

25.6 $(\mathbf{x}'\mathbf{x})^2 \leq \mathbf{x}'\mathbf{V}\mathbf{x} \cdot \mathbf{x}'\mathbf{V}^{-1}\mathbf{x}$, i.e., $\dfrac{(\mathbf{x}'\mathbf{x})^2}{\mathbf{x}'\mathbf{V}\mathbf{x} \cdot \mathbf{x}'\mathbf{V}^{-1}\mathbf{x}} \leq 1,$

where the equality holds (assuming $\mathbf{x} \neq \mathbf{0}$) iff $\mathbf{x}$ is an eigenvector of $\mathbf{V}$: $\mathbf{V}\mathbf{x} = \lambda\mathbf{x}$ for some $\lambda \in \mathbb{R}$.

25.7 $(\mathbf{x}'\mathbf{V}\mathbf{x})^{-1} \leq \mathbf{x}'\mathbf{V}^{-1}\mathbf{x}$, where $\mathbf{x}'\mathbf{x} = 1$ equality iff $\mathbf{V}\mathbf{x} = \lambda\mathbf{x}$ for some λ

25.8 Let $\mathbf{V} \in \mathrm{NND}_n$ with $\mathbf{V}^{\{p\}}$ defined as

$$\begin{aligned} \mathbf{V}^{\{p\}} &= \mathbf{V}^{p}; \quad p = 1, 2, \dots, \\ &= \mathbf{P}_{\mathbf{V}}; \quad p = 0, \\ &= (\mathbf{V}^{+})^{|p|}; \quad p = -1, -2, \dots \end{aligned}$$

Then: $(\mathbf{x}'\mathbf{V}^{\{(h+k)/2\}}\mathbf{y})^2 \leq (\mathbf{x}'\mathbf{V}^{\{h\}}\mathbf{x})(\mathbf{y}'\mathbf{V}^{\{k\}}\mathbf{y})$ for $h, k = \dots, -1, 0, 1, 2, \dots,$ with equality iff $\mathbf{V}\mathbf{x} \propto \mathbf{V}^{\{1+(k-h)/2\}}\mathbf{y}$.

KI $\tau_1^2 := \dfrac{4\lambda_1\lambda_n}{(\lambda_1 + \lambda_n)^2} \leq \dfrac{(\mathbf{x}'\mathbf{x})^2}{\mathbf{x}'\mathbf{V}\mathbf{x} \cdot \mathbf{x}'\mathbf{V}^{-1}\mathbf{x}}$ Kantorovich inequality $\lambda_i = \mathrm{ch}_i(\mathbf{V})$, $\mathbf{V} \in \mathrm{PD}_n$

25.9 Equality holds in Kantorovich inequality KI when $\mathbf{x}$ is proportional to $\mathbf{t}_1 \pm \mathbf{t}_n$, where $\mathbf{t}_1$ and $\mathbf{t}_n$ are orthonormal eigenvectors of $\mathbf{V}$ corresponding to λ_1 and λ_n; when the eigenvalues λ_1 and λ_n are both simple (i.e., each has multiplicity 1) then this condition is also necessary.

25.10 $\tau_1 = \sqrt{\lambda_1\lambda_n}/[(\lambda_1+\lambda_n)/2] =$ the first antieigenvalue of $\mathbf{V}$

25.11 $$\tau_1^{-2} = \frac{(\lambda_1+\lambda_n)^2}{4\lambda_1\lambda_n} = \frac{(\lambda_1+\lambda_n)}{2}\cdot\frac{(1/\lambda_1+1/\lambda_n)}{2} = \left[\frac{\frac{1}{2}(\lambda_1+\lambda_n)}{\sqrt{\lambda_1\lambda_n}}\right]^2 = \frac{1}{1-\left(\frac{\lambda_1-\lambda_n}{\lambda_1+\lambda_n}\right)^2}$$

25.12 $\mathbf{x}'\mathbf{V}\mathbf{x} - \dfrac{1}{\mathbf{x}'\mathbf{V}^{-1}\mathbf{x}} \le (\sqrt{\lambda_1}-\sqrt{\lambda_n})^2 \quad (\mathbf{x}'\mathbf{x}=1)$

WI Wielandt inequality. Consider $\mathbf{V} \in \mathrm{NND}_n$ with $\mathrm{r}(\mathbf{V}) = v$ and let λ_i be the ith largest eigenvalue of $\mathbf{V}$ and $\mathbf{t}_i$ the corresponding eigenvector. Let $\mathbf{x} \in \mathscr{C}(\mathbf{V})$ and $\mathbf{y}$ be nonnull vectors satisfying the condition $\mathbf{x}'\mathbf{y} = 0$. Then

$$\frac{(\mathbf{x}'\mathbf{V}\mathbf{y})^2}{\mathbf{x}'\mathbf{V}\mathbf{x}\cdot\mathbf{y}'\mathbf{V}\mathbf{y}} \le \left(\frac{\lambda_1-\lambda_v}{\lambda_1+\lambda_v}\right)^2 = 1 - \frac{4\lambda_1\lambda_v}{(\lambda_1+\lambda_v)^2} := \nu_v^2.$$

The upper bound is attained when $\mathbf{x} = \mathbf{t}_1 + \mathbf{t}_v$ and $\mathbf{y} = \mathbf{t}_1 - \mathbf{t}_v$.

25.13 If $\mathbf{X} \in \mathbb{R}^{n\times p}$ and $\mathbf{Y} \in \mathbb{R}^{n\times q}$ then

$$\mathbf{0} \le_{\mathsf{L}} \mathbf{Y}'(\mathbf{I}_n - \mathbf{P}_{\mathbf{X}})\mathbf{Y} \le_{\mathsf{L}} (\mathbf{Y}-\mathbf{X}\mathbf{B})'(\mathbf{Y}-\mathbf{X}\mathbf{B}) \quad \text{for all } \mathbf{B}_{p\times q}.$$

25.14 $|\mathbf{X}'\mathbf{Y}|^2 \le |\mathbf{X}'\mathbf{X}|\cdot|\mathbf{Y}'\mathbf{Y}|$ for all $\mathbf{X}_{n\times p}$ and $\mathbf{Y}_{n\times p}$

25.15 Let $\mathbf{V} \in \mathrm{NND}_n$ with $\mathbf{V}^{\{p\}}$ defined as in 25.8 (p. 112). Then

$$\mathbf{X}'\mathbf{V}^{\{(h+k)/2\}}\mathbf{Y}(\mathbf{Y}'\mathbf{V}^{\{k\}}\mathbf{Y})^-\mathbf{Y}'\mathbf{V}^{\{(h+k)/2\}}\mathbf{X} \le_{\mathsf{L}} \mathbf{X}'\mathbf{V}^{\{h\}}\mathbf{X} \quad \text{for all } \mathbf{X} \text{ and } \mathbf{Y},$$

with equality holding iff $\mathscr{C}(\mathbf{V}^{\{h/2\}}\mathbf{X}) \subset \mathscr{C}(\mathbf{V}^{\{k/2\}}\mathbf{Y})$.

25.16 $\mathbf{X}'\mathbf{P}_{\mathbf{V}}\mathbf{X}(\mathbf{X}'\mathbf{V}\mathbf{X})^-\mathbf{X}'\mathbf{P}_{\mathbf{V}}\mathbf{X} \le_{\mathsf{L}} \mathbf{X}'\mathbf{V}^+\mathbf{X}$ equality iff $\mathscr{C}(\mathbf{V}\mathbf{X}) = \mathscr{C}(\mathbf{P}_{\mathbf{V}}\mathbf{X})$

25.17 $\mathbf{X}'\mathbf{V}^+\mathbf{X} \le_{\mathsf{L}} \dfrac{(\lambda_1+\lambda_v)^2}{4\lambda_1\lambda_v}\mathbf{X}'\mathbf{P}_{\mathbf{V}}\mathbf{X}(\mathbf{X}'\mathbf{V}\mathbf{X})^-\mathbf{X}'\mathbf{P}_{\mathbf{V}}\mathbf{X}$ $\quad \mathrm{r}(\mathbf{V}) = v$

25.18 $\mathbf{X}'\mathbf{V}\mathbf{X} \le_{\mathsf{L}} \dfrac{(\lambda_1+\lambda_v)^2}{4\lambda_1\lambda_v}\mathbf{X}'\mathbf{P}_{\mathbf{V}}\mathbf{X}(\mathbf{X}'\mathbf{V}^+\mathbf{X})^-\mathbf{X}'\mathbf{P}_{\mathbf{V}}\mathbf{X}$ $\quad \mathrm{r}(\mathbf{V}) = v$

25.19 $\mathbf{X}'\mathbf{V}^{-1}\mathbf{X} \le_{\mathsf{L}} \dfrac{(\lambda_1+\lambda_n)^2}{4\lambda_1\lambda_n}\mathbf{X}'\mathbf{X}(\mathbf{X}'\mathbf{V}\mathbf{X})^{-1}\mathbf{X}'\mathbf{X}$ $\quad \mathbf{V}$ pd

25.20 If $\mathbf{X}'\mathbf{P}_{\mathbf{V}}\mathbf{X}$ is idempotent, then $(\mathbf{X}'\mathbf{V}\mathbf{X})^+ \le_{\mathsf{L}} \mathbf{X}'\mathbf{V}^+\mathbf{X}$, where the equality holds iff $\mathbf{P}_{\mathbf{V}}\mathbf{X}\mathbf{X}'\mathbf{V}\mathbf{X} = \mathbf{V}\mathbf{X}$.

25.21 $(\mathbf{X}'\mathbf{V}\mathbf{X})^+ \le_{\mathsf{L}} \mathbf{X}'\mathbf{V}^{-1}\mathbf{X}$, if $\mathbf{V}$ is pd and $\mathbf{X}\mathbf{X}'\mathbf{X} = \mathbf{X}$

25.22 Under the full rank model $\{\mathbf{y}, \mathbf{X}\boldsymbol{\beta}, \mathbf{V}\}$ we have

(a) $\operatorname{cov}(\tilde{\boldsymbol{\beta}}) \leq_{\mathsf{L}} \operatorname{cov}(\hat{\boldsymbol{\beta}}) \leq_{\mathsf{L}} \dfrac{(\lambda_1 + \lambda_n)^2}{4\lambda_1\lambda_n} \operatorname{cov}(\tilde{\boldsymbol{\beta}})$,

(b) $(\mathbf{X}'\mathbf{V}^{-1}\mathbf{X})^{-1} \leq_{\mathsf{L}} (\mathbf{X}'\mathbf{X})^{-1}\mathbf{X}'\mathbf{V}\mathbf{X}(\mathbf{X}'\mathbf{X})^{-1} \leq_{\mathsf{L}} \dfrac{(\lambda_1 + \lambda_n)^2}{4\lambda_1\lambda_n}(\mathbf{X}'\mathbf{V}^{-1}\mathbf{X})^{-1}$,

(c) $$\begin{aligned}\operatorname{cov}(\hat{\boldsymbol{\beta}}) - \operatorname{cov}(\tilde{\boldsymbol{\beta}}) &= \mathbf{X}'\mathbf{V}\mathbf{X} - (\mathbf{X}'\mathbf{V}^{-1}\mathbf{X})^{-1} \\ &\leq_{\mathsf{L}} (\sqrt{\lambda_1} - \sqrt{\lambda_n})^2\mathbf{I}_p, \quad \mathbf{X}'\mathbf{X} = \mathbf{I}_p.\end{aligned}$$

25.23 Consider $\mathbf{V} \in \mathrm{NND}_n$, $\mathbf{X}_{n\times p}$ and $\mathbf{Y}_{n\times q}$, satisfying $\mathbf{X}'\mathbf{P}_\mathbf{V}\mathbf{Y} = \mathbf{0}$. Then

(a) $\mathbf{X}'\mathbf{V}\mathbf{Y}(\mathbf{Y}'\mathbf{V}\mathbf{Y})^-\mathbf{Y}'\mathbf{V}\mathbf{X} \leq_{\mathsf{L}} \left(\dfrac{\lambda_1 - \lambda_v}{\lambda_1 + \lambda_v}\right)^2 \mathbf{X}'\mathbf{V}\mathbf{X}, \qquad \mathrm{r}(\mathbf{V}) = v$

(b) $\mathbf{V} - \mathbf{P}_\mathbf{V}\mathbf{X}(\mathbf{X}'\mathbf{V}^+\mathbf{X})^-\mathbf{X}'\mathbf{P}_\mathbf{V} \geq_{\mathsf{L}} \mathbf{V}\mathbf{Y}(\mathbf{Y}'\mathbf{V}\mathbf{Y})^-\mathbf{Y}'\mathbf{V}$,

(c) $\mathbf{X}'\mathbf{V}\mathbf{X} - \mathbf{X}'\mathbf{P}_\mathbf{V}\mathbf{X}(\mathbf{X}'\mathbf{V}^+\mathbf{X})^-\mathbf{X}'\mathbf{P}_\mathbf{V}\mathbf{X} \geq_{\mathsf{L}} \mathbf{X}'\mathbf{V}\mathbf{Y}(\mathbf{Y}'\mathbf{V}\mathbf{Y})^-\mathbf{Y}'\mathbf{V}\mathbf{X}$,
where the equality holds iff $\mathrm{r}(\mathbf{V}) = \mathrm{r}(\mathbf{V}\mathbf{X}) + \mathrm{r}(\mathbf{V}\mathbf{Y})$.

25.24 Samuelson's inequality. Consider the $n \times k$ data matrix $\mathbf{X}_0 = (\mathbf{x}_{(1)} : \ldots : \mathbf{x}_{(n)})'$ and denote $\bar{\mathbf{x}} = (\bar{x}_1, \ldots, \bar{x}_k)'$, and $\mathbf{S}_{\mathbf{xx}} = \frac{1}{n-1}\mathbf{X}_0'\mathbf{C}\mathbf{X}_0$. Then

$$\frac{(n-1)^2}{n}\mathbf{S}_{\mathbf{xx}} - (\mathbf{x}_{(j)} - \bar{\mathbf{x}})(\mathbf{x}_{(j)} - \bar{\mathbf{x}})' \geq_{\mathsf{L}} \mathbf{0},$$

or equivalently,

$$\begin{aligned}(\mathbf{x}_{(j)} - \bar{\mathbf{x}})'\mathbf{S}_{\mathbf{xx}}^{-1}(\mathbf{x}_{(j)} - \bar{\mathbf{x}}) &= \mathrm{MHLN}^2(\mathbf{x}_{(j)}, \bar{\mathbf{x}}, \mathbf{S}_{\mathbf{xx}}) \\ &\leq \frac{(n-1)^2}{n}, \quad j = 1, \ldots, n.\end{aligned}$$

The equality above holds iff all $\mathbf{x}_{(i)}$ different from $\mathbf{x}_{(j)}$ coincide with their mean.

25.25 Let $\mathbf{V}_{p\times p}$ be nnd. Then $\mathbf{1}'\mathbf{V}\mathbf{1} \geq \frac{p}{p-1}[\mathbf{1}'\mathbf{V}\mathbf{1} - \operatorname{tr}(\mathbf{V})]$, i.e., assuming $\mathbf{1}'\mathbf{V}\mathbf{1} \neq 0$,

$$1 \geq \frac{p}{p-1}\left(1 - \frac{\operatorname{tr}(\mathbf{V})}{\mathbf{1}'\mathbf{V}\mathbf{1}}\right) := \alpha(\mathbf{1}).$$

where the equality is obtained iff $\mathbf{V} = \gamma^2\mathbf{1}\mathbf{1}'$ for some $\gamma \in \mathbb{R}$. The term $\alpha(\mathbf{1})$ can be interpreted as Cronbach's *alpha*.

25.26 Using the notation above, and assuming that $\operatorname{diag}(\mathbf{V}) = \mathbf{V}_\delta$ is pd [see 22.26 (p. 102)]

$$\alpha(\mathbf{a}) = \frac{p}{p-1}\left(1 - \frac{\mathbf{a}'\mathbf{V}_\delta\mathbf{a}}{\mathbf{a}'\mathbf{V}\mathbf{a}}\right) \leq \frac{p}{p-1}\left(1 - \frac{1}{\mathrm{ch}_1(\mathbf{R}_\mathbf{V})}\right) \leq 1$$

for all $\mathbf{a} \in \mathbb{R}^p$, where $\mathbf{R}_V$ is a correlation matrix calculated from the covariance matrix $\mathbf{V}$. Moreover, the equality $\alpha(\mathbf{a}) = 1$ is obtained iff $\mathbf{V} = \gamma^2\mathbf{q}\mathbf{q}'$ for some $\gamma \in \mathbb{R}$ and some $\mathbf{q} = (q_1, \dots, q_p)'$, and $\mathbf{a}$ is a multiple of $\mathbf{a}_* = (1/q_1, \dots, 1/q_p)'$.

26 Kronecker product, some matrix derivatives

26.1 The Kronecker product of $\mathbf{A}_{n\times m} = (\mathbf{a}_1 : \ldots : \mathbf{a}_m)$ and $\mathbf{B}_{p\times q}$ and $\mathrm{vec}(\mathbf{A})$ are defined as

$$\mathbf{A} \otimes \mathbf{B} = \begin{pmatrix} a_{11}\mathbf{B} & a_{12}\mathbf{B} & \dots & a_{1n}\mathbf{B} \\ a_{21}\mathbf{B} & a_{22}\mathbf{B} & \dots & a_{2n}\mathbf{B} \\ \vdots & \vdots & & \vdots \\ a_{n1}\mathbf{B} & a_{n2}\mathbf{B} & \dots & a_{nm}\mathbf{B} \end{pmatrix} \in \mathbb{R}^{np\times mq}, \quad \mathrm{vec}(\mathbf{A}) = \begin{pmatrix} \mathbf{a}_1 \\ \vdots \\ \mathbf{a}_m \end{pmatrix}.$$

26.2 $(\mathbf{A} \otimes \mathbf{B})' = \mathbf{A}' \otimes \mathbf{B}', \quad \mathbf{a}' \otimes \mathbf{b} = \mathbf{b}\mathbf{a}' = \mathbf{b} \otimes \mathbf{a}', \quad \lambda \otimes \mathbf{A} = \lambda\mathbf{A} = \mathbf{A} \otimes \lambda$

26.3 $(\mathbf{F} : \mathbf{G}) \otimes \mathbf{B} = (\mathbf{F} \otimes \mathbf{B} : \mathbf{G} \otimes \mathbf{B}), \quad (\mathbf{A} \otimes \mathbf{B})(\mathbf{C} \otimes \mathbf{D}) = \mathbf{AC} \otimes \mathbf{BD}$

26.4 $(\mathbf{A} \otimes \mathbf{b}')(\mathbf{c} \otimes \mathbf{D}) = (\mathbf{b}' \otimes \mathbf{A})(\mathbf{D} \otimes \mathbf{c}) = \mathbf{A}\mathbf{c}\mathbf{b}'\mathbf{D}$

26.5 $\mathbf{A}_{n\times m} \otimes \mathbf{B}_{p\times k} = (\mathbf{A} \otimes \mathbf{I}_p)(\mathbf{I}_m \otimes \mathbf{B}) = (\mathbf{I}_n \otimes \mathbf{B})(\mathbf{A} \otimes \mathbf{I}_k)$

26.6 $(\mathbf{A} \otimes \mathbf{B})^{-1} = \mathbf{A}^{-1} \otimes \mathbf{B}^{-1}, \quad (\mathbf{A} \otimes \mathbf{B})^+ = \mathbf{A}^+ \otimes \mathbf{B}^+$

26.7 $\mathbf{P}_{\mathbf{A}\otimes\mathbf{B}} = \mathbf{P}_\mathbf{A} \otimes \mathbf{P}_\mathbf{B}, \quad \mathrm{r}(\mathbf{A} \otimes \mathbf{B}) = \mathrm{r}(\mathbf{A}) \cdot \mathrm{r}(\mathbf{B})$

26.8 $\mathrm{tr}(\mathbf{A} \otimes \mathbf{B}) = \mathrm{tr}(\mathbf{A}) \cdot \mathrm{tr}(\mathbf{B}), \quad \|\mathbf{A} \otimes \mathbf{B}\|_F = \|\mathbf{A}\|_F \cdot \|\mathbf{B}\|_F$

26.9 Let $\mathbf{A} \geq_\mathsf{L} \mathbf{0}$ and $\mathbf{B} \geq_\mathsf{L} \mathbf{0}$. Then $\mathbf{A} \otimes \mathbf{B} \geq_\mathsf{L} \mathbf{0}$.

26.10 Let $\mathrm{ch}(\mathbf{A}_{n\times n}) = \{\lambda_1, \dots, \lambda_n\}$ and $\mathrm{ch}(\mathbf{B}_{m\times m}) = \{\mu_1, \dots, \mu_m\}$. Then $\mathrm{ch}(\mathbf{A} \otimes \mathbf{B}) = \{\lambda_i\mu_j\}$ and $\mathrm{ch}(\mathbf{A} \otimes \mathbf{I}_m + \mathbf{I}_n \otimes \mathbf{B}) = \{\lambda_i + \mu_j\}$, where $i = 1, \dots, n$, $j = 1, \dots, m$.

26.11 $\mathrm{vec}(\alpha\mathbf{A} + \beta\mathbf{B}) = \alpha\,\mathrm{vec}(\mathbf{A}) + \beta\,\mathrm{vec}(\mathbf{B}), \quad \alpha, \beta \in \mathbb{R}$

26.12 $\mathrm{vec}(\mathbf{ABC}) = (\mathbf{I} \otimes \mathbf{AB})\,\mathrm{vec}(\mathbf{C}) = (\mathbf{C}' \otimes \mathbf{A})\,\mathrm{vec}(\mathbf{B}) = (\mathbf{C}'\mathbf{B}' \otimes \mathbf{I})\,\mathrm{vec}(\mathbf{A})$

26.13 $\mathrm{vec}(\mathbf{A}^{-1}) = [(\mathbf{A}^{-1})' \otimes \mathbf{A}^{-1}]\,\mathrm{vec}(\mathbf{A})$

26.14 $\mathrm{tr}(\mathbf{AB}) = [\mathrm{vec}(\mathbf{A}')]'\,\mathrm{vec}(\mathbf{B})$

26.15 Below are some matrix derivatives.

(a) $\dfrac{\partial \mathbf{Ax}}{\partial \mathbf{x}} = \mathbf{A}'$

(b) $\dfrac{\partial \mathbf{x}'\mathbf{Ax}}{\partial \mathbf{x}} = (\mathbf{A} + \mathbf{A}')\mathbf{x}; \quad 2\mathbf{Ax}$ when $\mathbf{A}$ symmetric

(c) $\dfrac{\partial \operatorname{vec}(\mathbf{AX})}{\partial \operatorname{vec}(\mathbf{X})'} = \mathbf{I} \otimes \mathbf{A}$

(d) $\dfrac{\partial \operatorname{vec}(\mathbf{AXB})}{\partial \operatorname{vec}(\mathbf{X})'} = \mathbf{B}' \otimes \mathbf{A}$

(e) $\dfrac{\partial \operatorname{tr}(\mathbf{AX})}{\partial \mathbf{X}} = \mathbf{A}'$

(f) $\dfrac{\partial \operatorname{tr}(\mathbf{X}'\mathbf{AX})}{\partial \mathbf{X}} = 2\mathbf{AX}$ for symmetric $\mathbf{A}$

(g) $\dfrac{\partial \log|\mathbf{X}'\mathbf{AX}|}{\partial \mathbf{X}} = 2\mathbf{AX}(\mathbf{X}'\mathbf{AX})^{-1}$ for symmetric $\mathbf{A}$

Index

A

Abadir, Karim M. vi
added variable plot 18
adjoint matrix 83
admissibility 65
Albert's theorem 110
ANOVA 37
  another parametrization 39
  conditional mean and variance 38
  estimability 48
  in matrix terms 38
antieigenvalues 58, 61
approximating
  the centered data matrix 75
AR(1)-structure *see* autocorrelation
autocorrelation
  $\mathbf{V} = \{\varrho^{|i-j|}\}$ 30
  BLP 31
  in linear model 32
  in prediction 67
  regression coefficients 32
  testing 32

B

Baksalary, Oskar Maria vi
Banachiewicz–Schur form 89
Bernoulli distribution 19
Bernstein, Dennis S. vi
best linear predictor *see* BLP
best linear unbiased estimator *see* BLUE
best linear unbiased predictor *see* BLUP, 65
best predictor *see* BP
bias 27
binomial distribution 19
Blanck, Alice vi
block diagonalization 28, 89
Bloomfield–Watson efficiency 62
Bloomfield–Watson–Knott inequality 58
BLP
  $\mathrm{BLP}(\mathbf{y};\mathbf{x})$ 22, 28
  $\mathrm{BLP}(\mathbf{z};\mathbf{A}'\mathbf{z})$ 29
  $\mathrm{BLP}(y_n;\mathbf{y}_{(n-1)})$ 31
  $\boldsymbol{\mu}_{\mathrm{y}} + \boldsymbol{\Sigma}_{\mathrm{yx}}\boldsymbol{\Sigma}_{\mathrm{xx}}^{-1}(\mathbf{x} - \boldsymbol{\mu}_{\mathrm{x}})$ 22
  $\boldsymbol{\mu}_{\mathrm{y}} + \boldsymbol{\Sigma}_{\mathrm{yx}}\boldsymbol{\Sigma}_{\mathrm{xx}}^{-}(\mathbf{x} - \boldsymbol{\mu}_{\mathrm{x}})$ 28
  autocorrelation 31
BLUE
  $\tilde{\boldsymbol{\beta}}$, representations 51
  $\tilde{\boldsymbol{\beta}}_2 = (\mathbf{X}_2'\dot{\mathbf{M}}_{1\mathrm{W}}\mathbf{X}_2)^{-1}\mathbf{X}_2'\dot{\mathbf{M}}_{1\mathrm{W}}\mathbf{y}$ 53
  $\tilde{\boldsymbol{\beta}}_2 = (\mathbf{X}_2'\dot{\mathbf{M}}_1\mathbf{X}_2)^{-1}\mathbf{X}_2'\dot{\mathbf{M}}_1\mathbf{y}$ 53
  $\tilde{\boldsymbol{\beta}}_i$ in the general case 53
  $\mathrm{BLUE}(\mathbf{K}'\boldsymbol{\beta})$ 49
  $\mathrm{BLUE}(\mathbf{X}\boldsymbol{\beta})$ 48
  definition 48
  fundamental equation 49
  $\mathrm{cov}(\mathbf{X}\tilde{\boldsymbol{\beta}})$ as a shorted matrix 111
  $\mathbf{X}\tilde{\boldsymbol{\beta}}$, general representations 49
  $\mathbf{X}\tilde{\boldsymbol{\beta}}$, representations 50
  equality of OLSE and BLUE, several conditions 54
  restricted BLUE 37
  without the ith observation 43
BLUP
  fundamental equation 66
  in mixed model 68
  of y_* 14
  of $\boldsymbol{\gamma}$ 68
  of $\mathbf{y}_f$ 66
  Pandora's Box 66
  representations for $\mathrm{BLUP}(\mathbf{y}_f)$ 66
BP
  $\mathrm{BP}(\mathbf{y};\mathbf{x})$ 23

S. Puntanen et al., *Formulas Useful for Linear Regression Analysis and Related Matrix Theory*, SpringerBriefs in Statistics,
DOI: 10.1007/978-3-642-32931-9,

N

O